# Enhancing Your Cloud Security with a CNAPP Solution

Unlock the full potential of Microsoft Defender for Cloud to fortify your cloud security

**Yuri Diogenes**

‹packt›

# Enhancing Your Cloud Security with a CNAPP Solution

**Senior Publishing Product Manager**: Reshma Raman

**Acquisition Editor – Peer Reviews**: Jane Dsouza

**Project Editor**: K. Loganathan

**Content Development Editor**: Deepayan Bhattacharjee

**Copy Editor**: Safis Editing

**Technical Editor**: Kushal Sharma

**Proofreader**: Safis Editing

**Indexer**: Rekha Nair

**Presentation Designer**: Pranit Padwal

**Developer Relations Marketing Executive**: Maran Fernandes

First published: October 2024

Production reference: 1241024

Published by Packt Publishing Ltd.

Grosvenor House

11 St Paul's Square

Birmingham

B3 1RB, UK.

ISBN 978-1-83620-487-9

www.packt.com

*To my loving friends and family*

*– Yuri Diogenes*

# Contributors

## About the authors

**Yuri Diogenes** has worked at Microsoft since 2006, and currently is the Principal PM Manager for the CxE Defender for Cloud Team. He is a professor at EC-Council University and at Trine University. Yuri has a master's degree in Cybersecurity from Utica University. He is currently working on a PhD in Cybersecurity Leadership (Capitol Technology University). He is also the author of 33 published books. Yuri holds many industry certifications, such as CISSP, CND, CEH, CSA, CHFI, and CASP.

> *I would like to thank my wife and daughters for their continuous support, my mother for always believing in me, the entire Defender for Cloud Team for the great partnership since 2015, and my coworkers. Also, thanks to all my mentees for pushing me to be a better professional, to the entire team of Packt for the continuous partnership.*

**Shay Amar** has over 20 years of experience in Cloud Security and customer success, and he specializes in driving digital transformation and aligning technology with business goals. As a Senior Product Manager for Cloud Security CxE at Microsoft, he leads initiatives that bridge the gap between product teams and enterprise customers, delivering impactful security solutions. He holds a degree in Computer Science and multiple certifications, including CCNA, MCSE, and Azure Architect, and regularly speaks at industry events. His focus is on innovation, automation, and securing growth for organizations in the cloud

# About the reviewer

**Dominik Hoefling** is a Security Cloud Solution Architect with a robust background in cybersecurity. He earned his bachelor's degree in business informatics from IU International University of Applied Sciences, where he developed a strong foundation in both business administration and IT. Dominik specializes in deploying security solutions for customers, particularly focusing on Microsoft security solutions. He holds the **Certified Cloud Security Professional (CCSP)** certification from ISC2 and numerous other Microsoft security certificates. Before joining Microsoft, he was recognized as a **Microsoft Most Valuable Professional (MVP)** and served as a Principal Consultant at a Microsoft partner, leading numerous successful projects. As a seasoned cybersecurity expert, Dominik focuses on cloud security, **Extended Detection and Response (XDR)**, and **Security Information and Event Management (SIEM)**. He is a sought-after speaker at global conferences, where he shares his knowledge and insights on modern cybersecurity threats and solutions, helping organizations enhance their security posture. Dominik is passionate about guiding businesses through the complexities of today's cybersecurity landscape, ensuring they are well equipped to handle emerging threats.

*In producing this book, I would like to extend my heartfelt thanks to my wife, Kathrin, and our son, Levi, who was born just a week before the work on this book began. Kathrin's support and understanding have been invaluable throughout this journey.*

# Join our community on Discord

Read this book alongside other users. Ask questions, provide solutions to other readers, and much more.

Scan the QR code or visit the link to join the community.

https://packt.link/SecNet

# Table of Contents

# Preface

With the growth of multicloud adoption, and the constant need to protect resources from code to runtime, the old approach of working in silos is not effective anymore. To overcome this challenge the use of a **Cloud Native Application Protection Platform** (**CNAPP**) becomes imperative for organizations that want to continue to elevate their cloud security posture, while prioritizing what is important to be remediated based on the risk factors tailored for their own environment. This book covers end-to-end CNAPP adoption, from setting the context of the need for a CNAPP and planning an agnostic approach to adopt CNAPP, all the way to the use of Microsoft Defender for Cloud as a CNAPP solution. I've been working with cloud security since 2012, when the emphasis was on Private Cloud security, and I've been part of the Defender for Cloud team since its conception in 2015, when it was still called Azure Security Center. I wrote this book based on all these years of experience, talking to hundreds of customers over the years and helping them to adopt the product. The book was reviewed by Dominik Hoefling, a great specialist in Defender for Cloud whom I have known and mentored for years. He did an amazing job performing the tech review and adding his own insights to improve the final project. *Chapter 17* was written by Shay Amar, a specialist in Security Exposure Management who has been working in the team that created the product since it was in incubation. Shay adds a lot of value to this project.

I hope you enjoy reading it!

## Who this book is for

This book is recommended for Cloud Security Administrators, Cloud Solution Architects, DevSecOps Engineers, Security Operations Engineer, members of the Cloud Security Posture Management Team, and any IT/Security professional that wants to learn more about CNAPP.

# What this book covers

*Chapter 1, Why CNAPP?*, covers the roots of cloud security with CSPM and CWP, the traditional CSPM lifecycle, the use of secure score to track progress over time, and the challenges introduced with multicloud and shift left. You will also learn about the main aspects of CWP, the use of MITRE ATT&CK framework to map alerts to different workloads, and the need to have agents for some types of workloads. Lastly, you will learn how CNAPP was idealized and the main advantages of using a CNAPP, which include attack disruption, agentless approach, proactive hunting, and SOC enrichment.

*Chapter 2, Assessing Your Environment's Security Posture*, goes further into the adoption of Defender for Cloud Foundational CSPM to start assessing your environment security posture. You will learn how to use security recommendations to visualize areas of improvement for your workloads, how to remediate security recommendations, and about the importance of using secure score to improve the overall security posture. Lastly, you will learn how to track your secure score overtime, about the use of MCSB to have a different view of your recommendations and adhere to compliance standards, and how to access your cloud inventory using Defender for Cloud.

*Chapter 3, CNAPP Design Considerations*, focuses on the importance of first establishing design principles four your CNAPP adoption, which includes the use of Zero Trust, Shift-left security, data protection, visibility and monitoring, dynamic threat detection and response, and compliance/governance. These design principles will be agnostic of your platform and will be the foundation of your implementation. You will also learn about the design considerations for your CNAPP adoption, which include considerations for posture management, considerations for DevOps security, and for workload protection. Lastly, you will learn some important questions to ask while going through these considerations, which should also be complemented by specific needs and constraints of your organization.

*Chapter 4, Creating an Adoption Plan*, details the overall approach to forming a Microsoft CNAPP (Defender for Cloud) adoption plan. You will learn about the different aspects of planning your posture management adoption by enabling Defender CSPM. We discuss the planning aspects of workload protection by covering each individual Defender for Cloud plan. Lastly, you will learn about the important aspects of how to create a proof of concept plan to validate the main use case scenarios that your organization needs.

*Chapter 5, Elevating Your Workload's Security Posture*, shows how to onboard Defender CSPM. You will also learn about the use of the attack path to disrupt potential attacks and how to use the risk-based recommendations to prioritize what is important for your environment. Lastly, you will learn about data security posture, how the discovery takes place, how to customize data sensitivity, and how to use the data security dashboard.

*Chapter 6, Multicloud*, demonstrates how to connect with AWS. You will learn about the pre-requisites, the architecture of the solution, and how to configure the AWS Connector. You will also learn how to leverage the risk-based recommendations to prioritize what it is important to remediate in the AWS environment. Lastly, you will learn about how to connect with GCP, and how to configure the GCP connector.

*Chapter 7, DevOps Security Capabilities*, covers the DevOps security capabilities available in Defender CSPM. You will also learn about the prerequisites and how to connect Defender for Cloud with GitHub, Azure DevOps, and GitLab. Lastly, you will learn about how recommendations from GitHub and ADO appear in Defender for Cloud, and how developers will experience these recommendations on their platform.

*Chapter 8, Governance and Continuous Improvement*, goes into how to use the Governance feature in Defender CSPM to assign ownership to security recommendations. You will also learn how to integrate the governance feature with ServiceNow. Lastly, you will learn how to create exemptions according to your needs and how to visualize all resources that have exemptions.

*Chapter 9, Proactive Hunting*, shows how to leverage the insights collected by Defender for Cloud to perform proactive hunting. You will also learn how to use the Cloud Security Explorer to create queries and how to use the available templates. In addition to that, you will learn how to use Azure Resource Graph (ARG) to create queries using KQL and how to access ARG via Defender for Cloud. Lastly, you will learn about the advantages of using threat intel while doing proactive hunting.

*Chapter 10, Implementing Workload Protection*, explains why it is important to have a tailored approach to workload protection, how threat detection works in Defender for Cloud, and the different types of detection. You will also learn about the Security Alert dashboard, alert correlation, sample alerts, and how alert suppression works. Lastly, you will learn about the different Defender for Cloud plans available.

*Chapter 11, Protecting Compute Resources (Servers and Containers)*, explains how to protect Containers with Defender for Containers and machines with Defender for Servers. You will also learn about the supported scenarios in Defender for Containers, the capabilities and constraints, how to enable Defender for Containers, how to visualize vulnerabilities in your containers, and how to configure binary drift detection. Lastly, you will learn about Defender for Servers capabilities such as agentless malware scanning, file integrity monitoring, just-in-time VM access, and vulnerability assessment powered by Microsoft Defender Vulnerability Management.

*Chapter 12, Protecting Storage and Databases*, discusses the importance of protecting the data located in storage and databases. You will learn about the most common threat vectors for storage and databases and how Defender for Storage and Defender for Databases can help protect this type of workload. You will also learn how to enable Defender for Storage in your subscription, and how malware scanning can be used to protect your storage accounts from getting compromised with malware. Lastly, you will learn how to enable Defender for Databases, about the different types of databases that are supported by this plan, and how to use the vulnerability assessment feature to improve your database security posture.

*Chapter 13, Protecting APIs*, looks at the importance of protecting APIs, what is necessary to do to prepare the environment before enabling Defender for APIs, and where Defender for APIs sits within the network architecture in a multilayered protection approach. You will also learn how to enable Defender for APIs in your Azure subscription, how to operationalize Defender for APIs, how to leverage the insights that are added by Defender for APIs in Cloud Security Explorer, and about the templates available to perform proactive hunting. Lastly, you will learn how to manage the onboarded APIs, including how to offboard APIs.

*Chapter 14, Protecting Service Layer*, addresses the importance of protecting Azure Resource Manager and how Defender for Resource Manager monitors and detects threats against ARM. You will also learn how to enable Defender for Resource Manager in your Azure subscription. Lastly, you will learn about the importance of protecting Azure App Service, and how Defender for App Service can help to protect your platform.

*Chapter 15, Incident Response*, covers the use of Defender for Cloud for Incident Response and the different alert insights provided by Defender for Cloud to empower IR teams to do a better investigation. You will also learn about the integration of Defender for Cloud with Microsoft XDR, the Alert experience in Microsoft XDR portal, and get an introduction to the advanced hunting capability in Microsoft XDR. Lastly, you will learn how to configure the Defender for Cloud connector in Microsoft Sentinel to enable the ingestion of Defender for Cloud security alerts in Microsoft Sentinel.

*Chapter 16, Leveraging AI to Improve Your Security Posture*, looks at Defender for Cloud integration with Copilot for Security, how to perform risk exploration by using Copilot for Security embedded experience in Defender for Cloud, and how to summarize recommendations and ask Copilot for Security to generate remediation script. Lastly, you will learn about AI posture management, the importance of having security recommendations tailored for AI scenarios, how Defender CSPM takes into consideration Azure AI as part of the attack path, and the AI queries available in Cloud Security Explorer.

*Chapter 17, Security Exposure Management*, explains more about Microsoft Security Exposure Management and how to enable it in your organization to manage the security posture of your workloads proactively. You will also learn about the importance of incorporating *Key Initiatives, Top Metrics,* and *Security Recommendations* into a cohesive strategy. In addition, you will learn about the importance of using *Security Events* to track how specific incidents and score drops affect an organization's security landscape. Lastly, you will learn about *Attack Surface Maps* and *Attack Paths* to visualize your environment and identify potential vulnerabilities.

# To get the most out of this book

- Ensure that you have an Azure subscription available to test the scenarios. You can get an Azure trial subscription at `https://azure.microsoft.com/en-us/pricing/purchase-options/azure-account`

- You can use Defender for Cloud free trial for 30 days per plan. You can build your own environment using the resources from `https://aka.ms/MDCLabs`

- For *Chapter 6, Multicloud*, you should also have access to AWS and GCP environments to test the connectors. Both (AWS and GCP) also have trial accounts available. Visit `https://aws.amazon.com/free/compute` for AWS and `https://cloud.google.com/free` for GCP.

# Download the color images

We also provide a PDF file that has color images of the screenshots/diagrams used in this book. You can download it here: `https://packt.link/gbp/9781836204879`.

## Conventions used

There are a number of text conventions used throughout this book.

`CodeInText`: Indicates code words in text, database table names, folder names, filenames, file extensions, pathnames, dummy URLs, user input, and Twitter handles. For example: "By the time this book was published, the GCP data collectors that were using a fixed scan time of 1 hour were `ComputeInstance`, `ArtifactRegistryRepositoryPolicy`."

**Bold**: Indicates a new term, an important word, or words that you see on the screen. For instance, words in menus or dialog boxes appear in the text like this. For example: "Open the Defender for Cloud dashboard and click **Environment settings**, under the **Management** section."

> Warnings or important notes appear like this.

> Tips and tricks appear like this.

## Get in touch

Feedback from our readers is always welcome.

**General feedback**: Email `feedback@packtpub.com` and mention the book's title in the subject of your message. If you have questions about any aspect of this book, please email us at `questions@packtpub.com`.

**Errata**: Although we have taken every care to ensure the accuracy of our content, mistakes do happen. If you have found a mistake in this book, we would be grateful if you reported this to us. Please visit `http://www.packtpub.com/submit-errata`, click **Submit Errata**, and fill in the form.

**Piracy**: If you come across any illegal copies of our works in any form on the internet, we would be grateful if you would provide us with the location address or website name. Please contact us at `copyright@packtpub.com` with a link to the material.

**If you are interested in becoming an author**: If there is a topic that you have expertise in and you are interested in either writing or contributing to a book, please visit `http://authors.packtpub.com`.

# Leave a Review!

Thank you for purchasing this book from Packt Publishing—we hope you enjoy it! Your feedback is invaluable and helps us improve and grow. Once you've completed reading it, please take a moment to leave an Amazon review; it will only take a minute, but it makes a big difference for readers like you.

*https://packt.link/r/1836204876*

Scan the QR code below to receive a free ebook of your choice.

*https://packt.link/NzOWQ*

# Download a free PDF copy of this book

Thanks for purchasing this book!

Do you like to read on the go but are unable to carry your print books everywhere?

Is your eBook purchase not compatible with the device of your choice?

Don't worry, now with every Packt book you get a DRM-free PDF version of that book at no cost.

Read anywhere, any place, on any device. Search, copy, and paste code from your favorite technical books directly into your application.

The perks don't stop there, you can get exclusive access to discounts, newsletters, and great free content in your inbox daily.

Follow these simple steps to get the benefits:

1. Scan the QR code or visit the link below:

*https://packt.link/free-ebook/9781836204879*

2. Submit your proof of purchase.
3. That's it! We'll send your free PDF and other benefits to your email directly.

# 1

# Why CNAPP?

For the past decade, cloud security has evolved according to the threat landscape in addition to the overall business needs of the companies that were migrating to the cloud. In the beginning, cloud security solutions provided very basic security hygiene based on a set of baselines and workload visibility, which, at the time, addressed the needs of most companies. As the market evolved around cloud security, companies started to demand specialized solutions to address specific challenges in cloud security, such as multicloud adoption and the shift-left initiatives. New solutions were developed to tackle these challenges; however, they were done in an isolated manner.

The evolution of attack methods, the growth of cloud automation such as the high usage of **Infrastructure as Code (IaC)**, the wide adoption of multicloud, and the need to have a better way to prioritize risk based on a contextual approach led the market to a new reality when it comes to cloud security. The best-of-breed approach to deciding which cloud security solution should be adopted wasn't working anymore. Customers demanded a better way to cross-reference the data consumed by different tools in a single place to enable them to make smarter decisions when it comes to risk prioritization.

It becomes imperative to not only improve the security posture but also identify how threat actors can exploit existing vulnerabilities and move laterally to potentially compromise highly sensitive assets. The solution for all this is called **Cloud Native Application Protection Platform (CNAPP)**.

This chapter covers:

- Cloud Security Posture Management
- Cloud Workload Protection
- Cloud Native Application Protection Platform

# Cloud Security Posture Management

The term **Cloud Security Posture Management (CSPM)** was introduced around 2018. It appeared as companies started to adopt more and more cloud computing, which led to the need to have tools to manage and secure their cloud environments. The term was coined by Gartner, a leading research and advisory company. Gartner introduced CSPM to describe a category of security tools designed to identify and manage security risks in cloud environments. The main objective of CSPM was to ensure that organizations were strengthening their cloud security posture across their workloads.

The core of CSPM was based on the discoverability of cloud workloads, and the assessment of these workloads according to cloud security best practices. These cloud security best practices were grounded in a mix of cloud solution providers' benchmarks and industry security standards, such as the **Center for Internet Security (CIS)**, the **International Organization for Standardization (ISO)**, and the **National Institute of Standards and Technology (NIST)**.

Over time, some CSPM solutions also started to offer regulatory compliance lenses on top of the data to help organizations validate if their workloads were compliant with certain standards, such as the **Health Insurance Portability and Accountability Act (HIPAA)** and **Payment Card Industry (PCI)**.

Regardless of the benchmark in use, the traditional CSPM lifecycle used in the beginning had the following phases:

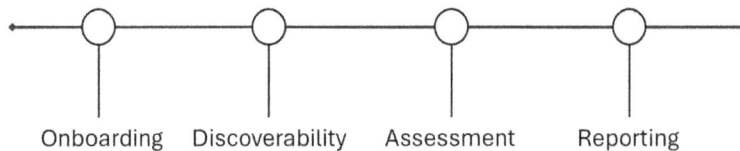

Onboarding    Discoverability    Assessment    Reporting

*Figure 1.1: Traditional CSPM lifecycle*

These phases are highlighted below:

- **Onboarding**: The first step was to onboard the CSPM solution to the cloud solution provider. For example, in an Azure environment, this step means enabling CSPM in the Azure subscription.
- **Discoverability**: Once the CSPM platform is enabled, it will perform the scan to discover all supported workloads in the cloud environment.

- **Assessment**: After discovering all supported workloads, and creating the initial inventory, it will perform the security assessment to evaluate if the workloads that were discovered are using security best practices based on industry standard benchmarks.

- **Reporting**: Security assessment is a continuous operation, but at the end of each assessment, a report will be created to present the current security state of the workloads. While some workloads may already be configured using security best practices, others may require additional steps to be compliant. For these scenarios, security recommendations will be presented to guide you through the steps on how to remediate the workload.

While these steps are generic and vendor agnostic, each CSPM solution available at that time (around 2018) was adding specific features to improve the overall user experience. For example, Azure Security Center, the Microsoft CSPM at that time, had the *secure score*, which was a measurement for organizations to identify their security posture by scoring recommendations based on the benchmarks' asset criticality. The advantage of using a metric such as secure score was that it gave organizations the capability to evaluate progress over time and a North Star to follow: reach 100% on their secure score.

However, the use of secure score also exposed another problem in the cloud security environment, which was the lack of governance. This problem was exposed with the constant fluctuation of the secure score, as shown in the diagram below:

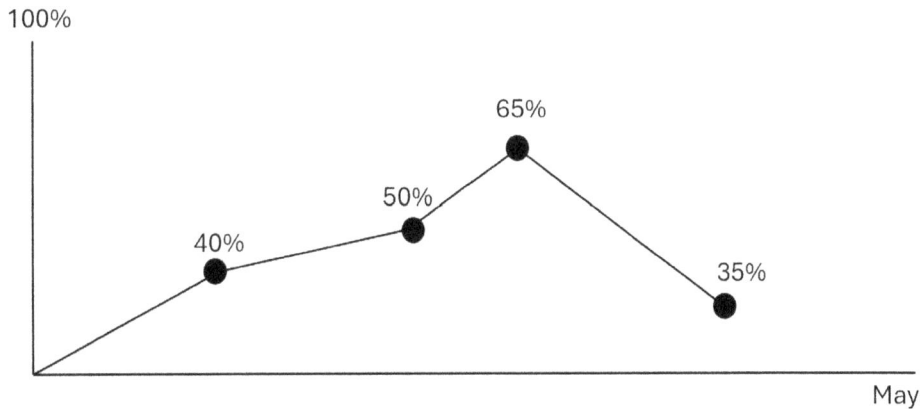

*Figure 1.2: Secure score over time*

In the fictitious sample diagram above, you have the secure score during the month of May. Notice that in the beginning, the progress was going in the right direction; the initial score was 40%, and it got better, all the way to 65%.

But then something happened, and it dropped to 35%. The question that many cloud administrators had at that time was: I didn't do anything to change the environment, why did my secure score drop so much?

The reason that those drops occurred, and are still occurring, is the lack of security guardrails at the beginning of the pipeline. In other words, when users provision new resources (for example, a new storage account) that are not using security best practices from the get-go, the number of security recommendations that will need to be applied to leverage that resource's security posture is high, and this will negatively affect the secure score. Every time the CSPM platform performs the assessment, it will either increase the score (if the resources are secure) or drop the score (when the resources are not secure). Of course, the score can stay the same in case the environment hasn't changed, or new resources haven't been provisioned, but in a cloud environment, the likelihood of having many resources getting created and deleted on a daily basis is very high. The lack of guardrails at the beginning of the pipeline led organizations to realize that CSPM was not the sole solution for cloud security. Governance became imperative to ensure that resources were created with security defaults.

Another buzzword that started to become more reality around that time was *shift-left*. The shift-left approach encouraged practices like early testing, continuous integration, and incorporating security considerations (often referred to as DevSecOps) from the very beginning of the development process. The shift-left approach also influenced how cloud workloads were provisioned with the proliferation of IaC. Amazon Web Services (AWS) introduced AWS CloudFormation in 2011, allowing users to define and manage their infrastructure using templates. In 2014, HashiCorp released Terraform, a tool that has since become one of the most widely used IaC solutions, allowing for the codification of infrastructure in a declarative manner. All these technologies contributed to ensuring that workloads were provisioned with security best practices from the beginning, and therefore contributed to a better overall security posture.

With this context in mind, we can all agree that security posture management is a preventative control, because it helps to improve the security posture, which reduces the likelihood of successful compromise of workloads. According to Microsoft Digital Defense Report 2022[1], effective security hygiene can protect against 98% of attacks. This is a very important number, because it means that if you have solid security posture management, are aligned with good governance, and are constantly improving your security hygiene, you are going to strengthen your cloud environment against most attacks.

Having said that, organizations also understand that it is important to operate with the *assume breach* mindset. The *assume breach* approach gained prominence around the early 2010s, although its exact origin as a term is not well-documented. This approach emerged as cybersecurity professionals began to recognize the limitations of traditional perimeter-based defenses and the inevitability of breaches. Microsoft has been a notable advocate of the *assume breach* approach, incorporating it into their security strategies and guidelines in the early 2010s. This advocacy has helped popularize the term within the industry. Around 2010-2011, the cybersecurity industry started to increasingly acknowledge that breaches were not just possible but likely. This shift in mindset was influenced by high-profile data breaches and **advanced persistent threats** (APTs).

With the assume breach mindset, it became imperative to not only have a strong posture management with CSPM but also to actively monitor cloud workloads and detect potential attempts to compromise them. Threat detection for cloud workloads becomes a reality with Cloud Workload Protection.

# Cloud Workload Protection

One of the major differences between Cloud Workload Protection (CWP) and other threat detection technologies such as **Intrusion Detection System** (IDS) is the variation in the threat landscape according to each type of cloud workload. For example, the threat landscape of a cloud container is not the same as the threat landscape of a cloud storage. Therefore, it becomes imperative that the analytics that are built to create detection for each workload are tailored for the needs of that specific workload.

CWP is a critical pillar in cloud security because it enables organizations to quickly identify potential attacks on their cloud workloads, while it equips **Security Operations Center** (SOC) teams to perform incident response. Rich threat detection aligned with a solid incident response can be the difference between identifying a threat at the beginning of the cyber kill chain (for example, during reconnaissance) to take measures that can stop the proliferation of the threat, and only identifying a threat after the threat actor was able to fully compromise the environment.

Over the years, cloud vendors started to align their threat detections with the MITRE ATT&CK (`https://attack.mitre.org/`) framework. This approach helps cloud administrators, security analysts, and incident responders understand which phase of the attack an alert is related to. The code below is extracted from a sample alert from Microsoft Defender for Cloud, specifically for the Defender for Containers threat detection. Notice that this alert has a field called `intent`, which has the value "InitialAccess".

This value represents the MITRE ATT&CK Initial Access (https://attack.mitre.org/tactics/TA0001) phase, which makes it easier for whoever is investigating this incident to understand the techniques that were potentially used in this attack.

```
Copied alert from Microsoft Defender for Cloud on 06/01/24, 09:07 AM (UTC-5)
https://portal.azure.com/#blade/Microsoft_Azure_Security_
AzureDefenderForData/AlertBlade/location/centralus/alertId/251685050089166
2950_058ad4df-ac35-4dd6-92ec-db17363e2062/referencedFrom/copyAlertButton
{
  "id": "/subscriptions/XXXXXXXXXXX/resourceGroups/Sample-RG/providers/
Microsoft.Security/locations/centralus/alerts/2516850500891662950_058ad4
df-ac35-4dd6-92ec-db17363e2062",
  "name": "2516850500891662950_058ad4df-ac35-4dd6-92ec-db17363e2062",
  "type": "Microsoft.Security/Locations/alerts",
  "properties": {
    "status": "Active",
    "timeGeneratedUtc": "2024-06-01T14:06:27.226Z",
    "processingEndTimeUtc": "2024-06-01T14:06:26.8337049Z",
    "version": "2022-01-01.0",
    "vendorName": "Microsoft",
    "productName": "Microsoft Defender for Cloud",
    "productComponentName": "Containers",
    "alertType": "SIMULATED_K8S_ExposedDashboard",
    "startTimeUtc": "2024-06-01T14:05:10.8337049Z",
    "endTimeUtc": "2024-06-01T14:05:10.8337049Z",
    "severity": "High",
    "isIncident": false,
    "systemAlertId": "2516850500891662950_058ad4df-ac35-4dd6-92ec-
db17363e2062",
    "intent": "InitialAccess",
    "resourceIdentifiers": [
      {
        "$id": "centralus_1",
        "azureResourceId": "/subscriptions/XXXXXXXX/resourceGroups/Sample-
RG/providers/Microsoft.Kubernetes/ConnectedClusters/Sample-Cluster",
        "type": "AzureResource",
        "azureResourceTenantId": "XXXXXXX-XXXXXXXXX"
```

```
      },
      {
        "$id": "centralus_2",
        "aadTenantId": " XXXXXXX-XXXXXXXXX ",
        "type": "AAD"
      }
    ],
    "compromisedEntity": "Sample-Cluster",
    "alertDisplayName": "[SAMPLE ALERT] Exposed Kubernetes dashboard
detected (Preview)",
    "description": "THIS IS A SAMPLE ALERT: Kubernetes audit log analysis
detected exposure of the Kubernetes Dashboard by a LoadBalancer service.\
nExposed dashboard allows an unauthenticated access to the cluster
management and poses a security threat.",
    "remediationSteps": [
      "Review the LoadBalancer service in the alert details. In case the
dashboard is exposed to the Internet, delete the LoadBalancer service
immediately and escalate the alert to the information security team."
    }
  }
```

CSPM and CWP are heavily utilized in the *protect, detect, and response* pillars. When you increase your security posture, you reduce the likelihood of successful compromise, which means you will likely have fewer threats to detect because your attack surface is more restricted. This will positively affect the SOC team, because they will have fewer alerts to triage, and they can invest more in proactive threat hunting in the environment.

Understanding this perspective that CSPM and CWP are different platforms, but should always work together, many vendors started to offer one single solution for CSPM and CWP. This was the case for Azure Security Center, which, since 2021, has been called **Microsoft Defender for Cloud (MDC)**. Since its origins back in 2015 when Azure Security Center was released in Public Preview, CSPM and CWP have always been part of the platform. The goal was always to improve the security posture while detecting threats against cloud workloads. Over time, the product became more mature and created a feedback loop that allows cloud administrators to learn from incidents and see which gaps must be filled in their security posture to avoid that same type of attack happening again.

Some workloads, such as VMs, may require a separate agent to be installed to be able to have deeper visibility, real-time threat detection, and response. While many organizations don't like to have an extra agent installed, the reality is that there are many functions that require an agent. For example, an agent can be used to analyze the behavior of applications and processes to identify anomalies that might indicate a compromise. This means that depending on the type of workload, the CWP platform may require the installation of an agent to provide better functionality and protection. For example, if VMs are very short-lived and may be reprovisioned every second day, and there is no publicly exposed workload running, an agentless approach might be sufficient. And that's why most CWP providers offer both solutions.

# Cloud Native Application Protection Platform

In less than a decade, cloud security technology grew from security posture management with CSPM to an amalgamation of many other platforms that were created to address specific issues within the cloud security space, such as **Cloud Infrastructure Entitlement Management (CIEM)**, which is focused on managing identities and their entitlements (permissions) within cloud environments. In addition to CIEM, other platforms started to proliferate, such as:

- **External Attack Surface Management (EASM)**: Focused on identifying, monitoring, and managing the external-facing digital assets of an organization.

- **Data Security Posture Management (DSPM)**: Focuses on managing and improving the security posture of an organization's data across various environments, including cloud workloads.

- **Vulnerability Assessment and Management (VAM)**: Focused on identifying, evaluating, prioritizing, and addressing security vulnerabilities within an organization's cloud or on-premises environment.

Organizations started to adopt these tools by using the rationale of adopting the best-of-breed strategy. While a best-of-breed strategy can provide some benefits in terms of performance and functionality, it also involves challenges such as increased complexity in integration, potential compatibility issues, and the need for skilled IT management to maintain and support a heterogeneous environment. In addition to that, the growth of multicloud adoption added even more challenges when it comes to managing all these tools in different dashboards, across different cloud providers.

In 2021, Gartner introduced the term **Cloud-Native Application Protection Platform (CNAPP)** to describe a new category of security platforms designed to provide comprehensive protection for cloud-native applications throughout their lifecycle.

The goal was to integrate various security functionalities, such as vulnerability management, compliance, runtime protection, and identity and access management, into a unified platform, aiming to address the complex security needs of cloud-native environments. In 2023, Gartner published the *Market Guide for Cloud-Native Application Protection Platforms*[2], which documents the architecture of a CNAPP solution, which includes elements shown in *Figure 1.3*:

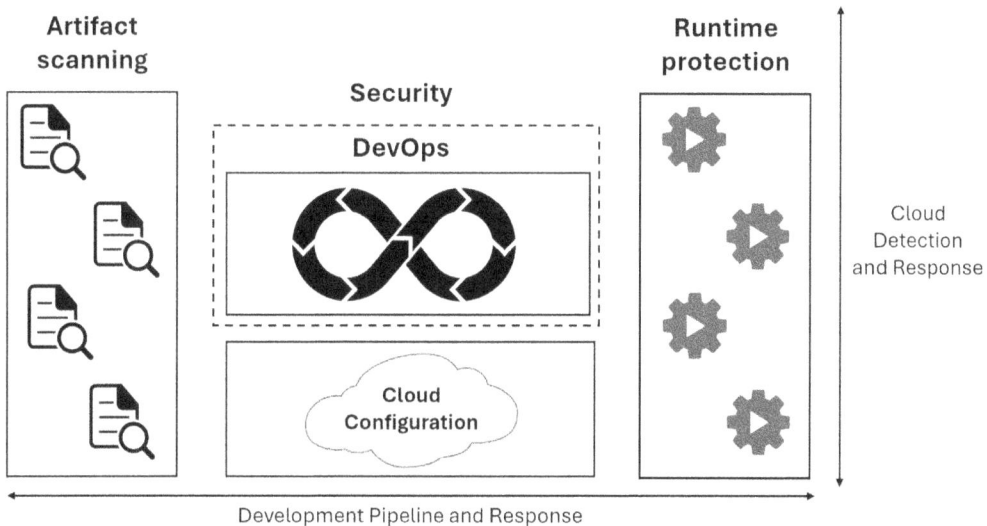

*Figure 1.3: CNAPP architecture*

As shown in *Figure 1.3*, a CNAPP solution must contain these major pillars, which start with *artifact scanning*. This component describes the platform's capability to scan different types of artifacts, including traditional workloads such as VMs, storage accounts, and containers, as well as code and **Application Program Interfaces (APIs)**. The insights generated by artifact scanning will help enhance the security posture of the DevOps lifecycle and take into account the different aspects of cloud configuration, which includes IaC. These components will also integrate with runtime protection, which contains CWP. As you evaluate which CNAPP vendor you will adopt, you must ensure that the vendor's solution is aligned with these components.

## Attack disruption

One of the main benefits of having the artifact scanning capability integrated with the other elements of this platform is the possibility of sharing and crossing information to allow a better understanding of the assets and using this information to prioritize risk mitigation.

The artifact scanning will generate a series of *insights* that can be leveraged by the platform. For example, the artifact scanning of a storage account may find the following *insights* about the storage account:

- **Access**: The storage account is widely accessible through the internet.
- **Permissions**: The storage account has a very permissive set of permissions.
- **Type of data**: The storage account contains sensitive information.

Upon having these insights, the CNAPP will perform an attack path analysis to identify potential areas of compromise, including the capability to move laterally across workloads. *Figure 1.4* has an example of what this looks like:

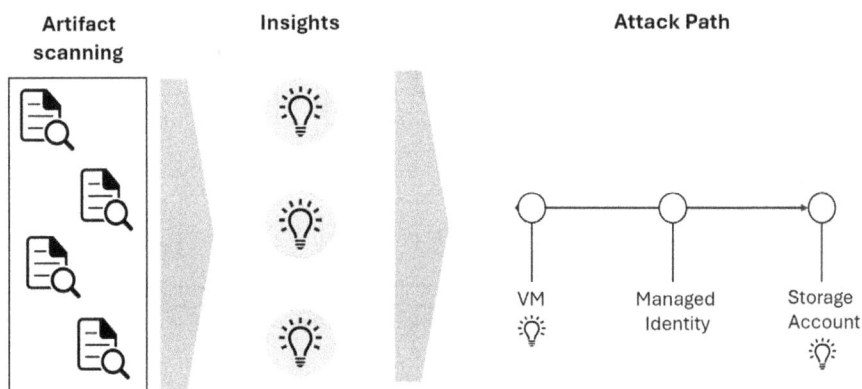

*Figure 1.4: Attack path*

The attack path shown in *Figure 1.4* has three workloads, a VM, a managed identity, and a storage account. The insights into these workloads will give more details about the potential attack. For example, in this case, the VM's insights show that this VM is exposed to the internet and has a series of vulnerabilities (CVEs) that were not patched and could be exploited by a threat actor. Once the threat actor gains access to this VM, it could authenticate to a managed identity that has permissions to a storage account. The insights from the storage account show that this storage account contains sensitive information.

Only CNAPP can provide this level of detail across multiple workloads (even if they are located in multiple cloud environments) due to the nature of the platform, which allows you to obtain intelligent insights from workloads that were scanned. CNAPP will analyze the correlation of these workloads with others, understand the potential attack, and show the results to you so you can take proactive measures to disrupt potential attacks.

This native CNAPP capability empowers cloud posture management teams to be more proactive and effective, and to prioritize what is truly important in their environment.

> The insights can also come from code, which means that the attack path can also highlight potential vulnerabilities in your code that can be used as the entry point for threat actors.

Attack path disruption also adds another KPI for organizations that want to track progress over time. Let's say that your company opened the CNAPP dashboard and saw 100 attack paths. Their ultimate goal is to drop the number of attack paths to zero. When tracking the attack path over time, you can also find out what the **time to resolution** (**TTR**) is for those attack paths. In other words, once the attack path appears, how long does it take to resolve?

This is an important metric to track because it will directly reflect on how effective the security posture of your cloud environment is. Here, you will have the opportunity to use the continuous improvement mindset to always chase a better metric and ensure that you are driving your team to be more effective when it comes to rapidly remediating critical issues. *Figure 1.5* has an example of an attack path over time, and the set of questions (A and B) that you must answer:

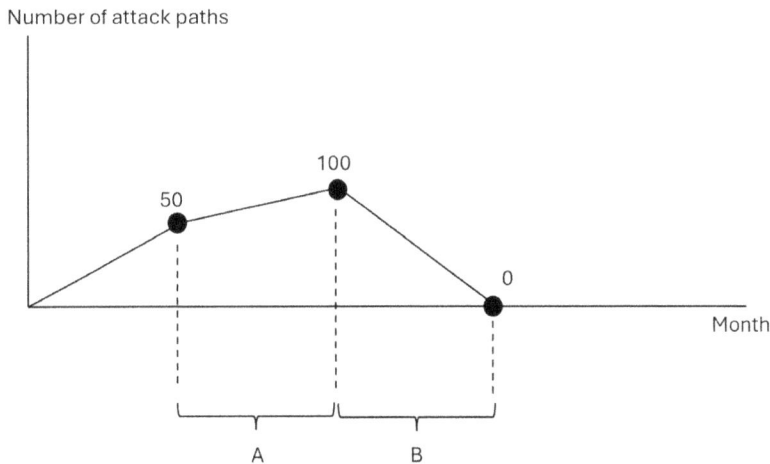

*Figure 1.5: Tracking an attack path*

In this case, some sample questions that could be asked by looking at this diagram are:

- **Set of questions A:**

  - How long did it take for the attack paths to grow from 50 to 100?

- Why did the attack path grow in this period instead of dropping?
- What were the lessons learned from this event?

- **Set of questions B**:
    - How long did it take for the attack paths to drop from 100 to 0?
    - What was the TTR for these attack paths?
    - How can we improve this TTR?

The answers to those questions will help your organization improve its security posture over time and keep fine-tuning the TTR for future attack paths.

> Attack disruption becomes even more critical when you are dealing with a multicloud scenario where threat actors could start their attack campaign in one cloud provider and pivot to another. This scenario is very difficult to identify without CNAPP.

# Agentless approach

The posture management side of CNAPP enables organizations to quickly obtain insights about workloads due to the agentless approach. When you onboard a cloud environment to use CNAPP, the onboarding process is frictionless because, by default, you will not need an agent just to obtain the initial security posture insights of a workload. This means that you don't need to wait until an agent is deployed to be able to obtain information about the workload's security posture. *Figure 1.6* summarizes the process:

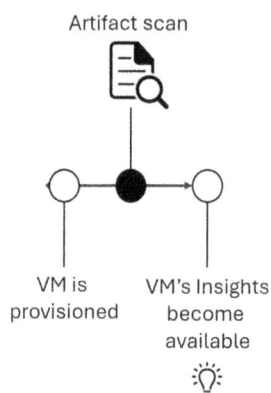

*Figure 1.6: Agentless process*

*Figure 1.6* shows a VM being provisioned and the artifact scan taking place to generate the insights. It is important to mention that different vendors may implement different methods to perform this artifact scanning for VMs upon provisioning. Regardless of the method running behind the scenes, the result is a faster onboarding process as you will have rapid access to key information from the VM, such as vulnerability assessment, software inventory, secret scanning, and potential malware.

> It is important to emphasize that despite the advantage of having an agentless approach for posture management, when you need deeper threat detection, you may need to install an agent according to the type of workload.

## Proactive hunting

While the term hunting is more often used in the context of threat hunting, which is more of a task done by the SOC team, CNAPP enables you to perform proactive hunting based on security posture information available about your workloads.

Once you have all workloads scanned and all insights created, you will not only have access to potential paths of attack but also access to the *big data* that was collected. This data contains the full inventory of your workloads and the security posture information of those workloads. With this information available, you can create queries that will give you even more information about different scenarios.

How this query will be executed depends on the CNAPP vendor. The Microsoft CNAPP solution, Defender for Cloud, enables you to perform visual queries using a feature called Cloud Security Explorer. *Figure 1.7* has an example of a query that returns all VMs that are vulnerable to the Log4Shell vulnerability and have an identity attached with permissions to a storage account.

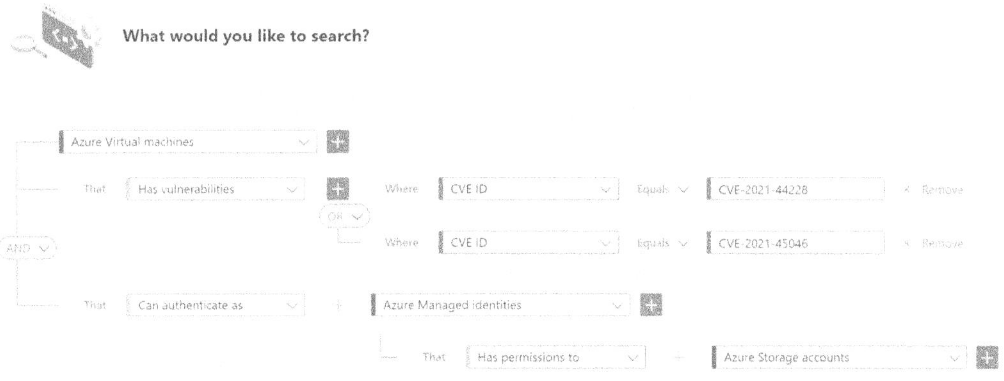

*Figure 1.7: Cloud Security Explorer*

Cloud Security Explorer is a functionality that will be covered in more detail in *Chapter 9* of this book.

## Alert enrichment

Although many of the capabilities that really highlight the power of CNAPP are related to posture management, there is a lot of value added when it comes to enhancements in CWP. Mainly because now, you can analyze the data from different angles. For example, you may see an attack path that has a VM that is exposed to the internet and has unpatched vulnerabilities and this VM has been attacked already. Notice that, in the same sentence, I included proactive elements (attack path) with reactive elements (has been attacked already). This is possible because the CWP is integrated with all other security posture modules of the overall CNAPP solution.

This data enrichment can also benefit the SOC team when they are triaging alerts, as they will have additional information that can help them prioritize how fast they need to respond. The data can be streamed to the **Security Information and Event Management (SIEM)** platform and the investigation will take place on the SIEM level where data ingestion from multiple data sources is taking place.

## Summary

In this chapter, we discussed the roots of cloud security with CSPM and CWP. We covered the traditional CSPM lifecycle, the use of secure score to track progress over time, and the challenges introduced with multicloud and shift-left. We also discussed the main aspects of CWP, the use of the MITRE ATT&CK framework to map alerts to different workloads, and the need to have agents for some types of workloads. The foundational knowledge obtained throughout this chapter will help you connect the dots about CNAPP and how Defender for Cloud implements those core capabilities.

We discussed how CNAPP was idealized, and the main advantages of using a CNAPP, which included attack disruption, agentless approach, proactive hunting, and SOC enrichment.

The next chapter is about accessing your environment security posture.

## Notes

1.  You can download this report from `https://query.prod.cms.rt.microsoft.com/cms/api/am/binary/RE5bUvv?culture=en-us&country=us`.

2.  You can download a copy of this report at `https://start.paloaltonetworks.com/gartner-market-guide-cnapp`.

# Additional resources

- Webinar with author Yuri Diogenes about an agnostic approach to CNAPP: `https://bit.ly/CNAPPBook1`

- Podcast with author Yuri Diogenes where he talks about CNAPP: `https://bit.ly/CNAPPBook2`

- Episode of *Defender for Cloud in the Field* with author Yuri Diogenes talking about CNAPP: `https://bit.ly/CNAPPBook3`

# Join our community on Discord

Read this book alongside other users. Ask questions, provide solutions to other readers, and much more.

Scan the QR code or visit the link to join the community.

`https://packt.link/SecNet`

# 2

# Assessing Your Environment's Security Posture

In every cloud migration, security considerations should be part of the planning from day one. If you introduce security considerations after your workloads migrate to the cloud, it's already too late and you most likely have workloads that are vulnerable to potential attacks.

But you can't protect what you don't know you have. This is the simplest principle of all when it comes to security hygiene. To elevate the security posture of your cloud environment, you first need to create an inventory of all your assets, and then evaluate their current security state. Based on this assessment's result, you will have a better understanding not only of the current areas of improvement, you will also be able to start prioritizing what needs to be resolved first.

By using the Foundational CSPM tier from Microsoft Defender for Cloud, you will be able to perform this initial security assessment without additional cost and establish initial metrics for security posture improvements over time.

This chapter covers:

- Planning your security posture assessment
- Adopting Foundational CSPM
- Improving your security posture

# Planning your security posture assessment

Now that you understand the benefits of adopting a CNAPP solution, you can start planning your CNAPP adoption. However, we all understand that in today's economy, saving costs while improving overall security is the goal to aim for. With that in mind, this book will first guide you through the security assessment of your environment by leveraging the free tier of Microsoft Defender for Cloud, called Foundational CSPM. By performing this initial assessment using the free tier of Defender for Cloud, you will be able to:

- Create an inventory of your cloud workloads
- Understand the security state of these workloads
- Evaluate the security state of your cloud workloads and establish metrics to improve this state

Once you have this information, you should be able to answer the following questions that will be important as you prepare for your CNAPP adoption:

- What types of workloads do I have in my cloud environment?
- What's the security state of these workloads?
- What's the security state of my cloud environment?
- What do I need to do to improve the security state of my cloud environment?
- Are there any modifications to the benchmarks or even exceptions needed?

These questions are critical to get started, and the adoption of Defender for Cloud Foundational CSPM will help you to answer those questions. Another benefit of performing these tasks before even starting to plan CNAPP adoption is that you can elevate your overall security posture from the beginning while taking notes of important considerations for the upcoming CNAPP adoption.

CNAPP adoption is performed in multiple phases. Each phase will have multiple tasks, and the major milestones of this adoption are shown in *Figure 2.1*:

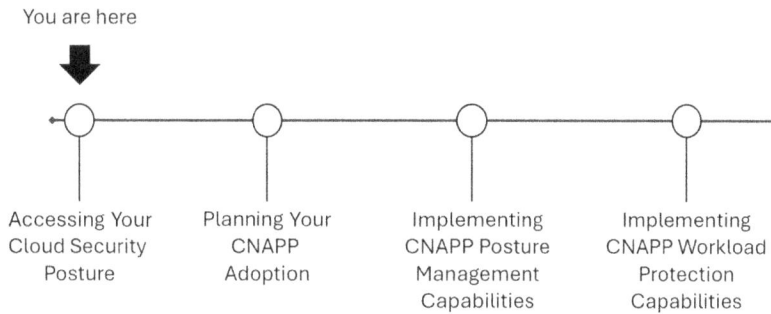

*Figure 2.1: CNAPP adoption milestones*

Although there are four major milestones, there will be many tasks in between these milestones. These tasks are covered in different chapters of the book, as below:

- **Accessing your cloud security posture**: This milestone is covered in this chapter.

- **Planning your CNAPP adoption**: The planning phase is covered in *Chapters 3* and *4*.

- **Implementing CNAPP posture management capabilities**: These capabilities are covered in *Chapters 5* to *8*.

  *Chapters 14* to *17* and *19* will also cover another set of posture management capabilities for CNAPP.

- **Implementing CNAPP workload protection capabilities**: These capabilities are covered in *Chapters 9* to *14*.

*Chapter 15, Incident Response*, and *Chapter 17, Extended Security Posture Management*, will cover additional operations that can be added once the CNAPP implementation is completed.

# Adopting Foundational CSPM

When you provision a new Azure subscription using the Azure portal, Defender for Cloud will automatically enable the free tier plan called Foundational CSPM. You can access the Defender for Cloud dashboard by accessing the Azure portal (`portal.azure.com`) and typing `Defender for Cloud` in the search box.

Once you open the Defender for Cloud dashboard, click **Environment settings** under the **Management** section, expand the management group, and click on the subscription, as shown in the example in *Figure 2.2*:

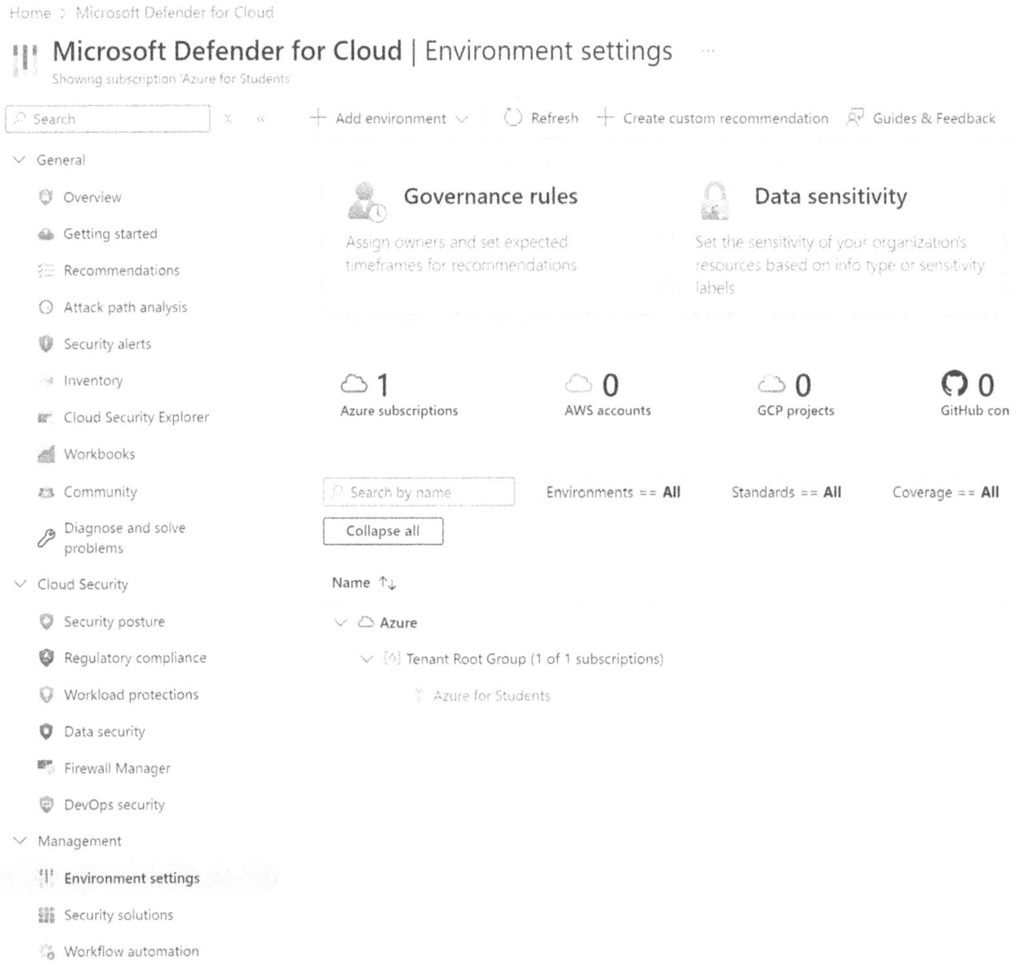

*Figure 2.2: Environment settings page*

The **Defender plans** page appears as shown in *Figure 2.3*, and there, you can see that Foundational CSPM is already enabled, showing **Full** under the **Monitoring coverage** column.

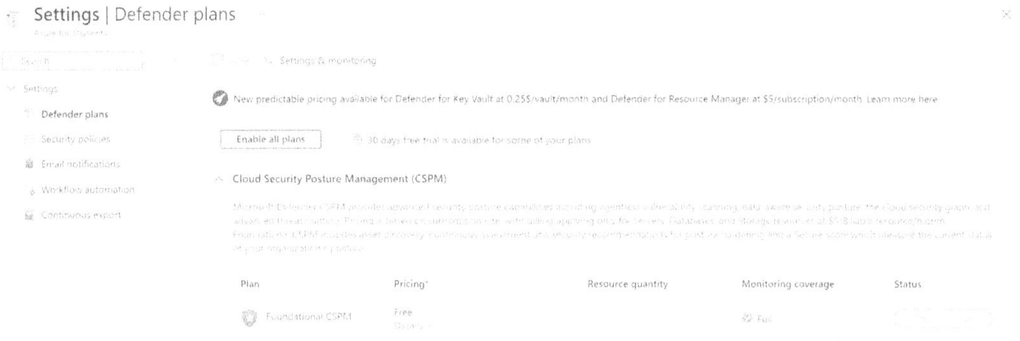

*Figure 2.3: Defender plans page (for better visualization, refer to https://packt.link/ gbp/9781836204879)*

To perform the security assessment of the environment, Defender for Cloud will utilize the **Microsoft Cloud Security Benchmark (MCSB)**[1], which is a multicloud security framework based on CIS, PCI, and NIST controls. Defender for Cloud continuously evaluates your hybrid cloud environment against these controls, which are part of the MCSB.

> Defender for Cloud's security standards are derived from Azure Policy initiatives or the Defender for Cloud native platform. Currently, Azure adheres to Azure Policy standards, while AWS and GCP follow Defender for Cloud standards.

After the initial assessment, Defender for Cloud will generate a series of security recommendations for you to address. The advanced feature of proactively prioritizing security recommendations based on the risk factors requires the enablement of the paid plan called Defender CSPM. However, as a starting point in your CNAPP adoption, these free recommendations can help you get started with understanding the gaps in your cloud environment.

To view these security recommendations, click the **Recommendations** option under the **General** section on the Defender for Cloud dashboard, and the **Recommendations** page will appear as shown in *Figure 2.4*:

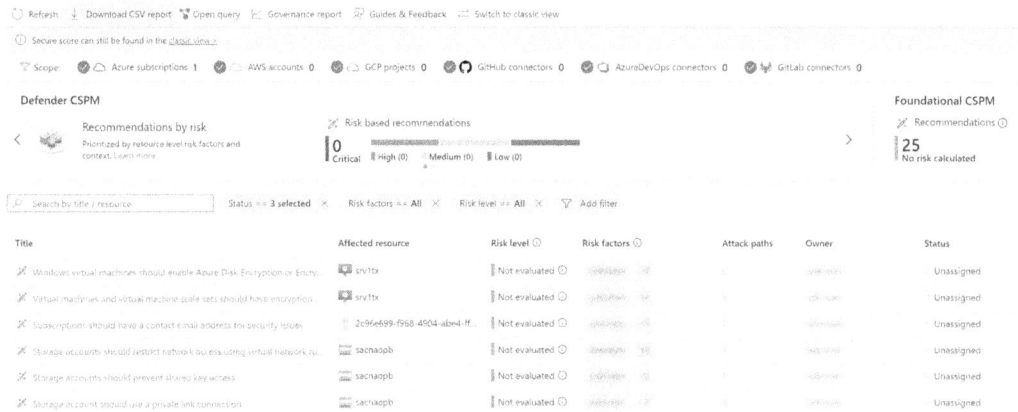

*Figure 2.4: Recommendations page (for better visualization, refer to https://packt.link/ gbp/9781836204879)*

Notice that some columns on this page, such as **Risk factors**, are blurry, and that's expected because you do not have the Defender CSPM plan enabled. On the top pane, you can also see that all Defender CSPM metrics are zeroed out, while Foundational CSPM shows the number of recommendations (in this case, 25).

You will learn more about risk factors and other elements of this page in *Chapter 5*, which covers Defender CSPM enablement and usage.

To better understand what needs to be done to improve your security posture using Foundational CSPM, click the **Switch to classic view** button on the top navigation pane and the classic **Recommendations** page will appear, as shown in *Figure 2.5*:

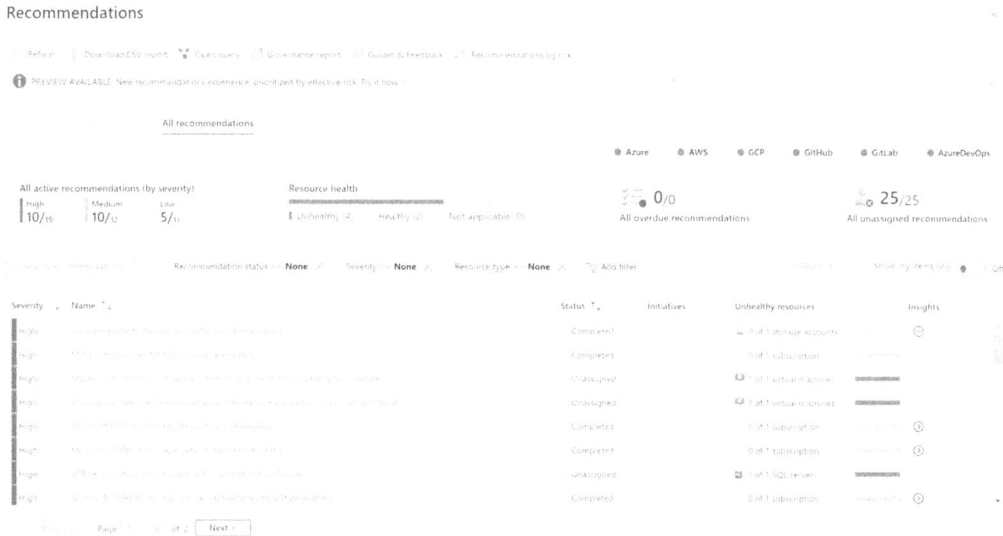

*Figure 2.5: Classic recommendations page (for better visualization, refer to https://packt.link/gbp/9781836204879)*

The recommendations now appear in top-down order based on the level of criticality (High, Medium, and Low). Keep in mind that the level of criticality in Foundational CSPM doesn't take into consideration the risk factors of the workloads. This advanced approach to evaluate risk is a feature that is only available when Defender CSPM is enabled in the subscription.

Here is where things can get a bit overwhelming, because you may have an environment with hundreds of high-severity recommendations, and without the proper risk context, it can become challenging to decide which one you should address first. But remember, this is exactly why you initially use this free tier plan: to identify your gaps and better understand your environment. In other words, this is not your final operational state; this is the pre-CNAPP adoption assessment of your cloud environment.

To help you decide which high-severity recommendations you should address first within the static context provided by Foundational CSPM, you can follow different approaches, as shown in the table below:

| Approach | Steps | Rationale |
|---|---|---|
| Number of affected resources | <ul><li>Briefly review (no need to open the recommendation at this point) the titles of each recommendation.</li><li>Look at the *unhealthy resources* column to understand how many resources are affected by this recommendation.</li></ul> | In this case, the goal is to prioritize the high-severity recommendations that are affecting more resources. |
| MITRE ATT&CK Framework | <ul><li>On the *Recommendations* page, add a new filter for *Tactics* (see *Figure 2.6*).</li><li>Add only the tactics that you want to address first. For example, if you want to remediate all security recommendations that are related to the Initial Access phase of the MITRE ATT&CK Framework, then select only *Initial Access* in the *Value* field.</li></ul> | You can create different campaigns to address security recommendations based on the MITRE ATT&CK framework, focusing on the left-to-right approach where the goal is to first prioritize the security recommendations from the beginning of the cyber kill chain (starting with initial access). |
| Resource type | <ul><li>On the *Recommendations* page, add a new filter for *Resource type*.</li><li>Select the resource type that you want to prioritize.</li></ul> | In this case, you may want to prioritize all resources that can store data, such as Storage Account and SQL databases. |

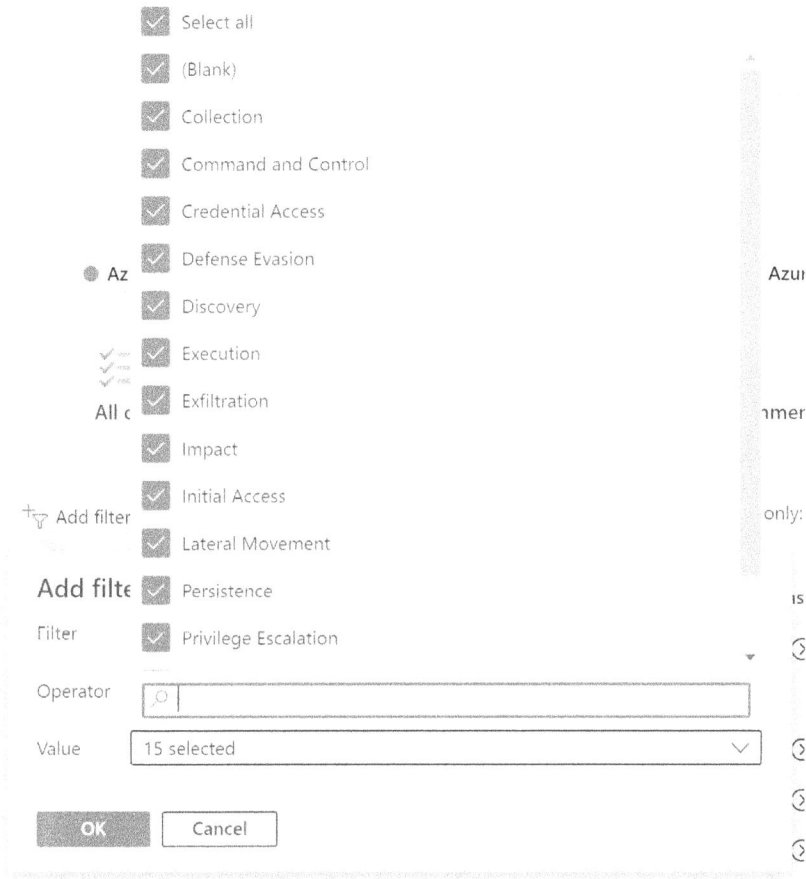

*Figure 2.6: Filtering by MITRE ATT&CK tactics*

After you identify which approach is better for you, you can start opening each recommendation to see more details about the remediation steps. It is important to mention that some recommendations can be quickly remediated from the Defender for Cloud dashboard if you have the right level of privilege to do so.

During this exercise of identifying which recommendations you should address first, make sure to reflect upon the following points:

- Why is this recommendation getting triggered?
- Is this recommendation relevant to my environment?

- Is it possible to create guardrails to avoid this type of resource being provisioned if this security recommendation is present? In other words: provision the resource using secure configuration by default.

    - This is an important reflection because there are many security recommendations that can be prevented from being triggered by applying the proper Azure policy to disallow a user from provisioning a resource (deny action) if not using security standards.

    - However, this discussion will likely involve different stakeholders because cloud governance is often performed by a different group and this group may have its own reasons to allow this type of unsecured provisioning to occur.

    - Although this is a debatable choice, since as a security practitioner you will always advocate for a more secure setting, there are legacy systems that may be in production that do not support a more secure setting.

    - Make sure to not enforce deny policies before a common agreement among the different groups has been reached, as this may cause downtime for the end users.

- Who are the stakeholders that I need to involve to remediate this recommendation?

    - This reflection may trigger a series of different discussions with different groups within your organization because, most likely, you as a cloud security administrator will not have the right level of privilege across all workloads to perform this remediation. This means you need to identify the resource owner and assign ownership of this recommendation to the proper owner.

As you can see, there is a lot of reflection that can be done in this initial review of the security recommendations. This is an ongoing task that gets refined every time that you go through this exercise.

# Remediation

Each security recommendation is different from another in many ways because there are different variables that are unique for each workload. Let's use as an example the recommendation "Azure SQL Database should be running TLS version 1.2 or newer," shown in *Figure 2.7*:

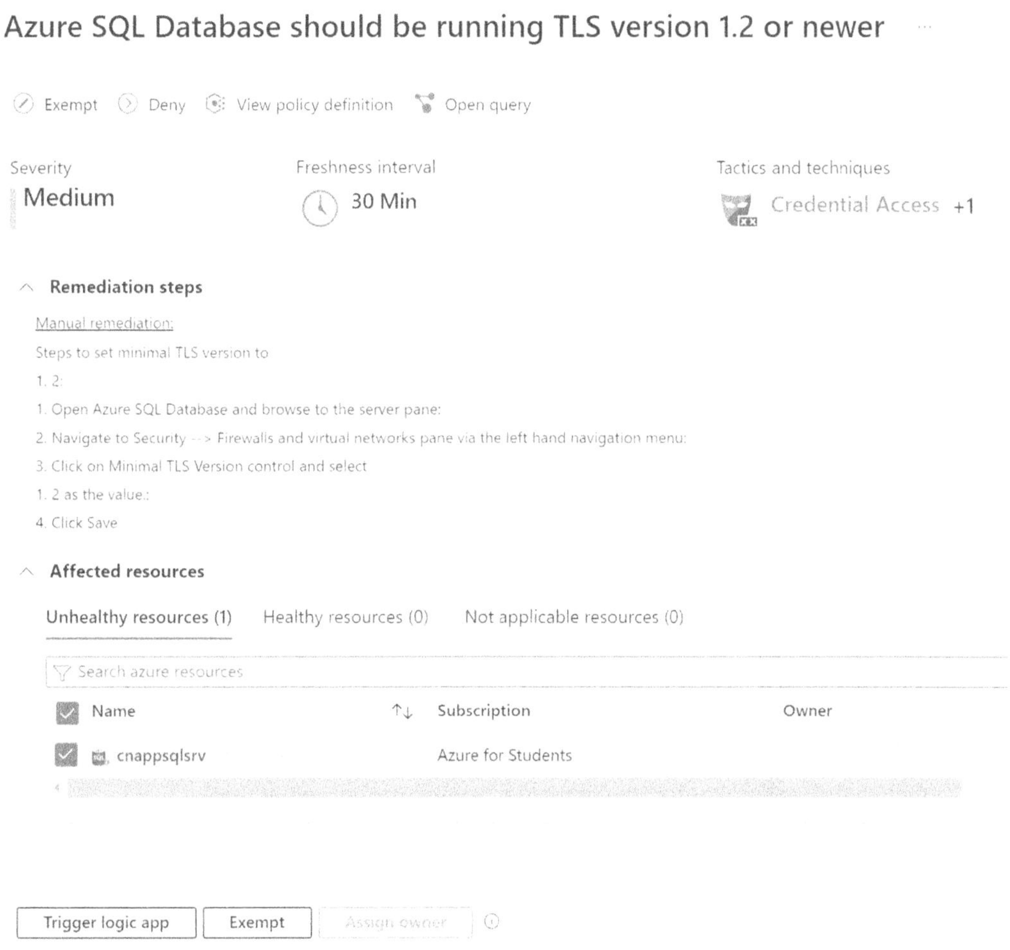

## Azure SQL Database should be running TLS version 1.2 or newer   ...

⊘ Exempt    ⊘ Deny    ⊙ View policy definition    ⋎ Open query

Severity                   Freshness interval                        Tactics and techniques
**Medium**                 ⏱ 30 Min                                  🐾 Credential Access  **+1**

∧   **Remediation steps**

Manual remediation:

Steps to set minimal TLS version to

1. 2:

1. Open Azure SQL Database and browse to the server pane:

2. Navigate to Security --> Firewalls and virtual networks pane via the left hand navigation menu:

3. Click on Minimal TLS Version control and select

1. 2 as the value.:

4. Click Save

∧   **Affected resources**

Unhealthy resources (1)    Healthy resources (0)    Not applicable resources (0)

▽ Search azure resources

| ☑ | Name | ↑↓ | Subscription | Owner |
|---|------|-----|--------------|-------|
| ☑ | 🗄 cnappsqlsrv | | Azure for Students | |

| Trigger logic app | Exempt | Assign owner | ⓘ

*Figure 2.7: Medium severity recommendation*

Let's start with the options available at the top of this recommendation, which will also vary according to the security recommendation and workload. The options available in the first two top bars are:

- **Exempt:** This button enables you to create an Azure Policy exemption for this recommendation applying to the resource that is selected (in this case, the *cnappsqlsrv* resource at the bottom). You will learn more about resource exemption in *Chapter 8*.

- **Deny:** This button enables you to create a deny effect in Azure Policy for this recommendation. The deny effect prevents the creation of resources that don't satisfy the recommendation. The deny effect can be in the scope of the subscription or management group. As mentioned previously, do not create a deny effect in Azure Policy before having a common agreement with the team responsible for Azure governance.

- **View policy definition**: This button will open the policy definition in JSON format (see *Figure 2.8*) so that you can see more about how the assessment is being done in detail.

- **Open query**: This button will lead you to Azure Resource Graph, which allows you to perform open queries using the **Kusto Query Language** (**KQL**) format.

- **Severity**: This option reflects the current severity of this recommendation.

- **Freshness interval**: The number that appears in this field reflects the refresh cycle of this recommendation. For this example, the freshness interval is 30 minutes, which means that every 30 minutes there will be a refresh of this recommendation. This is an important piece of information because when you remediate a security recommendation, you can have the right expectation about when the recommendation will appear as resolved, which in this case is when it disappears from the list.

- **Tactics and techniques**: This field reflects the mapping of MITRE ATT&CK framework tactics and techniques for this particular security recommendation.

Azure SQL Database should be running TLS version 1.2 or newer

*Figure 2.8: Policy definition (for better visualization, refer to https://packt.link/ gbp/9781836204879)*

The body of the recommendation will also vary, but the core fields are always the same, which are:

- **Description:** This is a more detailed explanation about the security recommendation.

- **Remediation steps:** It is the process to remediate the recommendation step by step. Sometimes, there will be no step-by-step remediation because the remediation needs to be done outside of the box, for example, by visiting a vendor's page and applying a patch.

- **Affected resources:** A list of the resources that are affected by this security recommendation.

In the case of this recommendation (*Figure 2.7*), you don't have the option to fix it directly from this dashboard (see the example shown in *Figure 2.9* of a recommendation that has the **Fix** button available). However, if you have created an Azure Logic Apps automation to remediate this recommendation, you can click the **Trigger logic app** button to call the logic app that you created. In this case, the **Assign owner** button is not available because it uses the Governance feature, and this feature is only available when Defender CSPM is enabled in the subscription.

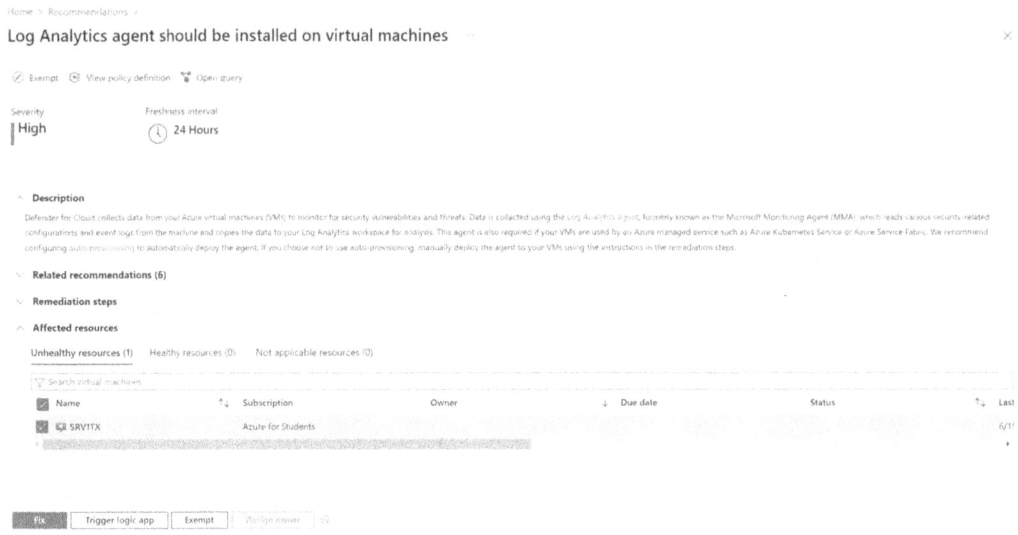

*Figure 2.9: Recommendation with the Fix button available (for better visualization, refer to https://packt.link/gbp/9781836204879)*

## Secure score

After reviewing the initial assessment, addressing some of the recommendations, and reflecting on important points to follow up, you can now look at your secure score to have better visibility of the overall security posture of your cloud environment. To access your secure score from the classic **Recommendations** page, click the **Secure score recommendations** tab and the page shown in *Figure 2.10* will appear:

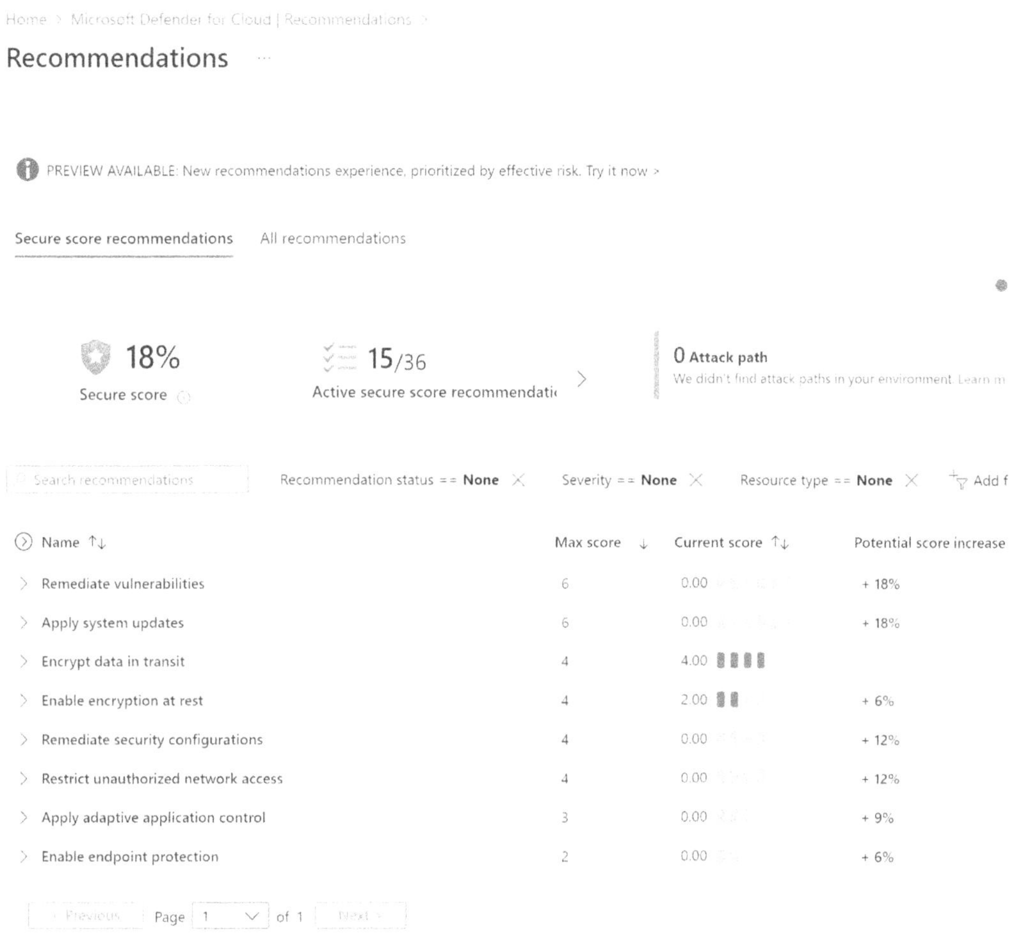

Home > Microsoft Defender for Cloud | Recommendations >

## Recommendations

PREVIEW AVAILABLE: New recommendations experience, prioritized by effective risk. Try it now >

**Secure score recommendations**    All recommendations

🛡 **18%**
Secure score

**15**/36
Active secure score recommendati

**0** Attack path
We didn't find attack paths in your environment. Learn m

Search recommendations | Recommendation status == **None** ✕ | Severity == **None** ✕ | Resource type == **None** ✕ | Add f

| Name ↑↓ | Max score ↓ | Current score ↑↓ | Potential score increase |
|---|---|---|---|
| Remediate vulnerabilities | 6 | 0.00 | + 18% |
| Apply system updates | 6 | 0.00 | + 18% |
| Encrypt data in transit | 4 | 4.00 | |
| Enable encryption at rest | 4 | 2.00 | + 6% |
| Remediate security configurations | 4 | 0.00 | + 12% |
| Restrict unauthorized network access | 4 | 0.00 | + 12% |
| Apply adaptive application control | 3 | 0.00 | + 9% |
| Enable endpoint protection | 2 | 0.00 | + 6% |

Previous | Page 1 ∨ of 1 | Next >

*Figure 2.10: Secure score*

In the example shown in *Figure 2.10*, the secure score is 18%, and the goal is always to reach 100%. The secure score is not a test grade, where if you have 70% it means you pass. As a matter of fact, 70% means that 30% of your cloud environment is still vulnerable in one way or another. Keep in mind that "vulnerable" in this context doesn't mean open software vulnerabilities, because a vulnerability in this context may mean a specific setting that is not configured in a secure manner.

Notice in *Figure 2.10* that there is a banner showing **0 Attack path**. This does not mean that you don't have an attack path in your environment. Attack Path is a Defender CSPM feature, and since this plan is not enabled in the subscription, you will not see an attack path.

To increase your secure score, you need to focus on the security controls that are listed on the **secure score** page. The organization of these security controls is done in a top-down approach, where the top controls will have a bigger impact on your secure score. Let's use, as an example, the first security control shown in *Figure 2.10* and expanded in *Figure 2.11*:

| Name ↑↓ | Max score ↓ | Current score ↑↓ | Potential score increase ↑↓ |
|---|---|---|---|
| ⌄ Remediate vulnerabilities | 6 | 0.00 | + 18% |
| Machines should have a vulnerability assessment solution | | | |
| Machines should have vulnerability findings resolved | | | |

*Figure 2.11: Remediate vulnerabilities security control (for better visualization, refer to https:// packt.link/gbp/9781836204879)*

This security control has two security recommendations, and once you remediate these recommendations and completely address this security control, you will have a potential increase of 18% in your secure score. In other words, to have a positive impact on your secure score, you must remediate all recommendations that belong to each security control in top-down order.

When a security control is partially resolved, such as in the example of *Figure 2.12*, you will see a number assigned to the **Current score** column. In the example below, the number is 4. You will also see the status as **Completed**, showing that this assessment passed and is compliant.

| ⌄ Encrypt data in transit | 4 | 4.00 | ∘ Completed |
|---|---|---|---|
| Secure transfer to storage accounts should be enabled | | | ∘ Completed |
| Azure SQL Database should be running TLS version 1.2 or newer | | | ∘ Unassigned |

*Figure 2.12: Remediate vulnerabilities security control (for better visualization, refer to https:// packt.link/gbp/9781836204879)*

As you continue to remediate your security recommendations, you should progressively see an improvement in your secure score. However, don't expect these changes to immediately reflect in your secure score. As mentioned previously, each recommendation has its own refresh interval, which will dictate when the recommendation will be considered resolved/completed, which directly affects the freshness of your secure score. Once the secure score is updated, you will also be able to see it on the **Overview** page in the main Defender for Cloud dashboard, under the **Security posture** tile, as shown in *Figure 2.13*:

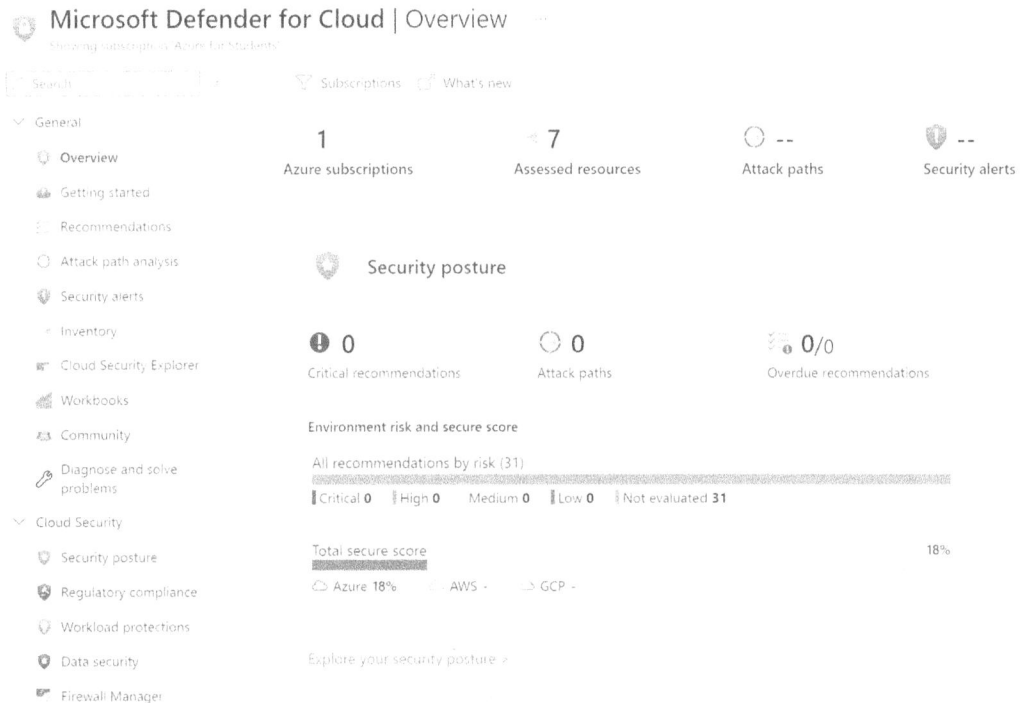

*Figure 2.13: Security posture tile*

Notice in the above secure score that there are some metrics that are zeroed out, and as mentioned before, this happens because these metrics are collected when Defender CSPM is enabled.

When you click on this tile, you will see more details, as shown in *Figure 2.14*:

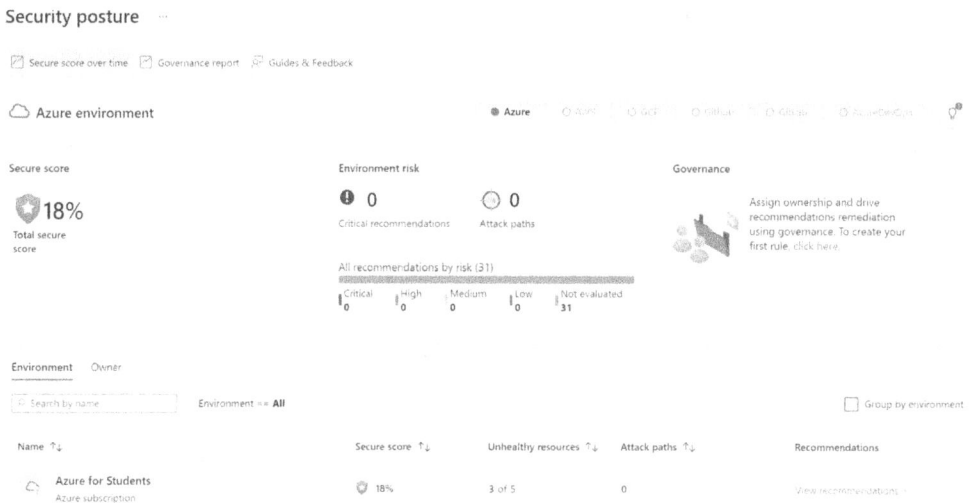

*Figure 2.14: More details about your security posture (for better visualization, refer to https:// packt.link/gbp/9781836204879)*

# Improving your security posture

By now, you already understand that to improve your overall security posture using Foundational CSPM, you will need to remediate the recommendations by focusing on the security controls available and using a top-down approach. By doing so, you will see the continuous improvement of your secure score.

When working on security posture improvement over time, it is important to keep track of the secure score progress. Defender for Cloud has a built-in workbook called *Secure score over time*, which is accessible via the **Security posture** page (see the button in the top-left corner in *Figure 2.14*). However, this workbook depends on the collection of secure score snapshots over time and the storage of these snapshots in a Log Analytics workspace. These settings are not configured by default because it incurs extra charges to store the data in the Log Analytics workspace. However, if you want to track progress, which is highly recommended, then you need to configure this option using the **Continuous export** feature. This feature is available from **Environment settings**, **Subscription properties**, and then **Continuous export**, as shown in *Figure 2.15*:

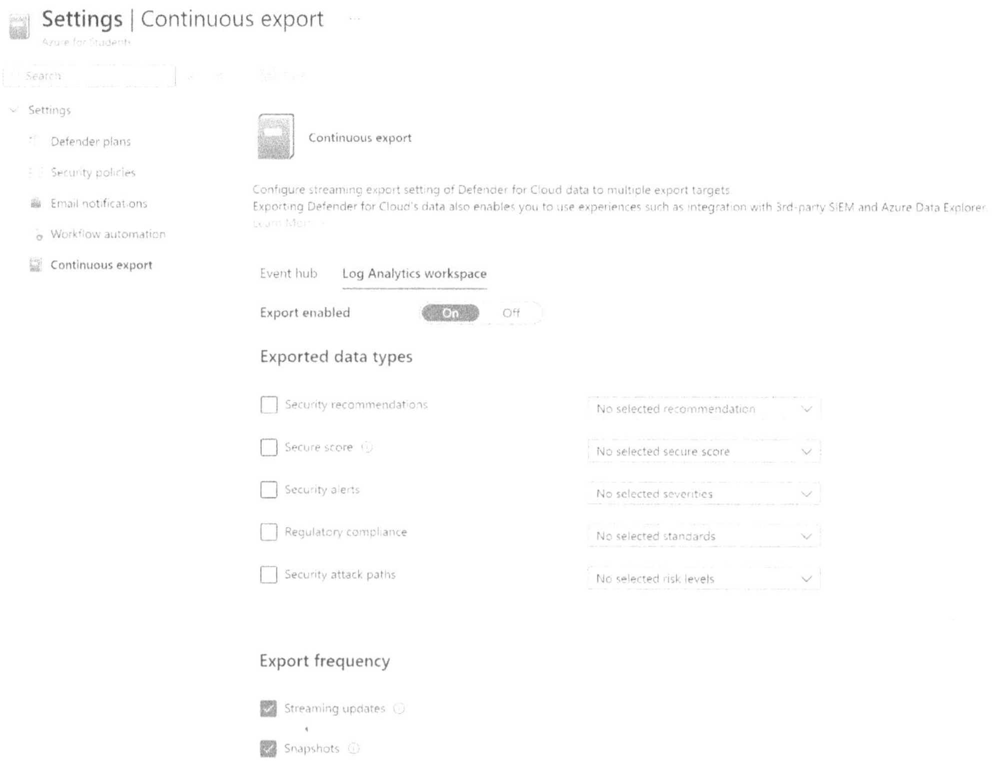

*Figure 2.15: Continuous export (for better visualization, refer to https://packt.link/ gbp/9781836204879)*

As shown in *Figure 2.15*, you can export different data types, but in this case, you can start by only selecting the secure score with the options shown in *Figure 2.16*:

*Figure 2.16: Exporting the secure score*

In the second part of the screen, shown in *Figure 2.17*, you have the option to select the resource group, the subscription, and the target Log Analytics workspace that will receive the data.

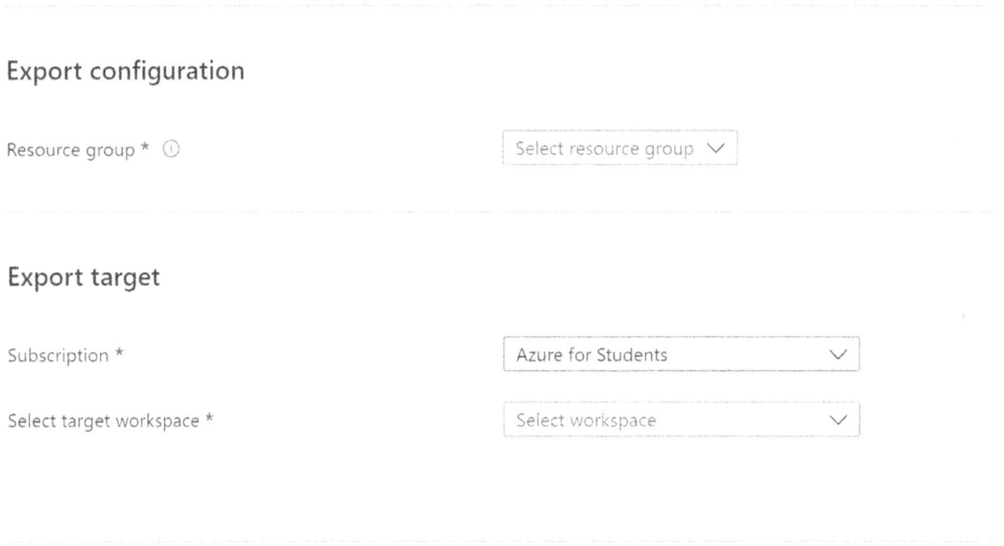

## Export configuration

Resource group * ⓘ                                       Select resource group ⌄

## Export target

Subscription *                                           Azure for Students            ⌄

Select target workspace *                                Select workspace              ⌄

*Figure 2.17: Additional configuration*

Once you save these settings, the *SecurityCenterFree* solution will be enabled on the target workspace in order to be able to export data to it.

After this configuration is enabled, you will be able to track your secure score progress. Some of the metrics available in this workbook once the data is populated include trends (shown in *Figure 2.18*) and secure score value over time (shown in *Figure 2.19*):

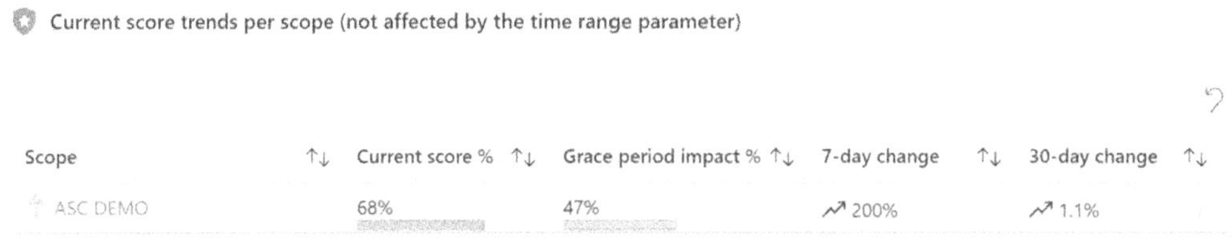

⭐ Current score trends per scope (not affected by the time range parameter)

| Scope | ↑↓ | Current score % ↑↓ | Grace period impact % ↑↓ | 7-day change | ↑↓ | 30-day change | ↑↓ |
|-------|----|--------------------|--------------------------|--------------|----|---------------|----|
| ⚲ ASC DEMO | | 68% | 47% | ↗ 200% | | ↗ 1.1% | |

*Figure 2.18: Secure score trend analysis*

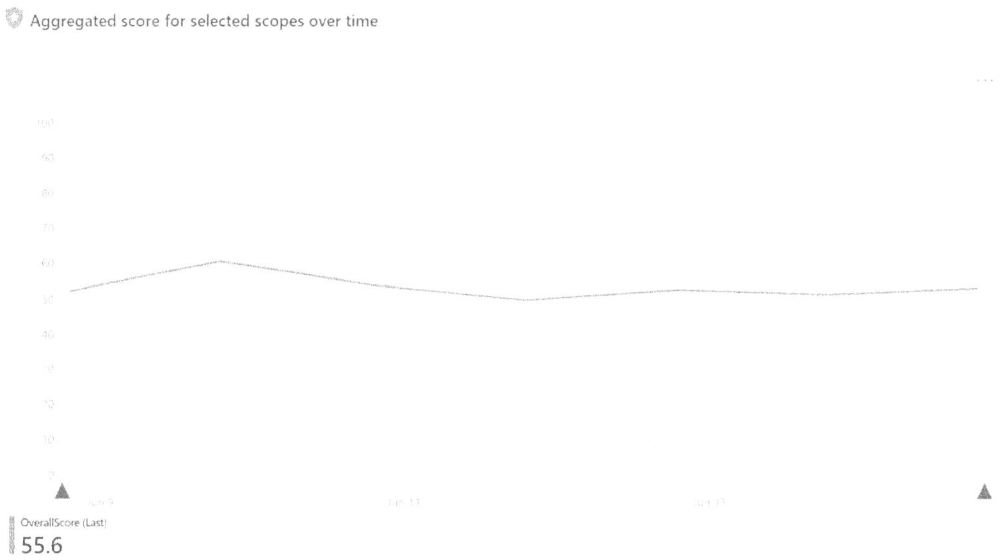

Aggregated score for selected scopes over time

OverallScore (Last)
55.6

*Figure 2.19: Secure score over time (for better visualization, refer to https://packt.link/ gbp/9781836204879)*

You should use this data to evaluate your progress and to better understand areas of improvement. For example, in *Figure 2.19*, you can see that there is an increase in the score between January 9 and January 10, and then a drop around January 11. This can help you to investigate what happened on January 10 that led to this drop.

## Microsoft Cloud Security Benchmark (MCSB)

As mentioned earlier in this chapter, Defender for Cloud recommendations are based on the MCSB. While the previous approaches shown in this chapter are ideal for prioritizing which recommendations should be remediated first and having a direct reflection on secure score improvement, you can also look at your recommendations through a different set of lenses.

By leveraging the MCSB visualization available in the **Regulatory compliance** dashboard, you can view the recommendations based on MCSB controls. Usually, this approach is used for compliance purposes, when you need to visualize which items are compliant according to MCSB standards. In other words, this may not be something that you as a Cloud Administrator will use, but you may have a compliance team in your organization that will benefit from viewing the data using this approach.

To access this visualization, click the **Regulatory compliance** option, under the **Cloud Security** section in the Defender for Cloud dashboard, and the **Regulatory compliance** page will appear as shown in *Figure 2.20*:

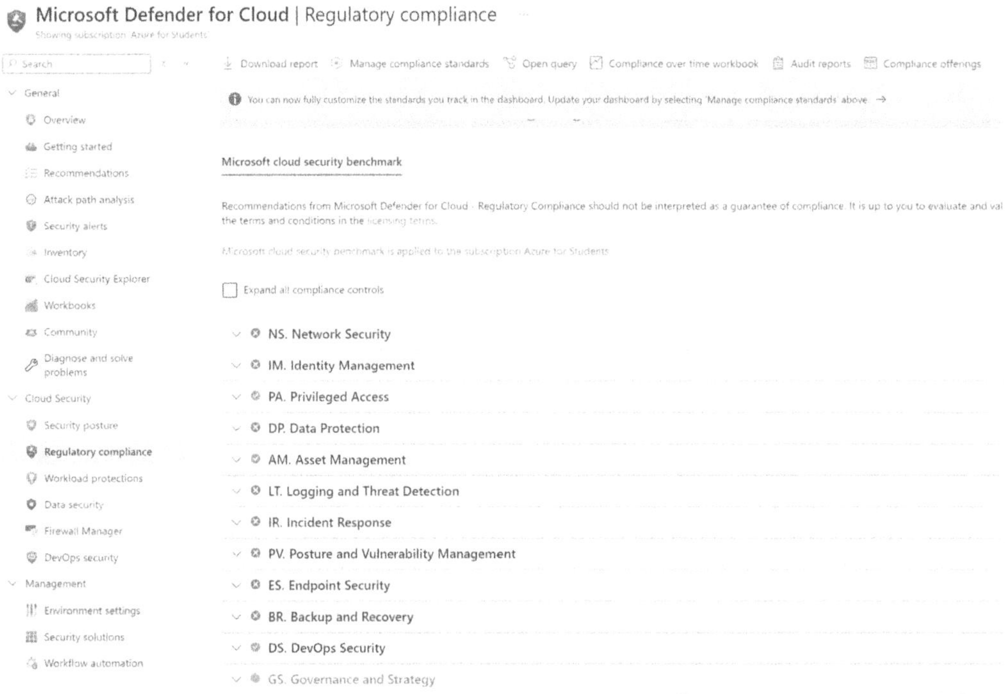

*Figure 2.20: MCSB visualization (for better visualization, refer to https://packt.link/ gbp/9781836204879)*

The MCSB presents 12 controls, and each control aggregates a set of security recommendations that are relevant to that control. For example, the **Network Security** control contains security recommendations that evaluate how secure and protected networks are, which includes safeguarding virtual networks, establishing private connections, preventing and mitigating external attacks, and securing DNS.

If you expand one of these controls, you will see the enumeration of the controls (NS 1, NS 2, and so on), and at the end of each line, you may see **MS** (which means this control is Microsoft's responsibility) or **C** (which means it is the customer's responsibility). There will be cases where you will see both, as shown in "NS-1. Establish network segmentation boundaries" (see *Figure 2.21*), which means that this control is a shared responsibility between Microsoft and the customer.

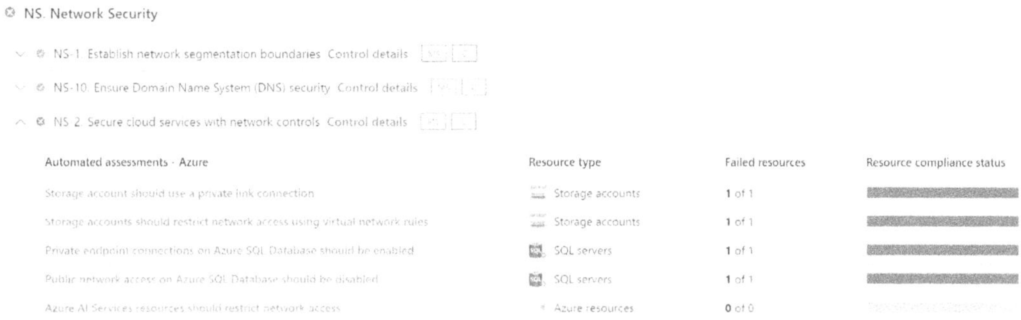

NS. Network Security

NS-1. Establish network segmentation boundaries  Control details

NS-10. Ensure Domain Name System (DNS) security  Control details

NS-2. Secure cloud services with network controls  Control details

| Automated assessments - Azure | Resource type | Failed resources | Resource compliance status |
|---|---|---|---|
| Storage account should use a private link connection | Storage accounts | 1 of 1 | |
| Storage accounts should restrict network access using virtual network rules | Storage accounts | 1 of 1 | |
| Private endpoint connections on Azure SQL Database should be enabled | SQL servers | 1 of 1 | |
| Public network access on Azure SQL Database should be disabled | SQL servers | 1 of 1 | |
| Azure AI Services resources should restrict network access | Azure resources | 0 of 0 | |

*Figure 2.21: Individual controls (for better visualization, refer to https://packt.link/ gbp/9781836204879)*

If you click on the **Control details** hyperlink available on each control, you will see more details about what is considered your responsibility, as shown in the example in *Figure 2.22*:

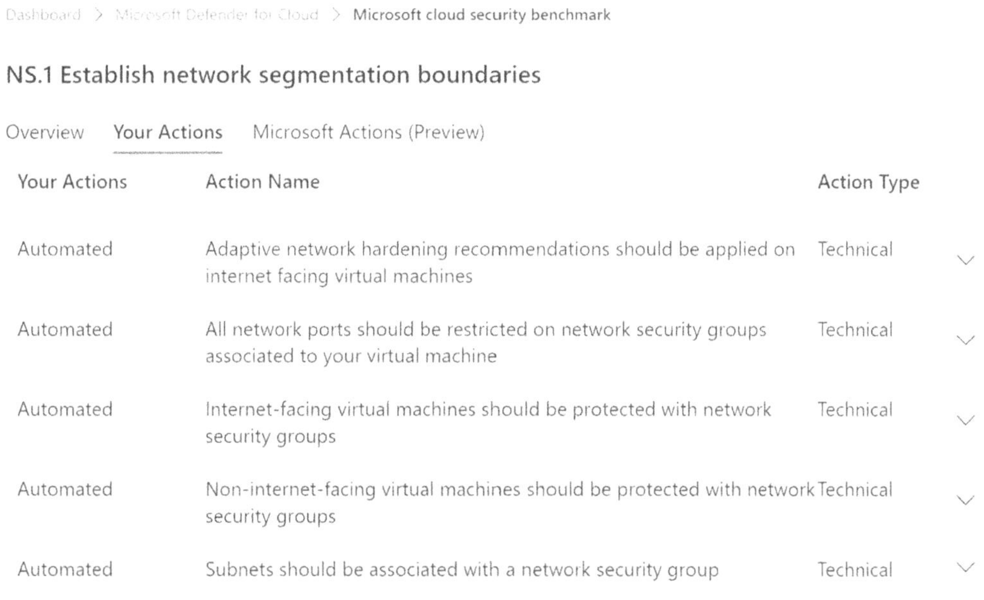

Dashboard > Microsoft Defender for Cloud > Microsoft cloud security benchmark

## NS.1 Establish network segmentation boundaries

Overview    **Your Actions**    Microsoft Actions (Preview)

| Your Actions | Action Name | Action Type | |
|---|---|---|---|
| Automated | Adaptive network hardening recommendations should be applied on internet facing virtual machines | Technical | ∨ |
| Automated | All network ports should be restricted on network security groups associated to your virtual machine | Technical | ∨ |
| Automated | Internet-facing virtual machines should be protected with network security groups | Technical | ∨ |
| Automated | Non-internet-facing virtual machines should be protected with network security groups | Technical | ∨ |
| Automated | Subnets should be associated with a network security group | Technical | ∨ |

*Figure 2.22: Customer's responsibility*

You can also download this report in PDF format (see the sample in *Figure 2.23*) by using the **Download report** button available in the **Regulatory compliance** dashboard (see the button in *Figure 2.20*):

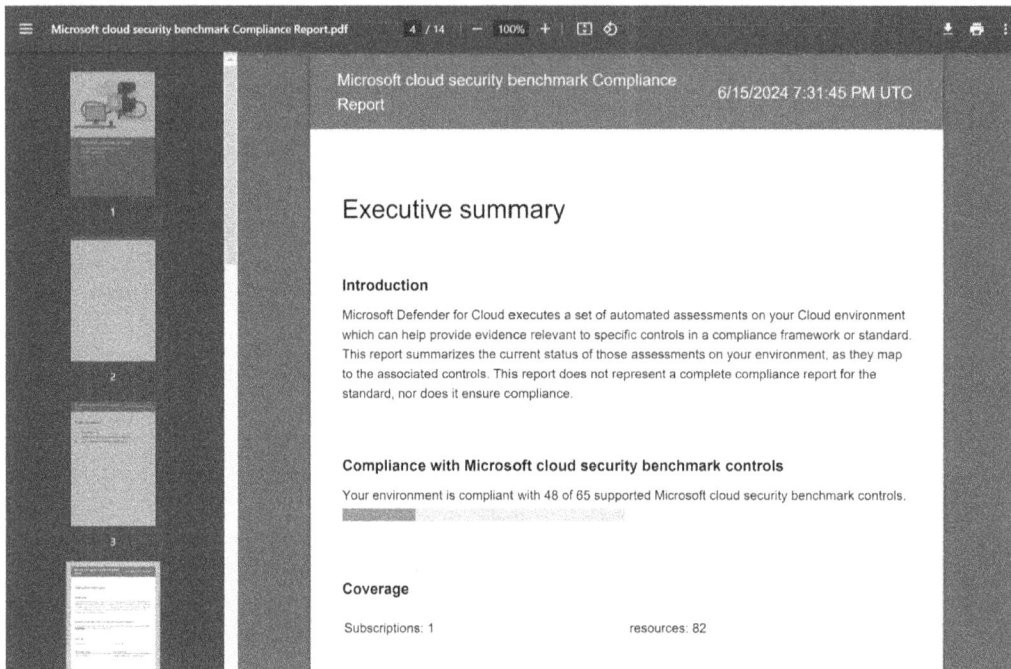

*Figure 2.23: Regulatory Compliance Report*

> To see other industry standards such as GDPR, PCI, NIST, HIPAA, and others mapped into the **Regulatory compliance** dashboard, you will need to have at least one paid plan enabled in your subscription.

## Inventory

After going through all security recommendations, prioritizing what is important, and addressing recommendations to improve your secure score, you should have a better understanding of your environment. This is the goal of this initial exercise using the free tier Foundational CSPM. You could even call this phase zero of your CNAPP adoption.

As of now, you have the complete inventory of your assets, which you can also view using the **Inventory** dashboard, available under the **General** section of the Defender for Cloud dashboard (see the example in *Figure 2.24*).

Microsoft Defender for Cloud | Inventory

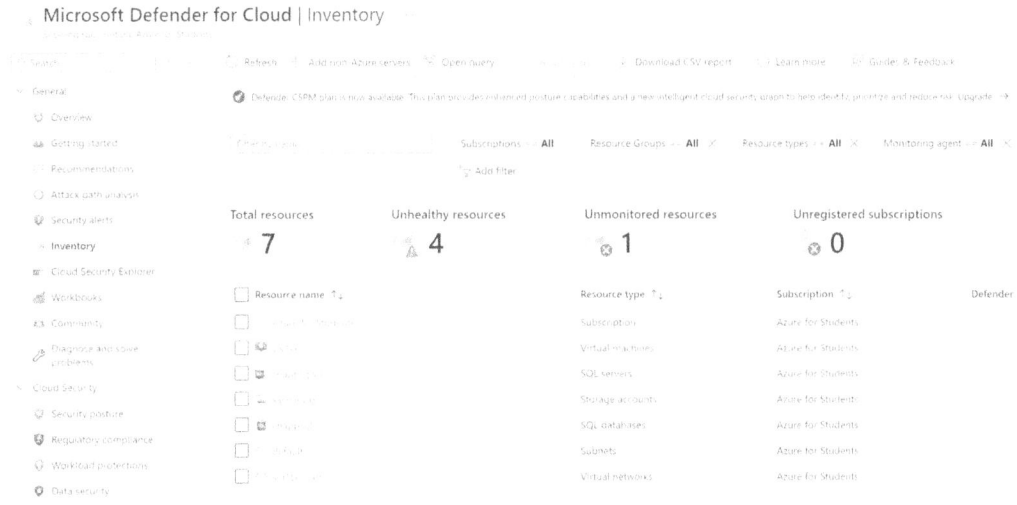

*Figure 2.24: Inventory dashboard (for better visualization, refer to https://packt.link/ gbp/9781836204879)*

You can use this inventory visualization as part of your final reflection before moving on to the CNAPP planning phase. For example, before moving on, you may revisit the inventory to take a deeper look at each individual resource. Let's say that you want to have a deeper look only at the SQL database that appears in this inventory. Once you click on it, you will open the **Resource health** page (see the example in *Figure 2.25*), which shows all the security recommendations that are linked to this particular resource.

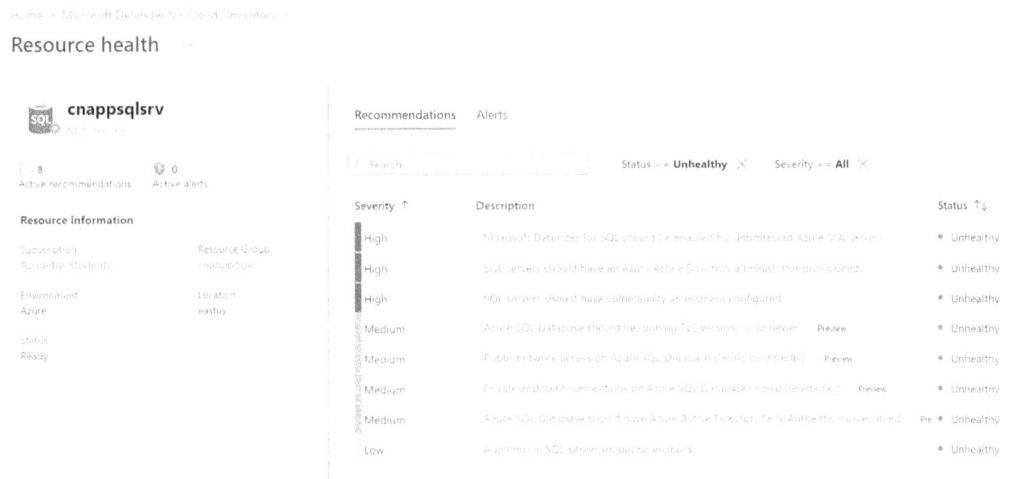

*Figure 2.25: Resource health page (for better visualization, refer to https://packt.link/ gbp/9781836204879)*

This final look can be a good opportunity to better understand the security state of each resource, and revisit important questions such as who the owner of this resource is. This is always an important question when it comes to security posture improvement—this is a team effort. You can't enhance the security posture of your cloud environment on your own; you must engage others, delegate assignments, and establish metrics to track progress.

An important realization at this point is that Foundational CSPM is a great start, not only because it is free but also because it gives a rich set of data for you to understand your current cloud security state. In other words: always use Foundational CSPM across all your cloud environments! However, as mentioned in *Chapter 1,* CNAPP is way beyond just a static security assessment using industry standards. While the actions executed in this chapter are a good start, you need to continue to evolve, and the next step is to plan your CNAPP adoption.

## Summary

In this chapter, we discussed the adoption of Defender for Cloud Foundational CSPM to start assessing your environment's security posture. You learned how to use security recommendations to visualize areas of improvement for your workloads, how to remediate security recommendations, and the importance of using the secure score to improve the overall security posture.

We also discussed how to track the secure score over time, the use of the MCSB to have a different view of your recommendations and adhere to compliance standards, and how to access your cloud inventory using Defender for Cloud.

In the next chapter, you will learn more about key CNAPP considerations.

## Notes

1.  For more information about the MCSB, visit `https://bit.ly/CNAPPBook4`.

## Additional resources

- Learn more about secure score calculation: `https://bit.ly/CNAPPBook5`
- Check out this episode of *Defender for Cloud in the Field* with author Yuri Diogenes talking about security posture improvements in Defender for Cloud: `https://bit.ly/CNAPPBook6`
- Author Yuri Diogenes explains the use of Secure Score in Azure Security Center (the previous name for Defender for Cloud): `https://bit.ly/CNAPPBook7`

# Join our community on Discord

Read this book alongside other users. Ask questions, provide solutions to other readers, and much more.

Scan the QR code or visit the link to join the community.

`https://packt.link/SecNet`

# 3

# CNAPP Design Considerations

After understanding the need for CNAPP and performing the initial assessment of your environment, you are ready for the next step in your CNAPP adoption journey. Before diving into more details on CNAPP planning, you need to create your own design considerations guide, which is a document that lists critical factors, best practices, and recommendations to be considered before implementing CNAPP.

The CNAPP design process is well-informed, efficient, and aligned with the desired outcomes. Throughout this process you will define the goals, objectives, and requirements of the project. You will also review the scope, which includes any limitations or constraints.

Throughout this process, you will establish guiding principles for the overall design and architecture of your CNAPP adoption. The design considerations also include the capability to support integration with other systems, as well as scalability to accommodate future growth and increased demand.

This chapter covers:

- Establishing design principles
- Design considerations for posture management
- Design considerations for DevOps security
- Design considerations for workload protection

# Establishing designing principles

Before you start planning your CNAPP adoption, it is important to establish several design principles that will guide the implementation to ensure effective security, scalability, and maintainability. These principles will be based on the CNAPP capabilities that you learned in *Chapter 1*, and the understanding of your environment that you gained in *Chapter 2*.

## Zero Trust

While this is not mandatory per se, it is always a good practice to include Zero Trust as part of your design principle for every technology that you want to adopt. By ensuring that the principles of Zero Trust are included in the foundation of every CNAPP adoption phase, you help to improve the overall security posture and protection of your workloads.

Consider adopting Zero Trust principles across all technologies that are part of CNAPP. These principles are:

- **Verify explicitly**: Make sure to always validate all available data points, which includes user identity and location, device health, data classification, and workload context.
- **Use a least privilege access approach**: Ensure that you always grant the minimum levels of access or permissions necessary.
- **Assume breach**: By assuming a breach has already occurred, your organization can better prepare for and respond to security incidents, reducing the impact and recovery time.

## Shift-left security

Shift-left security is a proactive approach that integrates security practices early in the **software development lifecycle (SDLC)**. This is a design principle for CNAPP, since the goal is to identify and address security issues as early as possible, which contributes to vulnerability reduction and reduces the cost of remediations.

Consider the use of automated security testing at every stage of the development process, including unit testing, integration testing, and system testing. Evaluate if your CNAPP solution has **static application security testing (SAST)**, **dynamic application security testing (DAST)**, and **interactive application security testing (IAST)** to identify vulnerabilities early. If these capabilities are not natively available, evaluate the possibility of integrating with third-party solutions that offer these capabilities and are able to share insights with the CNAPP solution.

# Data protection

Another critical design principle for CNAPP is data protection. While the term "data protection" can be broad, there are some important fundamentals that must be part of the design principles, such as:

- **Discoverability**: data discoverability is a big problem because you can't protect the data you don't even know you have. Ensuring that data can be seamlessly discovered across multiple cloud providers is imperative.

- **Data classification**: while it is not expected that CNAPP will replace dedicated data classification solutions, it is a design principle for CNAPP to be able to identify different data types, perform simple classification, and use this insight to help prioritize risks.

- **Exposure**: ensure that network configuration, access permissions, and configured data flows associated with data resources are visible to the CNAPP solution.

While the three previous principles are more relevant to data security posture management, it is also important to understand that data can be stored in different types of workloads, such as databases and storage accounts. This means that you must ensure that the CNAPP solution also has threat detection available for the workloads that will store data.

> Keep in mind that data protection also varies, according to the data state and encryption mechanism that's used. In other words, you have to understand the data protection while data is at rest (on the end user's device), in transit, and at rest on the server side (or in storage).

# Comprehensive visibility and monitoring

According to *SANS 2023 Multicloud Survey: Navigating the Complexities of Multiple Cloud*[1], sponsored by Microsoft, 86% of the organizations interviewed said that they already used a multicloud approach for their workloads. When you have a scenario like this, centralized visibility in one single dashboard of all workloads distributed across multiple cloud providers is extremely important. Visibility becomes the first step because once you have resources available in a single dashboard, you then need to effectively monitor these workloads from this single dashboard.

Ensure that your CNAPP solution has the capability to easily onboard multicloud resources, providing the level of visibility necessary to manage all these resources from a single place.

# Dynamic threat detection and response

As covered earlier in the Zero Trust section, assuming breach is one of the Zero Trust principles, and once you assume breach, you need to have proper threat detection to quickly identify an attack and rapidly respond.

Make sure the CNAPP solution has threat detections that are tailored based on the type of workload you are protecting. The CWP that is embedded in the CNAPP must have analytics that are able to detect different types of attack, according to the workload's threat landscape. For example, the threat landscape of a virtual machine is not the same as a storage account.

# Compliance and governance

While the security posture management team of your organization may only be monitoring the security posture of cloud workloads, the compliance team may need to see how these workloads are compliant according to industry standards. This means that the CNAPP solution needs to be able to evaluate the data that was gathered about these workloads and map this assessment according to industry standards. Ensuring that the CNAPP solution supports compliance with relevant regulations and standards (e.g., GDPR, HIPAA, PCI-DSS) is imperative. You should also be able to implement governance frameworks to manage and enforce compliance requirements.

# Design considerations

Now that you have established your core design principles, you need to look inward to your organization's needs and how the CNAPP adoption will lead your organization to achieve its goals in the context of cloud security. Design considerations look at important factors and constraints that must be taken into account before planning CNAPP adoption. These considerations ensure that the final design meets its intended purpose effectively and efficiently while adhering to relevant requirements and constraints. The following section explores the design considerations for the three major CNAPP pillars.

Cloud security is a very broad area that will touch different workloads that are usually managed by different teams. For this reason, it is important to have stakeholders from these different teams as part of this design consideration exercise. Consider putting together a team that can have at least the following roles:

- **Project coordinator**: responsible for coordinating the execution of necessary tasks in your organization.

- **Technical leader**: responsible for the technical expertise of the different cloud providers used by your organization, and able to coordinate the technical tasks required for the project.

- **Stakeholders from different teams**: decision-makers and influencers from other teams that will have workloads managed by CNAPP and are able to provide input to success criteria.

The sections that follow will look at design considerations for the following CNAPP pillars and their subcomponents:

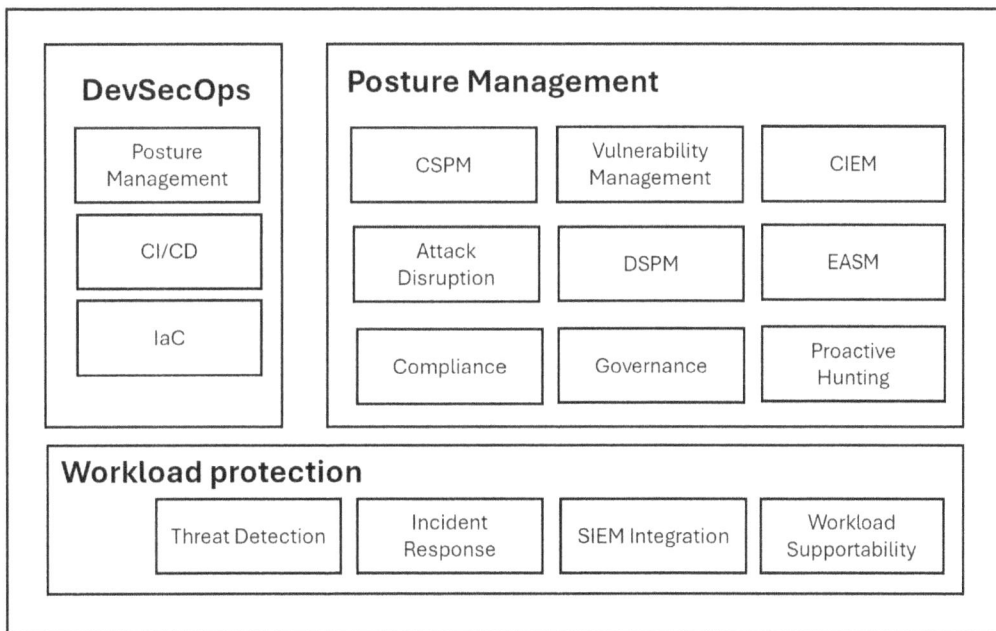

Figure 3.1: CNAPP pillars and their subcomponents

# Design considerations for posture management

With the team assembled, now you need to start from the foundational requirements to the more specific ones. If you consider cloud security posture foundational requirements as platform requirements, you first need to answer the following questions:

- What cloud providers does your organization use right now, and which one do they plan to use in the future?

- The answer to this question will determine if the CNAPP solution offered by the vendor will even make the cut or not. For example, if your organization has workloads in **Google Cloud Platform (GCP)** but the CNAPP vendor doesn't support GCP, then they are likely to be off the list of potential providers. The reason I say *likely* is because if the CNAPP provider doesn't immediately have this capability, but it is in their near-term roadmap, then you may want to consider it.

- What regulatory compliance standards do the workloads of your organization need to adhere to?

  - Regulatory compliance requirements vary according to the industry, and they also vary from country to country. If your organization has workloads that must be compliant with certain standards and they want to use CNAPP to monitor the level of compliance, you need to ensure that the required standard is supported by the CNAPP solution.

- How does your organization manage vulnerability remediation across different workloads, and who is responsible for this remediation?

  - Vulnerability assessment and management is a key part of cloud security posture management. In a cloud environment, where there are multiple workload owners, governance plays a big role in ensuring that these workloads have their vulnerabilities remediated promptly. For that, the CNAPP solution should have the capability to assign governance rules to workload owners and establish a **service-level agreement (SLA)** for remediation.

These three foundational considerations will dictate a big part of the posture management conversation around CNAPP adoption. Next, as you move to the top, you will establish specific posture management requirements. The list below has some specific requirements that you should consider during this design process:

- **Agentless**: to reduce onboarding friction and provide faster insights for supported workloads, the CNAPP solution must have agentless posture management capabilities. Ask the following questions:

  - What types of workloads support agentless capability?
  - How does the agentless capability collect the data from the workloads, and where is the data stored?

- **Vulnerability scanning**: as mentioned before, vulnerability assessment is an important part of CNAPP, and it should be available across all supported clouds and workloads. Also, the vulnerability assessment results must be natively used as insights to help prioritize risk remediation. Ask the following questions:

  - What types of workloads support vulnerability scanning?
  - How are these vulnerabilities exposed in the dashboard?

- **Attack disruption**: while an attack path is a pivotal part of CNAPP, it is important to consider the attack path experience provided by the CNAPP vendor. Ask the following questions:

  - How easy is it to use the attack path?
  - What's the level of insight provided in the attack path?
  - Is the attack path integrated with threat detection?

- **Query insights**: the data collected by the artifact scanning capability of CNAPP should be available for cloud administrators to open a query, to do, for example, proactive hunting. Ask the following questions:

  - How easy is it to query the data?
  - Does it require learning a new query language? (This is an important question for preparation purposes.)
  - What type of data can be queried?
  - Is it possible to query data across all cloud providers that the CNAPP is connected with from a single dashboard?

- **Risk prioritization**: the capability to evaluate all insights, including data security posture management, and create a list of security recommendations that need to be remediated, based on different risk factors. Ask the following questions:

  - What are the risk factors that are taken into consideration to prioritize a risk?
  - Is it possible to exempt security recommendations that are not applicable to my environment or are considered false positives?
  - Is it possible to automate the remediation of security recommendations?
  - Is it possible to integrate the security recommendation with the **IT Service Management (ITSM)** system used in our organization?

- **Data security posture management (DSPM) capability**: ensure that the CNAPP has the capability to perform data scanning, perform basic data classification, and use these insights to prioritize data risk across different workloads. Ask the following questions:

    - Data discoverability is supported on which types of workloads?

    - What type of classification is performed?

    - Is this classification customizable?

- **Native integration with Cloud Infrastructure Entitlement Management (CIEM)**: ensure that the CNAPP has the capability to leverage CIEM to help manage and control user access and entitlements in their cloud infrastructure. Ask the following questions:

    - What type of insights are generated by this CIEM integration?

    - Where are the insights shared?

    - Is this a true native integration, or does it require additional licensing?

- **Native integration with External Attack Surface Management (EASM)**: just like with CIEM, the integration with EASM should be embedded in the CNAPP solution, which helps to identify possible external attack vectors across your cloud asset. Ask the following questions:

    - What type of insights are generated by this EASM integration?

    - Where are the insights shared?

    - Is this a true native integration, or does it require additional licensing?

Once you have answered all these questions, you should get back to your team to evaluate the answers and see whether there are any constraints in your environment that will be affected by them. For example, if the answer to the question *"Data discoverability is supported on which types of workloads?"* is *"Microsoft SQL Server only"* and your organization has data in open-source SQL databases, then you have a constraint in your environment that will block this CNAPP solution from being adopted.

> Keep in mind that this can be a very extensive exercise, and it is critical that you do it accurately because it will determine the success of your CNAPP adoption.

# Design considerations for DevOps security

DevOps Security, or DevSecOps, is an essential part of CNAPP. But when it comes to design considerations, you will need to step back and think about how this will affect the culture and collaboration across teams. You must foster a culture where security is everyone's responsibility, from developers to operations to security teams. You need to encourage collaboration between these teams to ensure seamless integration of security practices. This also means providing ongoing security training for developers, operations, and security teams to keep them informed about the latest threats and security best practices.

The design considerations for DevOps security in the context of CNAPP must cover the following areas:

- **CI/CD integration**: defines how CNAPP integrates into **continuous integration and continuous deployment (CI/CD)** pipelines to automate security checks. The first step is to assess which CI/CD platforms are currently used by the organization and whether there are plans to either consolidate or adopt a new CI/CD platform. Ask the following questions:

    - What CI/CD platforms are used in the organization today?

    - Are there any plans to adopt a single platform? If so, which one?

    - Are there any plans to expand to other platforms? If so, which ones?

    Based on the answers to these questions, you can follow up with, what CI/CD platforms are natively supported by the CNAPP solution?

- **Infrastructure as Code (IaC)**: you must assess the environment to see if IaC is used to automate the deployment of secure infrastructure, and to ensure consistent security configurations. Ask the following questions:

    - Does the organization utilize IaC? If so, which ones (Terraform, AWS CloudFormation, etc.)?

    - Which IaC capabilities are available in the CNAPP solution?

    - How is IaC source control done today?

    - Source control in this context means how the organization stores IaC scripts in a version control system (e.g., Git) to track changes, enable collaboration, and maintain history.

    - How is sensitive data managed in IaC?

- The idea is to identify if the current practices allow hardcoding sensitive information like passwords or API keys in IaC scripts.

- How is IaC automation currently implemented?

The intent is to understand how the organization is automating IaC deployments using CI/CD pipelines to ensure consistent and repeatable deployments.

- **Visibility into DevOps security posture**: this is an area that you need to investigate if the level of visibility applies to both developers and security teams. Ask the following questions:

  - How do developers and security teams receive insights about the DevOps security posture?

  - Is there any tool in use that enables developers and security teams to have a centralized view of the DevOps security posture?

- **Security posture management controls**: an initial assessment to understand how the organization discovers and remediates risky misconfigurations in the DevOps platform is going to be necessary. Once you have this information, you can ask the following questions:

  - Does the CNAPP solution support access to scoped secrets?

  - Does the CNAPP solution prevent unauthorized executions and potential escalations?

  - Does the CNAPP solution provide branch protection?

While these are important questions to ask, it is commonplace that as you go through these questions, you may find peculiar aspects that apply to your environment only. Make sure to take notes and evaluate the constraints and the options available to address those constraints.

## Design considerations for workload protection

The first step in designing considerations for workload protection is to understand the workloads that you have in your cloud environment. During *Chapter 2*, when you did the cloud security assessment, you should have identified the different types of workloads that exist in your cloud environment. Ensure that you revisit this list, and then start asking the questions below for each type of workload. For example, if your environment has VMs, containers, and SQL databases, you will need to go over the questions below for each one of those workloads:

- **Threat detection**: defines the type of threat detection available in the CNAPP solution. Some threat detections are rule-based, while others are based on machine learning, which usually requires more time to learn about an environment before triggering any alert. Ask the following questions:

  - Were the threat detections for workload X (replace X with the type of workload that you are asking about) created based on the threat landscape for this type of workload?

  - What types of detections (rule-based, machine learning, AI-based, etc.) does the CNAPP solution have for workload X?

  - Is it possible to validate this threat detection (i.e., trigger an alert) to see the raw data generated by the alert?

  - Is it possible to suppress alerts that are considered false positives?

  - Does the platform perform automatic correlation of alerts across multiple protected workloads?

- **Incident response**: one of the main aspects of a CNAPP solution is to empower SOC teams to better respond to incidents; this means you will need to evaluate your current SOC operations and see how they can benefit from the CNAPP adoption. Ensure that, during this step, you have representatives from the SOC team to help you establish design considerations that are relevant to their team. In addition, ask the following questions:

  - Does the CNAPP solution have mechanisms to trigger an automated response if an alert is triggered?

  - Does the CNAPP solution allow the creation of playbooks to automate incident detection?

  - Does the CNAPP solution bridge the gap between incident response and posture management by offering lessons-learned insights from an incident, to improve the workload security posture and reduce the likelihood that the same type of incident will happen again in the future?

  - Does the CNAPP solution map alerts according to the MITRE ATT&CK framework?

  - Are the alerts that are triggered by a specific workload also visible in the attack path?

- **SIEM integration**: nowadays, organizations have a SIEM solution to aggregate data coming from different sources and consolidate it in a single place, helping them investigate security-related incidents. This is another area that will require you to work closely with the SOC team to establish design considerations that are relevant to their team, including identifying which SIEM solution the SOC team currently uses. Ask the following questions:

    - Can the alerts generated by the CNAPP solution be streamed to a SIEM platform?

    If the SIEM solution is a cloud-based solution, make sure to investigate where the data is located for privacy purposes.

    - Can the alerts generated by the CNAPP be synchronized with the SIEM to a point where when you close an alert in the CNAPP, it will also get closed in the SIEM (and vice versa)?

- **Workload supportability**: some workloads will vary according to the cloud provider. A container, for example, will have different nuances depending on which cloud provider you are using. At this stage, you need to take into consideration which workloads you have and with which cloud provider they reside; this will dictate the level of supportability. Ask the following questions:

    - Does the CNAPP solution offer threat detection for workload X (replace X with the type of workload) running in Y (replace Y with the cloud provider name)?

    - Sample question: does the CNAPP solution offer threat detection for containers running in Azure?

    - Does the threat detection for workload X require the installation of an agent?

When it comes to workload protection, you also need to take into consideration your organization's SOC team structure, such as whether they have a red team that is responsible for trying to penetrate the environment, and how the CNAPP solution will trigger alerts based on those tests.

It is important to understand that these are all sample questions; however, this is not the ultimate list of questions to ask, as some questions are very specific to an organization's needs and constraints. The intent of these questions is to give you an idea of what information needs to be gathered as you go through the CNAPP design process.

# Summary

In this chapter, we discussed the importance of first establishing design principles for your CNAPP adoption, which includes the use of Zero Trust, shift-left security, data protection, visibility and monitoring, dynamic threat detection and response, and compliance/governance. These design principles will be agnostic of your platform and the foundation of your implementation.

We discussed the design considerations for your CNAPP adoption, which include considerations for posture management, DevOps security, and workload protection. You learned some important questions to ask while going through these considerations, which should also be complemented by the specific needs and constraints of your organization.

In the next chapter, you will start planning the adoption of Microsoft Defender for Cloud as your CNAPP solution.

## Notes

1. For more information about this report, download the PDF at https://bit.ly/CNAPPBook8.

## Additional resources

- Additional design consideration questions are available as a Microsoft CNAPP eBook: https://aka.ms/MSCNAPP
- Check out this episode of *Defender for Cloud in the Field* with author Yuri Diogenes interviewing a cybersecurity specialist about lessons learned using Defender for Cloud: https://bit.ly/CNAPPBook9

## Join our community on Discord

Read this book alongside other users. Ask questions, provide solutions to other readers, and much more.

Scan the QR code or visit the link to join the community.

https://packt.link/SecNet

# 4

# Creating an Adoption Plan

Now that the design principles and considerations have been established, you need to create a CNAPP adoption plan. While the design principles and considerations are agnostic, the plan must be tailored for the CNAPP solution that you are going to adopt, which in the case of this book is Microsoft Defender for Cloud.

One of the biggest advantages of Defender for Cloud when compared to other CNAPP vendors is the depth and breadth of the product without requiring the addition of third-party solutions.

To create an adoption plan for Defender for Cloud, you need to understand the different components of the solution to ensure that the design principles and considerations that were pre-established can be met. The adoption plan ensures that you have the right set of steps that will lead to a successful implementation and integration into your organization's environment.

This chapter covers:

- Adoption plan
- Planning posture management adoption
- Planning workload protection adoption
- Creating a **Proof of Concept**

## Adoption plan

There are different approaches to adopt a CNAPP solution, and while the suggested phases shown in *Figure 4.1* is based on my experience working with Defender for Cloud since its creation in 2015 and helping hundreds of *Fortune 500* companies to enable this solution, you can make adjustments based on your organization's needs.

In other words: this is not a fixed plan; you can adapt it to your organization's reality.

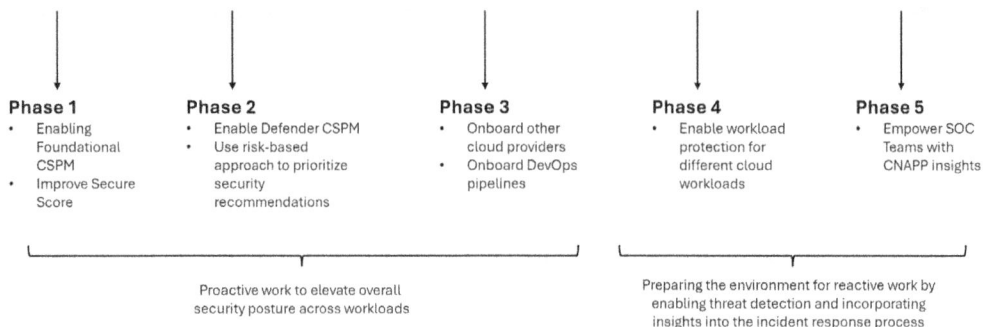

| Phase 1 | Phase 2 | Phase 3 | Phase 4 | Phase 5 |
|---------|---------|---------|---------|---------|
| • Enabling Foundational CSPM<br>• Improve Secure Score | • Enable Defender CSPM<br>• Use risk-based approach to prioritize security recommendations | • Onboard other cloud providers<br>• Onboard DevOps pipelines | • Enable workload protection for different cloud workloads | • Empower SOC Teams with CNAPP insights |

Proactive work to elevate overall security posture across workloads

Preparing the environment for reactive work by enabling threat detection and incorporating insights into the incident response process

*Figure 4.1: CNAPP adaption phases (for better visualization, refer to https://packt.link/ gbp/9781836204879)*

From this book's perspective, Phase 1 (shown in *Figure 4.1*) was done in *Chapter 2* when you enabled Foundational CSPM, performed the assessment, identified the types of resources available in your subscription, identified which security recommendations are high in the priority list, and started working to improve your secure score by remediating these recommendations.

In a large organization with thousands of cloud resources, this phase alone can take a long time to be completed. Of course, the time varies according to the number of people working on this project, how mature the company is in terms of cloud security posture management, and the readiness level of the team that will be responsible for this phase. In the sections that follow, you will learn about the different phases of this diagram.

> It is important to mention that the enablement of the different Defender for Cloud plans will not cause any downtime in the environment.

## Planning posture management adoption

In Phase 2 of the plan, you will enable Defender CSPM to help prioritize recommendations based on risk factors and to provide the necessary insights to disrupt future attacks. Compared to Foundational CSPM, Defender CSPM plan provides additional advanced security measures, governance, and compliance features.

Review the table below to understand the capabilities that you already have available in Foundational CSPM and the capabilities that will be available once you enable Defender CSPM. This is an important list to take into consideration when creating your adoption plan.

| Capability | Description | Foundational CSPM | Defender CSPM |
| --- | --- | --- | --- |
| Security recommendations | This is a list of security recommendations based on the **Microsoft Cloud Security Benchmark (MCSB)**. | X | X |
| Asset inventory | This enables you to visualize and query all your multicloud workloads from a single dashboard. | X | X |
| Secure score | This provides a score that represents your cloud environment current secure posture. You can use this as KPI for cloud security posture improvement overtime. | X | X |
| Data visualization and reporting with Azure Workbooks | This leverages Azure Workbook's capability to visualize Defender for Cloud data. | X | X |
| Data exporting | This enables you to export security alerts, recommendations, and compliance results to external storage, such as Log Analytics workspace. | X | X |
| Workflow automation | This enables you to create automation to trigger Azure Logic Apps for security recommendations and alerts. | X | X |
| Integration with MCSB | The security recommendations are based on MCSB. | X | X |
| AI security posture management | This provides the capability to discover generative **AI Bill of Materials (AI BOM)** and strengthen generative AI application security posture. | - | X |

| Capability | Description | Foundational CSPM | Defender CSPM |
| --- | --- | --- | --- |
| Agentless VM vulnerability scanning | This lets you enable visibility into installed software and software vulnerabilities on your workloads, extending vulnerability assessment coverage to server workloads without the need to install a vulnerability assessment agent. | - | X |
| Agentless VM secrets scanning | This capability can detect various types of secrets, including tokens, passwords, keys, and credentials, stored in different file types within the OS file system. | - | X |
| Attack path analysis | This creates a potential path that an attack can take across multiple workloads and enables you to disrupt this potential attack. | - | X |
| Risk prioritization | This adds contexts in security recommendations to help prioritize the level of criticality based on different risk factors. | - | X |
| Risk hunting with security explorer | This enables you to perform proactive hunting (free query) across all the data that was gathered to generate insight. | - | X |
| Code-to-cloud mapping for containers | This capability enables you to identify vulnerability in a container image stored in a container registry or running in a Kubernetes cluster, and trace back to the CI/CD pipeline that first built the container image. | - | X |
| Code-to-cloud mapping for IaC | This capability assists you in ensuring consistency, as well as secure and auditable infrastructure provisioning, by automatically mapping **Infrastructure as Code (IaC)** templates to cloud resources. | - | X |

| Capability | Description | Foundational CSPM | Defender CSPM |
|---|---|---|---|
| PR annotations | This capability helps developers prioritize critical code fixes. This is done by using pull request annotations and assigning developer ownership by triggering custom workflows feeding directly into the tool used by developers. | - | X |
| **External Attack Surface Management (EASM)** | This is a native integration with Defender EASM, which brings insights from Defender EASM even if you don't have a license for this product. | - | X |
| Permissions Management | This is a native integration with Microsoft Entra Permissions Management (Permissions Management) to obtain **Cloud Infrastructure Entitlement Management (CIEM)** - related insights. It is another native integration that doesn't require an extra license. | - | X |
| Regulatory compliance standards | Assess the workload's security posture in accordance with different industry standards and help organizations to improve workload security posture according to one or more industry standards. | - | X |
| ServiceNow integration | This provides the capability to create and monitor tickets in ServiceNow directly from Defender for Cloud. | - | X |
| Critical assets protection | This enables security administrators to identify and protect the most important assets. It leverages the critical assets engine by **Microsoft Security Exposure Management (MSEM)**. | - | X |

| Capability | Description | Foundational CSPM | Defender CSPM |
|---|---|---|---|
| Governance to drive remediation at scale | This lets you enable security administrators to create governance rules to assign ownership to security recommendations. | - | X |
| Data security posture management, sensitive data scanning | This capability helps you automatically discover sensitive data resources across multiple clouds, and evaluate data sensitivity and continuously uncover risks that might lead to data breaches. | - | X |
| Agentless discovery for Kubernetes | This provides agentless capability not only to discover Kubernetes but also for posture management improvement. | - | X |
| Agentless code-to-cloud containers vulnerability assessment | This provides agentless vulnerability assessment powered by Microsoft Defender Vulnerability Management. | - | X |

The capabilities listed in this table are natively available in Defender CSPM, and the majority of these capabilities do not require a deployment per se; since they are natively available, you just need to use them.

## Planning Defender CSPM

A fundamental part of planning Defender CSPM adoption is to understand the cost of using this plan in your cloud environment. Defender CSPM pricing is based on billable resources, which, as of the time the book was written, was US $0.007 per billable resource per hour[1]. Although Defender CSPM protects multicloud workloads, the billing is applied only on specific resources. There are billable resources on Azure subscriptions, AWS accounts, and GCP projects.

The types of resources that are eligible for billing will vary according to where the resource resides (cloud provider). For Azure, the resources are:

- **Compute**, including the following resource types:
    - `Microsoft.Compute/virtualMachines`
    - `Microsoft.Compute/virtualMachineScaleSets/virtualMachines`
    - `Microsoft.ClassicCompute/virtualMachines`

- **Storage**, including the following resource type:

    - `Microsoft.Storage/storageAccounts`

- **Databases**, including the following resource types:

    - `Microsoft.Sql/servers`

    - `Microsoft.DBforPostgreSQL/servers`

    - `Microsoft.DBforMySQL/servers`

    - `Microsoft.Sql/managedInstances`

    - `Microsoft.DBforMariaDB/servers`

    - `Microsoft.Synapse/workspaces`

> For compute resources, Defender CSPM doesn't bill deallocated VMs and Databricks VMs. For storage resources, Defender CSPM doesn't bill storage accounts without blob containers or file shares.

The same resource types (compute, storage, and database) apply to AWS and GCP. However, for AWS, the supported resources are EC2 instances (compute), S3 buckets (storage) and RDS instances (database). For GCP, the supported resource types are the following:

- Compute:

    - Google Compute instances

    - Google Instance Group

- Storage:

    - Storage buckets

- Databases

- Cloud SQL instances

As part of your planning, you can use the built-in *Cost Estimation* workbook to check how much it will cost to enable this plan in your subscription. This workbook is available in the Defender for Cloud main dashboard, in the *Workbooks* section, as shown in *Figure 4.2*.

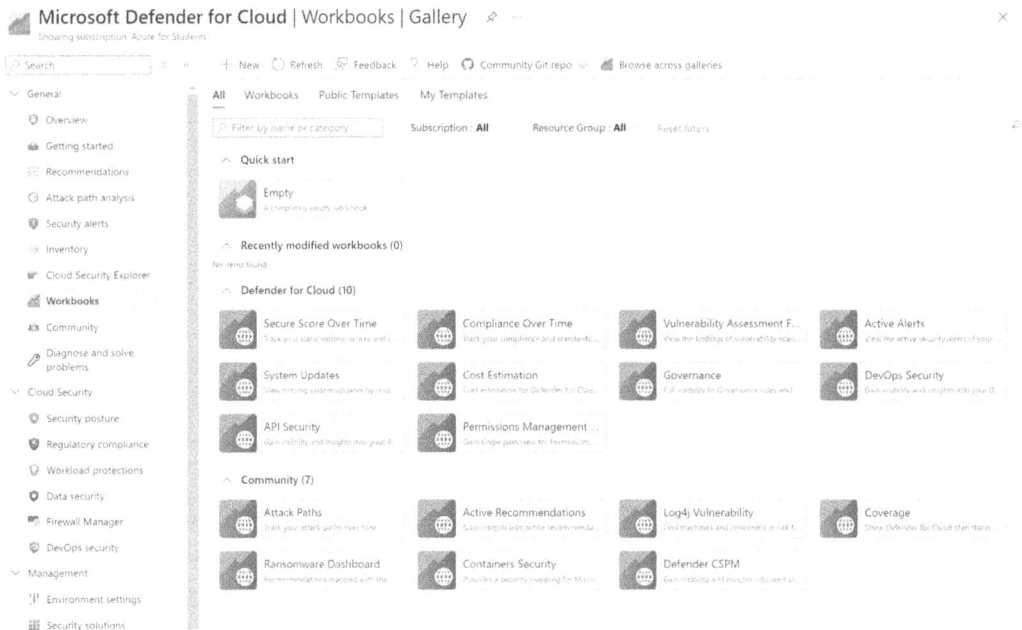

*Figure 4.2: The Cost Estimation workbook (for better visualization, refer to https://packt.link/gbp/9781836204879)*

Once you click on this workbook, you will be able to see the grand total for all plans, as shown in *Figure 4.3*:

*Figure 4.3: The Cost Estimation workbook's grand total (for better visualization, refer to https://packt.link/gbp/9781836204879)*

The advantage of using this built-in workbook is that you can obtain this information in the production environment without changing any settings. You don't need to be concerned about affecting the production environment, because you are not enabling anything, just assessing the environment to see how much it would cost to enable this plan in the subscription.

Don't make decisions about enabling the plan or not solely on the pricing. You should always perform a POC to experience the product and see if it addresses the use case scenarios that you established. Make sure to take advantage of the 30 days' free trial available for each Defender for Cloud plan.

While this is more of a strategical part of the planning, the sections that follow are going to talk about technical aspects that you should include in your plan.

## Privileges

Although Security Admin and Contributor have the privilege to enable the core components of Defender CSPM, you should ensure that you have the Owner privilege in the Azure subscription to be able to enable all Defender CSPM components. If you are onboarding Defender CSPM in an environment that has multicloud, you need to have the multicloud connector with AWS and/or GCP enabled first[2].

## Extensions

While some capabilities of Defender CSPM will be available from the get-go without the need to enable them, there are some settings that need to be enabled for you to take full advantage of the features available in this plan. When you are enabling this plan, which is something you will learn about in *Chapter 5*, you will have the option to enable the necessary Defender CSPM extensions, as shown in *Figure 4.4*:

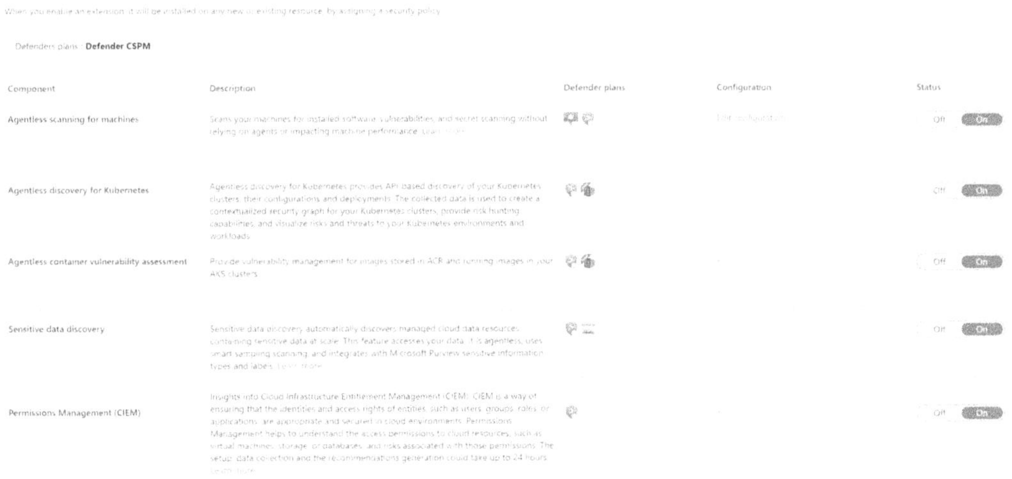

*Figure 4.4: Defender CSPM settings (for better visualization, refer to https://packt.link/ gbp/9781836204879)*

These extensions are a fundamental part of Defender CSPM, and if you don't enable them, some capabilities may not work properly. It is important to emphasize that these options will not incur any extra charges to your Defender CSPM billing. Another area that will be negatively affected if you do not enable these extensions is the Attack Path. Since the Attack Path is created based on the insights that were provided by the artifact scan, without these extensions, it will not be possible to extract the necessary data to build the insights.

## Governance

Defender CSPM enables you to create governance rules to assign security recommendations to workload owners and establish SLAs to remediate the recommendation. When planning your Defender CSPM adoption, ensure that you have already discussed the following points with your team:

- How will SLAs be established?
    - Per workload?
        - For example, workloads that contain data (database, storage) have a higher priority than others
    - Per severity?
        - For example, high severity recommendations should have a 5 business days SLA
    - Per workload and severity?
        - For example, high severity recommendations in workloads that contains data should have a 3 business days SLA

These are important questions to ask and answer during this planning phase. This will help expedite the creation of governance rules when it is time to deploy Defender CSPM.

## DevOps security

As part of the Defender CSPM adoption, you must also plan the onboarding of your DevOps platforms. At the time this book was written, Defender CSPM supported connections with GitHub, Azure DevOps, and GitLab. You will learn more about the onboarding process for these platforms in *Chapter 7*.

There is no additional cost when you onboard these DevOps platforms as they don't count as billable resources.

## Measure security posture management improvement

When you implemented Foundational CSPM in *Chapter 2*, the main KPI for measuring security posture management improvement was the secure score. While the secure score is still going to play an important role in measuring your security posture improvement, you should also plan the use of different metrics.

One important metric to consider is the number of attack paths and the time that it takes to resolve an attack path[3]. Let's take an example of the number of attack paths shown in *Figure 4.5*:

*Figure 4.5: Attack path (for better visualization, refer to https://packt.link/gbp/9781836204879)*

In this example, there are 120 attack paths and the goal should always be to see this number dropping to 0 (zero). This means that the first metric to measure is how many attack paths exist, and the second metric is how long it takes to resolve one attack path (**Time to Resolve (TTR)**).

Add to your plan the use of the continuous export feature (covered in *Chapter 2*), to export the attack path analytics. This will help you track these metrics over a period of time. Once you have this data ingested in a Log Analytics workspace, you can also use a workbook created by Microsoft to visualize this data. For more information about this workbook, visit `https://bit.ly/CNAPPBook10`.

## Planning workload protection adoption

The workload protection planning needs to be tailored to your cloud environment and based on the assessment that you've done using Foundational CSPM (*Chapter 2*). At this point, you already know which workloads you have, and you also know the security state of these workloads based on the initial assessment.

You can also use the **Coverage Workbook** to visualize which workloads are covered in your subscription. The sections that follow will cover the workload protection based on the available Defender for Cloud plans.

# Defender for Servers

One of the first resources that is usually migrated from on-premises to a cloud environment is the VM. In a production environment, that means many VMs. This means that you need to ensure that you are protecting these VMs as soon as they are provisioned in the cloud environment. This refers to not only the migrated VMs, but also all new ones that are provisioned from scratch.

To help you with that, Defender for Cloud has a plan to protect VMs, which is called Defender for Servers. As the name suggests, this plan is to protect servers, and there are two pricing offers with this plan:

- **Plan 1**

  - **Price**: US $0.007/Server/hour[5]

  - **Capabilities**:

    - Real-time scanning and protection and Microsoft Defender Antivirus.

    - Vulnerability assessment and mitigation provided by **Microsoft Defender Vulnerability Management (MDVM)**

    - Automatically provisions the Defender for Endpoint sensor on every supported machine

    - Threat detection for OS-level (require an agent to be installed on the VM)

- **Plan 2**

  - **Price**: US $0.02/Server/hour

  - **Capabilities**:

    - All capabilities available in P1

    - MDVM add-on

    - Security Policy and Regulatory Compliance

    - Adaptive application controls based on machine learning to identify safe applications and automatically create an allow list

- Azure Update Manager remediation of unhealthy resources and recommendations for Arc enabled machines
- Attack surface reduction with **Just-in-Time** (**JIT**) VM access
- Machine learning hardening of NSG rules with adaptive network hardening
- File integrity monitoring
- Docker host hardening using **Center for Internet Security** (**CIS**) Docker Benchmark
- Network map
- Agentless scanning

When planning Defender for Servers adoption, you must revisit the capabilities available for these plans to understand which business needs will lead you to select Plan 1 over Plan 2. If your organization only needs an EDR solution installed on the servers, then you clearly just need to enable Plan 1 (or P1). However, don't rush to make decisions before investigate the real business needs. Make sure to use the *Cost Estimation Workbook* that was presented earlier in this chapter to also obtain an estimation of how much this plan will cost based on the VMs that you have in your subscriptions.

> During the Defender for Servers planning phase, you also need to review the supportability matrix for this plan. Visit https://bit.ly/CNAPPBook11 for a complete list of Defender for Servers supported platforms.

Sometimes, you are led to select one plan over the other for budgetary reasons, but it was not considered the big picture, and how the capabilities of the plan will affect the **return of investment** (**ROI**). According to an article from *MIT Technology Review*[6], vendor consolidation "reduce[s] costs associated with IT support for vendor management." When adopting a solution like Defender for Servers P2, you may have the opportunity to save more money by decommissioning other solutions that were in use and will not be necessary anymore because there is one or more Defender for Servers capability that is going to fulfill that need.

As part of your planning, you should also establish how you will enable this plan. You can enable the plan at the subscription level (recommended option), which means that all VMs under that subscription (and new ones that will be created in the future) will be protected, or can exclude specific resources (VMs) by enabling the Defender for Servers plan at the resource level, using REST API, or at scale.

At the time this chapter was written, Microsoft was working on the deprecation of Log Analytics agent, and that's why I'm not covering the planning aspects of Log Analytics workspace. Revisit this article, `https://bit.ly/CNAPPBook13`, to understand the overall recommendations for this deprecation.

# Defender for Storage

This plan can also be enabled on the entire subscription to protect all storage accounts (recommended) or per storage account. Defender for Storage costs US $0.0134 per storage account per hour, and for storage accounts that exceed 73 million monthly transactions, a charge of US $0.1492 for every 1 million transactions that exceed the threshold will be added[7]. Enabling this plan will provide threat detection for storage accounts. The available capabilities vary according to the storage type, as shown below:

- **Blob Storage (Standard/Premium StorageV2, including Data Lake Gen2)**: supports activity monitoring, malware scanning and sensitive data discovery
- **Azure Files (over REST API and SMB)**: supports activity monitoring only

It is important to mention that although malware scanning is part of Defender for Storage, you will need to pay an additional price once you enable this capability. Malware scanning costs US $0.15 per GB of data scanned. Malware scanning is the only capability that is not included in the Defender for Storage free 30-day trial.

# Defender for Databases

This plan covers a variety of databases, some of which you may not even have in your environment. The assessment results gathered during the time that you enabled Foundational CSPM (*Chapter 2*) will be very important to identify what types of databases you have in your cloud environment. The pricing for each database will also vary according to the database type. *Figure 4.6* has the pricing (July 2024) for the supported databases:

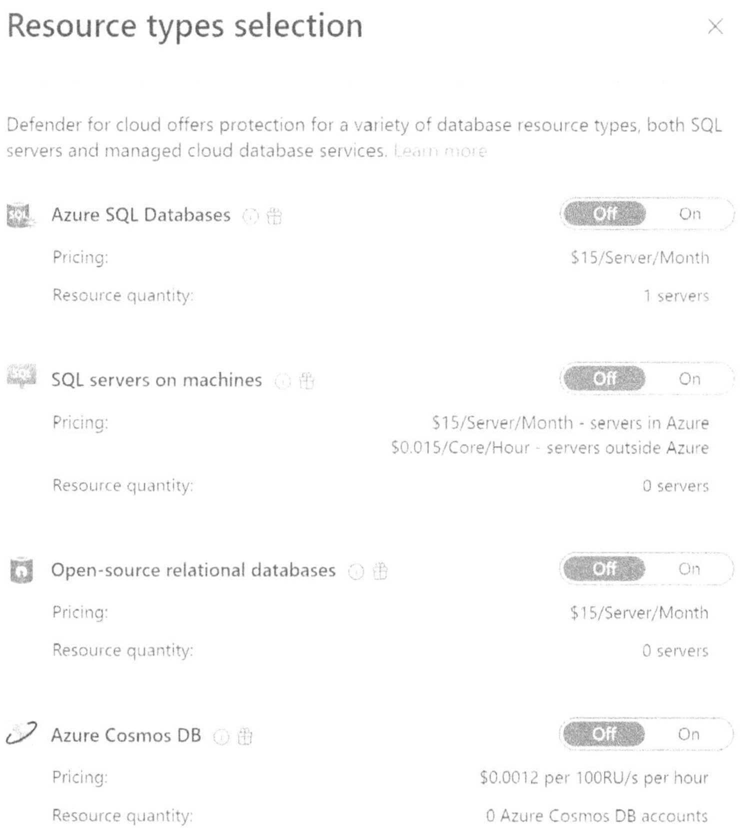

*Figure 4.6: Defender for Databases supported database type*

When planning your Defender for Database adoption, make sure to also use the *Cost Estimation Workbook* to evaluate how much it will cost to enable this plan at the subscription level. If you have cost constraints and need to enable the plan only on specific databases, you can also do that. For example, if you have Azure SQL in your subscription with 100 databases, but you can only afford to enable it on the most critical ones (let's say, 50), you can granularly select only the databases that you want to enable this plan for.

## Defender for Containers

This plan protects Kubernetes clusters running on Azure Kubernetes Service and Kubernetes on-premises/IaaS. It also offers vulnerability assessment for images stored in **Azure Container Registries (ACR)**, vulnerability assessment for images running in Azure Kubernetes Service, and run-time threat protection for nodes and clusters. This plan costs US $0.0095 per vCore per hour.

As part of your Defender for Containers planning, you also need to determine if you have on-premises Kubernetes clusters. If you do and you want to protect this type of resource, you first need to connect the Kubernetes cluster to Azure Arc. In addition, you need to ensure that the URLs below are configured for outbound access (port 443) so that the Defender sensor can connect to Microsoft Defender for Cloud.

## Defender for Key Vault

This plan doesn't have a lot of elements that need to be planned other than knowing the price, which is US $0.25 per Vault per month. It is important to mention that this plan cannot be enabled on certain resources only, which means you can only enable Defender for Key Vault at the subscription level.

## Defender for Resource Manager

Just like Defender for Key Vault, this plan doesn't have specific elements that need to be considered before enabling. This plan costs US $5.04 per subscription per month and can only be enabled at the subscription level. While it is easy to plan the usage of Defender Resource Manager, the value that this plan brings is unique. While the other plans will only be applicable if you have the workload in your cloud environment, this plan should always be enabled, since it protects the resource management operations (which is present in all Azure subscriptions).

## Defender for App Services

This plan is important if you have Azure App Services in your environment. This plan monitors requests and responses sent between App Service apps, the underlying sandboxes, and VMs and App Service internal logs. This plan can only be enabled at the subscription level and costs US $0.02 per instance per hour.

## Defender for APIs

API Security is a very new area for many organizations. I personally worked on the team that created this plan since it was in incubation, and during the interviews with customers, it was always clear that many organizations didn't have the right level of maturity to understand the risks of not securing APIs.

When planning your Defender for APIs adoption, you must ensure that the team that handles API management is involved in the discussion. At the time this book was written, Defender for APIs only supported APIs that were published via Azure **API Management** (**APIM**).

Defender for APIs' threat detections are based on OWASP Top 10 API Security Risks[8]. The plan can only be enabled at the subscription level; however, you do have the granularity to select the pricing according to your business needs, as shown in *Figure 4.7*:

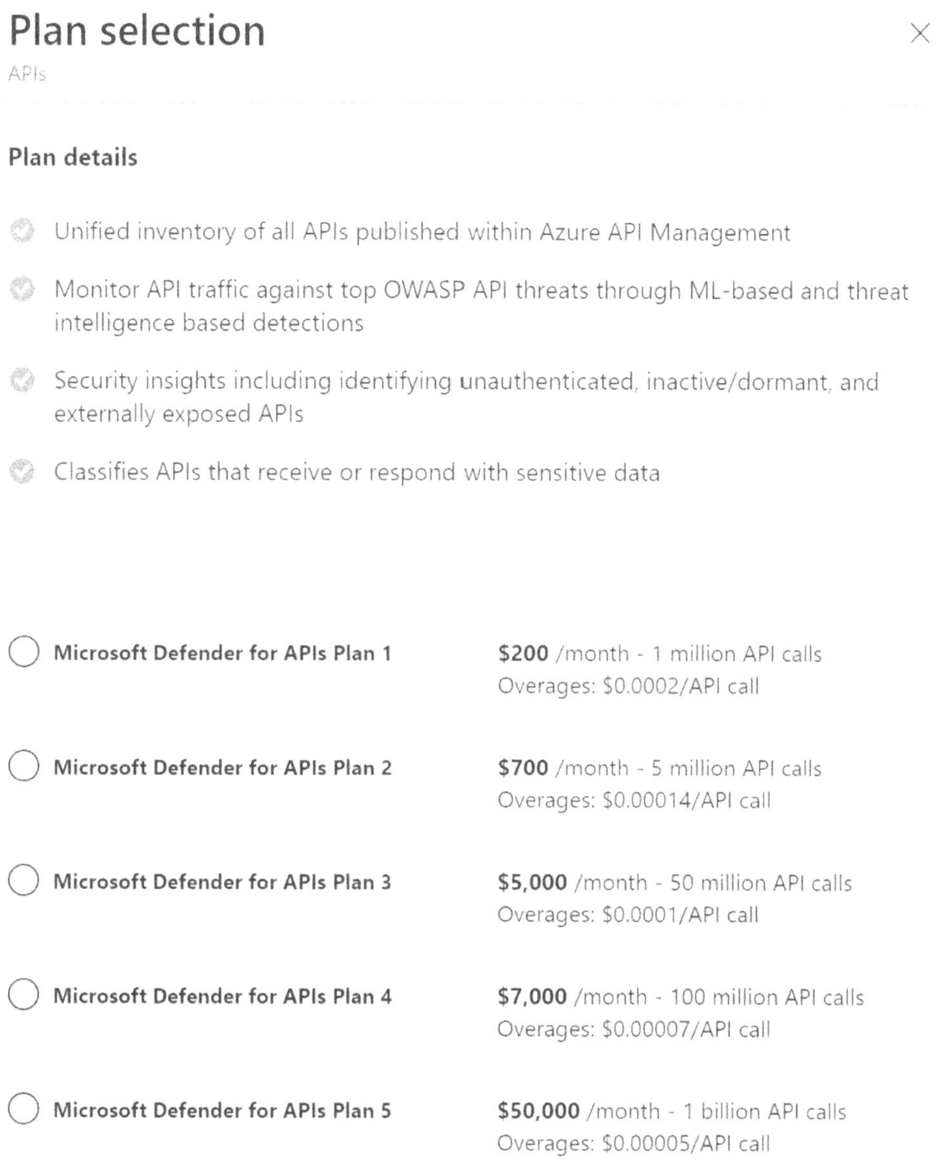

# Plan selection      ×

APIs

### Plan details

- Unified inventory of all APIs published within Azure API Management

- Monitor API traffic against top OWASP API threats through ML-based and threat intelligence based detections

- Security insights including identifying unauthenticated, inactive/dormant, and externally exposed APIs

- Classifies APIs that receive or respond with sensitive data

 

○ **Microsoft Defender for APIs Plan 1**      **$200** /month - 1 million API calls
Overages: $0.0002/API call

○ **Microsoft Defender for APIs Plan 2**      **$700** /month - 5 million API calls
Overages: $0.00014/API call

○ **Microsoft Defender for APIs Plan 3**      **$5,000** /month - 50 million API calls
Overages: $0.0001/API call

○ **Microsoft Defender for APIs Plan 4**      **$7,000** /month - 100 million API calls
Overages: $0.00007/API call

○ **Microsoft Defender for APIs Plan 5**      **$50,000** /month - 1 billion API calls
Overages: $0.00005/API call

*Figure 4.7: Defender for APIs plans*

You will need to work closely with the team that manages Azure APIM platform to measure the number of API calls and, based on that, decide which Defender for APIs plan is more recommended to your environment.

# Creating a Proof of Concept

After going over each Defender for Cloud plan and taking into account the particularities that each plan offers, it is time for you to create a **Proof Of Concept** (**POC**). To establish a successful POC, you need to do the following things:

- Determine what the scope of the POC is. You could do a POC that aggregates all plans (enabling all). This would be a very broad scope, but you could immediately see the advantage of having all these plans enabled and how the end-to-end experience would be. You could also have a smaller scope by having a POC for each plan, with a very specific scope.

- Part of the scope also includes where you will perform the POC. Will it be in the production environment or lab? Since Defender for Cloud doesn't cause downtime during the enablement, and only provides insights into what's happening, most organizations prefer to do a POC on the production environment so they can truly see how Defender for Cloud adds value to the business.

- Also, identify which workloads are more relevant for the POC. For example, if an organization has 80% of Kubernetes, it would make sense to start with Defender for Containers.

- **Use case scenarios:** to clearly understand what you need to validate, you need to establish the scenarios that you want to test. Here is an example:

  Be able to identify potential attacks against workloads (any workload that can store data) that have sensitive data —for this use case, you need to have the following plans enabled:

  - Defender CSPM: this is needed to have risk factor, as well as attack path and data posture management.

  - Defender for Database: You need to enable the database plan according to the type of database that you have in your environment. This will also enable threat detection for the database.

  - Defender for Storage: This is needed to be able to see potential threats against the storage account.

- **Timeline**: Establish the time frame for this POC, and if you want to perform the POC without paying for the Defender for Cloud plans, you just need to focus on getting the POC done during the 30-day free trial of the plans. In other words, only enable the plan when you are fully ready to start testing.

- **Success criteria**: How are you going to measure if this was a successful POC or not? You need to define these criteria with your team. The criteria could be something like *"successfully demonstrate how sensitive data can be discovered and how risk is analyzed"*.

# Summary

In this chapter, we discussed the overall approach to plan Microsoft CNAPP (Defender for Cloud) adoption plan. You learned about the different aspects of planning your posture management adoption by enabling Defender CSPM. We discussed the planning aspects of workload protection by covering each individual Defender for Cloud plan. Lastly, you learned important aspects of how to create a POC plan to validate the main use case scenarios that your organization needs.

This chapter concludes the last step before you start implementing Defender for Cloud as your CNAPP solution. Here, you learned important considerations regarding the creation of a concise adoption plan.

In the next chapter, you will learn how to deploy Defender CSPM.

# Notes

1.  Always consult the pricing page for the latest information on pricing; visit `https://azure.microsoft.com/en-us/pricing/details/defender-for-cloud/`.

2.  You will learn more about connecting to AWS and GCP in *Chapter 6*.

3.  You will learn more about Attack path in *Chapter 5*.

4.  The pricing is based on the day that this chapter was written in July 2024. For the latest pricing, visit the Defender for Cloud Pricing page at `https://azure.microsoft.com/en-us/pricing/details/defender-for-cloud/`.

5.  The price per hour is not applicable for deallocated VMs.

6.  You can read the MIT Technology Review article at `https://bit.ly/CNAPPBook12`.

7. As was mentioned in previous sections, any pricing referenced in this book is based on July 2024 pricing. To see the most updated pricing, visit `https://azure.microsoft.com/en-us/pricing/details/defender-for-cloud/`.

8. For more information about OWASP Top 10 API Security, visit `https://owasp.org/API-Security`.

## Additional resources

- Read the article written by the author Yuri Diogenes, wherein he covers the best practices to create a POC: `https://bit.ly/CNAPPBook14`

- You can access Defender for Cloud simulations at the Defender for Cloud GitHub repository. These simulations can be used during the POC to simulate alerts: `https://bit.ly/CNAPPBook15`

## Join our community on Discord

Read this book alongside other users. Ask questions, provide solutions to other readers, and much more.

Scan the QR code or visit the link to join the community.

`https://packt.link/SecNet`

# 5

# Elevating Your Workload's Security Posture

By now, it should be very clear to you how important it is to have a solid security posture in your cloud environment. It not only helps to reduce the attack footprint but also has been proven to prevent the vast majority of the potentially successful attacks. Having said that, you still need to have the right set of tools to help you prioritize what is important to remediate in your own cloud environment.

To take your security posture to the next level, you need insights that enable you to make intelligent decisions when it comes to remediation prioritization. Risk factors are an essential part of this decision-making process and Defender CSPM is the Defender for Cloud plan that empowers you to use risk factors to prioritize what is most important to remediate in your cloud environment.

Security posture improvement with Defender CSPM is a key part of your CNAPP adoption. The journey to full CNAPP adoption always starts with the elevation of your workload's security posture, which is part of the protect phase (from the protect, detect, and response lifecycle).

This chapter covers:

- Onboarding Defender CSPM
- Attack disruption
- Recommendation prioritization
- Data security posture

# Onboarding Defender CSPM

In the last two chapters, you learned about the prerequisites to enable Defender CSPM as part of the CNAPP planning phase. Assuming that you already have those prerequisites in place, the Defender CSPM onboarding process is very simple. All you need to do is to open the Defender for Cloud dashboard, click **Environment settings** in the **Management** section, click the subscription that you want to enable Defender CSPM, click the **Defender plans** option, and click **On** besides the Defender CSPM plan, as shown in *Figure 5.1*:

*Figure 5.1: Enabling Defender CSPM (for better visualization, refer to https://packt.link/ gbp/9781836204879)*

Notice that the **Settings** option (hyperlink) appears. You should click this option and double-check whether all extensions are enabled (set to **On**), as shown in *Figure 5.2*:

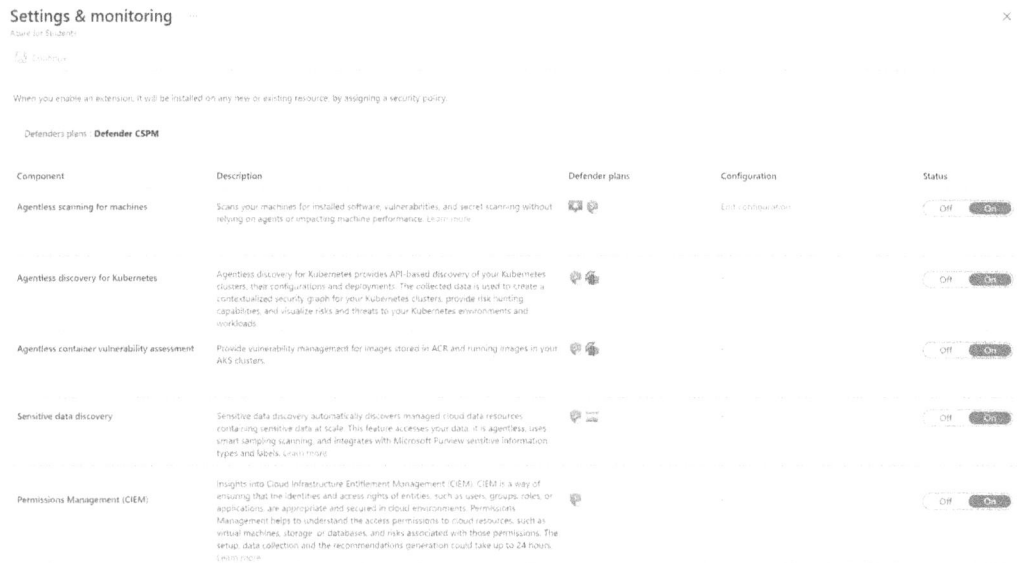

*Figure 5.2: Defender CSPM extensions (for better visualization, refer to https://packt.link/ gbp/9781836204879)*

To finish, click **Continue** on the extension page, and then click **Save** on the **Defender plans** page to commit the changes. After the plan is enabled, deeper analysis in your cloud environment will start taking place, which includes the calculation of risk factors based on existing recommendations, identification of potential attack paths, and data sensitivity discovery.

# Attack disruption

As you learned in *Chapter 1*, one of the key elements of CNAPP is the attack path analysis—the inherent capability that the platform offers to identify potential paths that could lead to an attack, and the potential lateral movement across workloads and even across cloud providers.

The attack path analysis is one of the features that is available when you enable Defender CSPM. If you just enabled Defender CSPM and you try to access the attack path, you may see a message stating that Defender CSPM was enabled recently, and it can take up to one day to fully populate the attack paths list (see *Figure 5.3*):

*Figure 5.3: Attack path page right after enabling Defender CSPM (for better visualization, refer to https://packt.link/gbp/9781836204879)*

After the necessary time to identify the potential attack paths, you will see this page fully populated, as shown in the example in *Figure 5.4*:

*Figure 5.4: Attack path page populated (for better visualization, refer to https://packt.link/gbp/9781836204879)*

Keep in mind that the number of attack paths will vary according to the environment. In your organization's production environment, the reality may be completely different from the example shown in *Figure 5.4*, and that's absolutely expected. Regardless of the number of attack paths, your approach to handling these attack paths should be the same: address the attack paths in top-down order. The attack paths are organized in such a way that the order reflects not only the level of criticality but also the highest level of risk, based on the risk level calculation.

To give you an example, I'm going to open the first attack path from the list of attacks shown in *Figure 5.4*, and the result is shown in *Figure 5.5*:

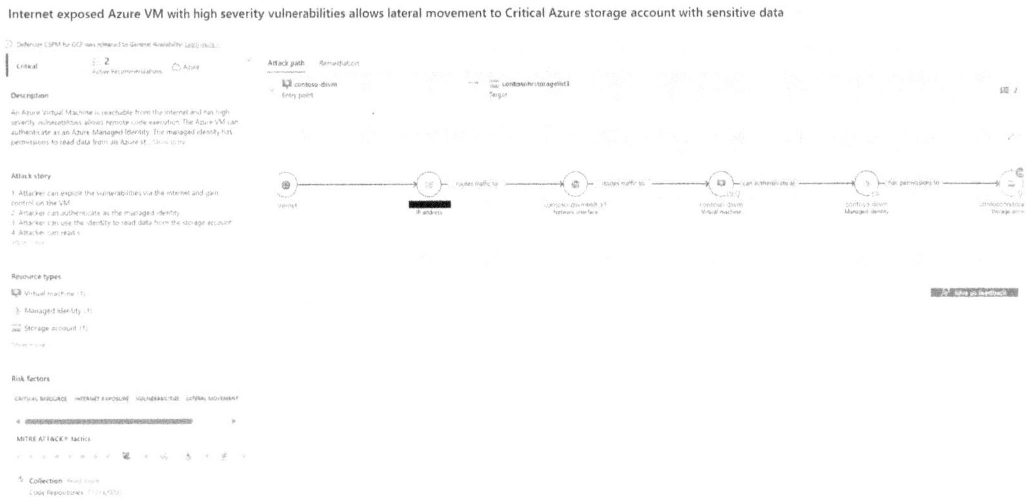

*Figure 5.5: Sample of an attack path (for better visualization, refer to https://packt.link/gbp/9781836204879)*

The attack path name gives you a good summary of what this attack is about. In this case, the attack path name is **Internet exposed Azure VM with high severity vulnerabilities allows lateral movement to Critical Azure storage account with sensitive data**. The name is very self-explanatory, but if you need additional comprehension, you can read the description. The example in *Figure 5.5* says: **An Azure Virtual Machine is reachable from the internet and has high severity vulnerabilities allows remote code execution. The Azure VM can authenticate as an Azure Managed Identity. The managed identity has permissions to read data from an Azure storage account. The Azure storage account stores sensitive data.**

To further your knowledge of how the threat actor can move laterally, the page also has the attack story. In the case of *Figure 5.5*, it says:

1. **Attacker can exploit the vulnerabilities via the internet and gain control on the VM**
2. **Attacker can authenticate as the managed identity**
3. **Attacker can use the identity to read data from the storage account**
4. **Attacker can read sensitive data from the Azure storage account**

On the same left side of the page, you also have the list of resources that can be affected if this attack path is exploited, the list of risk factors that were taken into account to rank this attack path's level of criticality, and the mapping for the MITRE ATT&CK tactics.

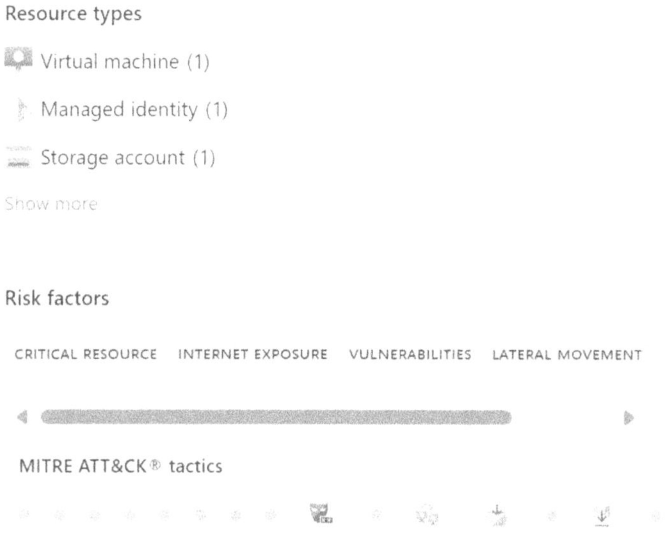

Resource types

Virtual machine (1)

Managed identity (1)

Storage account (1)

Show more

Risk factors

CRITICAL RESOURCE    INTERNET EXPOSURE    VULNERABILITIES    LATERAL MOVEMENT

MITRE ATT&CK® tactics

*Figure 5.6 Attack path components*

Now that you have a good understanding of what this attack path is about, you can move to the right side of the page to obtain more insights about each element of the attack path. To understand the details of each component involved in the attack path, you should click directly on it, and once you click, a blade opens on the right side.

Some of those components may have extra icons, such as the one shown in *Figure 5.6*, which has a bullet list icon, which means that once you click on this component, you will see the security recommendations that are open and need to be remediated. You will also see a light bulb, which means that there are insights available for this component.

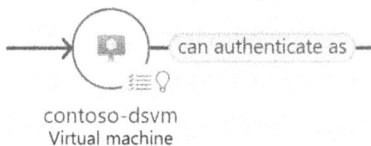

*Figure 5.7: Attack path icons*

For the component shown in *Figure 5.7* (Virtual machine), the insights that appear once you click on it are shown below:

contoso-dsvm
Virtual machine

| Info | Insights | Recommendations |

∨ **Insights - Exposed to the internet**

Description
The resource allows incoming network traffic from the internet

∨ **Insights - Vulnerable to remote code execution (Preview)**

Description
The resource has vulnerabilities allowing remote code execution

∨ **Insights - Has high severity vulnerabilities**

Description
The resource has high severity vulnerabilities

*Figure 5.8: Virtual machine insights*

These are important insights for you to understand the level of criticality that the vulnerabilities on this VM expose to the environment. As you continue to move from one component to the other (from left to right), you will arrive at the **crown jewels**, which in this case is a storage account that contains sensitive information.

contosohrstoragelist3
Storage account

*Figure 5.9: Attack path crown jewels*

Upon selecting this storage account, and looking at the insights, it is possible to see the type of sensitive information. This is another important aspect of Defender CSPM that brings data security posture management natively as part of the platform.

At this point, you have a deep understanding of the risks posed by this attack path, and now you need to learn how you can disrupt this attack by remediating the relevant recommendations. For that, you can click the **Remediation** tab, and you will see the list of recommendations that must be remediated to disrupt the attack:

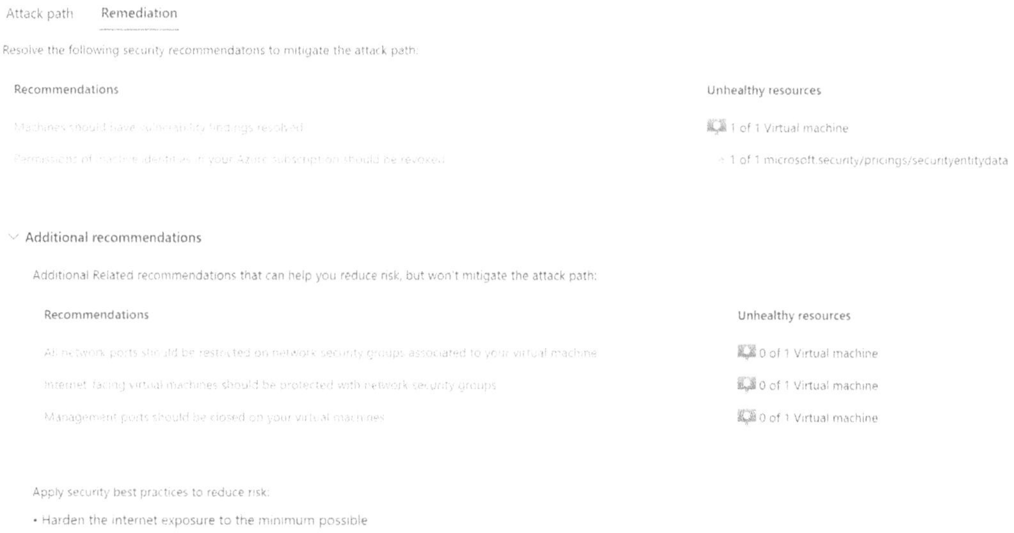

*Figure 5.10: Remediation plan (for better visualization, refer to https://packt.link/ gbp/9781836204879)*

While attack path analysis is mostly used to bring awareness about potential areas of compromise and allow you to remediate before threat actors can exploit them, there may be scenarios where a workload that belongs to the attack path is already under attack. When this happens, the workload appears in the attack path with an extra icon, which is a shield, as shown in *Figure 5.11*:

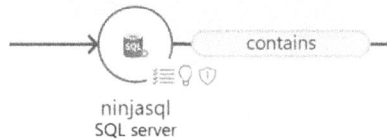

*Figure 5.11: Shield icon*

This shield means that this workload has been attacked, and there are security alerts available to give you more details. The number of security alerts appears at the bottom of the blade (see *Figure 5.12*), which becomes available once you click on the workload itself.

*Figure 5.12: Number of alerts available for this workload (for better visualization, refer to https://packt.link/gbp/9781836204879)*

This is an important piece of information because it helps you to justify even more why you need to prioritize the remediation of this attack path. A good business justification could be that this workload is already under attack, and the remediation needs to be done before the threat actor can get into the environment.

As explained in the last chapter, your goal is to drop the number of attack paths to zero, like the example shown in *Figure 5.13*.

Microsoft Defender for Cloud | Attack path analysis

Figure 5.13: Achieving and maintaining zero attack paths is the goal (for better visualization, refer to https://packt.link/gbp/9781836204879)

# Recommendation prioritization

In *Chapter 2*, when you enabled Foundational CSPM, you received a series of security recommendations that were based on the Microsoft Cloud Security Benchmark, however, as mentioned before, these recommendations don't consider the different risk factors of your own cloud environment. Now that you have Defender CSPM enabled, you will see a list of security recommendations organized by risk factors, as shown in *Figure 5.14*:

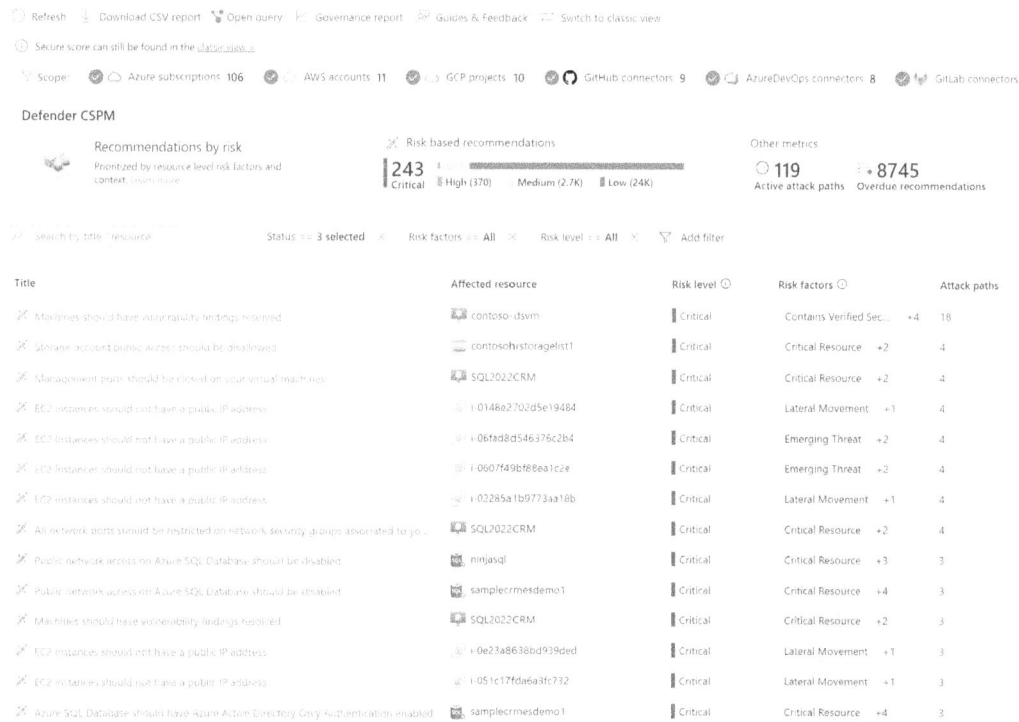

Figure 5.14: Recommendations organized by risk factors (for better visualization, refer to https://packt.link/gbp/9781836204879)

Just like the **Attack Path** page, the **Recommendations** page organizes the list in a top-down approach, where the recommendations at the top are the most critical ones. This order takes into consideration the risk level (critical, high, medium, and low), the risk factor (environmental factors that affect the resources, such as internet exposure, sensitive data, and lateral movement potential), and the number of attack paths that will be disrupted once the recommendation is remediated.

When you open the recommendation, you will see all the details about the security recommendation, as shown in *Figure 5.15*:

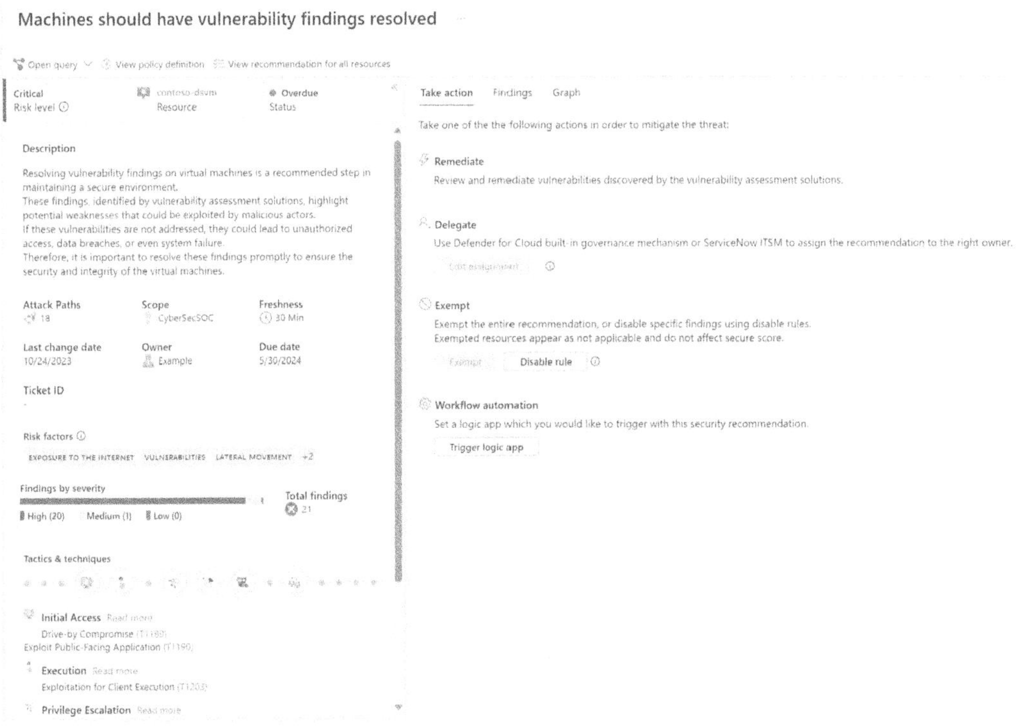

*Figure 5.15: Recommendations details (for better visualization, refer to https://packt.link/ gbp/9781836204879)*

This page is highly populated with lots of important information. To facilitate the absorption of this data, you should read the left pane first, which contains the following fields:

- **Critical**: Shows the criticality of this recommendation, which can be critical, high, medium, or low

- **Resource**: Contains the name of the affected resource

- **Status**: The correct status when you create governance rules to apply to recommendations (you will learn more about governance rules in *Chapter 8*)

- **Description**: Contains more details about the security recommendation

- **Attack paths:** Contains the number of attack paths that will be disrupted once this recommendation is remediated

- **Scope**: Contains the scope (environment in this context) of the recommendation, which in this case is an Azure subscription

- **Freshness**: Reflects the time that it will take for this recommendation to refresh—in this case, every 30 minutes

- **Last change date**: Date that the last change occurred

- **Owner**: In case this recommendation was assigned to someone using the governance feature, the name of the owner

- **Due date**: In case this recommendation was assigned to someone using the governance feature and a due date was established, the due date

- **Ticket ID**: In case this recommendation was integrated with ServiceNow, the ticket number

- **Risk factors**: List of risk factors for this recommendation, which in this case includes exposure to the internet, vulnerabilities, lateral movement, verified secret, and sensitive data

- **Findings by severity**: Summary of the different levels of criticality among the vulnerabilities that were discovered

- **Total findings**: Number of vulnerabilities found

- **Tactics & techniques**: List of MITRE ATT&CK framework tactics and techniques that this recommendation maps to

On the right pane, you may have three tabs, which the example shown in *Figure 5.14* has. The first one is **Take action**, which contains the following fields:

- **Remediate**: Callout to remediate the recommendation

- **Delegate**: Enables you to leverage the governance feature in Defender CSPM to create a governance rule for this recommendation

- **Exempt**: Allows you to exempt the recommendation in case you consider this a false positive (keep in mind that the **Exempt** button may not be available since not all recommendations can be exempted)

- **Workflow automation**: Enables you to call an Azure Logic App to trigger a series of actions to remediate the recommendation

The second tab in this example is **Findings**, which has the list of security vulnerabilities found, as shown in *Figure 5.16*:

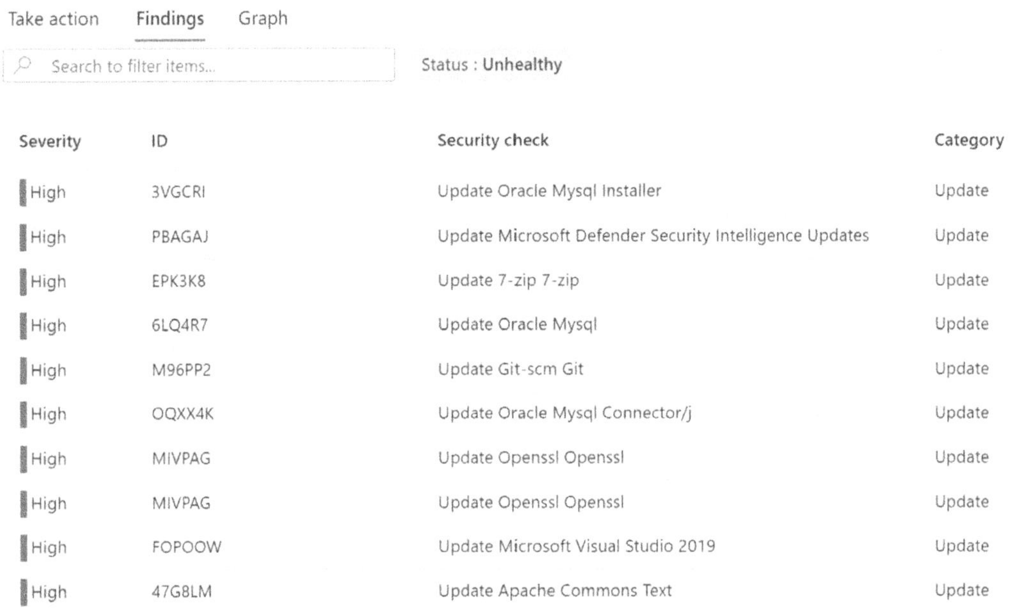

| Take action | Findings | Graph |
| --- | --- | --- |

| Search to filter items... | | Status : **Unhealthy** |

| Severity | ID | Security check | Category |
| --- | --- | --- | --- |
| High | 3VGCRI | Update Oracle Mysql Installer | Update |
| High | PBAGAJ | Update Microsoft Defender Security Intelligence Updates | Update |
| High | EPK3K8 | Update 7-zip 7-zip | Update |
| High | 6LQ4R7 | Update Oracle Mysql | Update |
| High | M96PP2 | Update Git-scm Git | Update |
| High | OQXX4K | Update Oracle Mysql Connector/j | Update |
| High | MIVPAG | Update Openssl Openssl | Update |
| High | MIVPAG | Update Openssl Openssl | Update |
| High | FOPOOW | Update Microsoft Visual Studio 2019 | Update |
| High | 47G8LM | Update Apache Commons Text | Update |

*Figure 5.16: Security vulnerabilities*

If you click on the security vulnerability itself, you will see more details about the vulnerability, as shown in *Figure 5.17*:

## Update Oracle Mysql Installer      ✕

⌃ **General information**

| | |
|---|---|
| ID | 3VGCRI |
| Category | Update |
| Solution | ⬢ Microsoft Defender vulnerability management |

⌃ **Remediation**

Update Mysql Installer (from Oracle) to the latest version.

⌃ **Weaknesses**

| | |
|---|---|
| Highest severity | ❶ High |
| Highest CVSS Score | 7.9 |
| CVSS Version | 3 |

| CVE ID | Severity | Applies to |
|---|---|---|
| CVE-2023-22094 | ❶ High | 1 of 1 resources |
| CVE-2022-39404 | ❶ Low | 1 of 1 resources |

Showing 1 - 2 of 2 results.

⌃ **Affected resources**

| Name | Subscription | Software version |
|---|---|---|
| 🖥 contoso-dsvm | CyberSecSOC | 1.4.3.0 |

Showing 1 - 1 of 1 results.

*Figure 5.17: Security vulnerabilities details*

The last tab is called **Graph**, which reflects the list of attack paths that will be disrupted once this recommendation is remediated. *Figure 5.18* has the example for this recommendation in this environment:

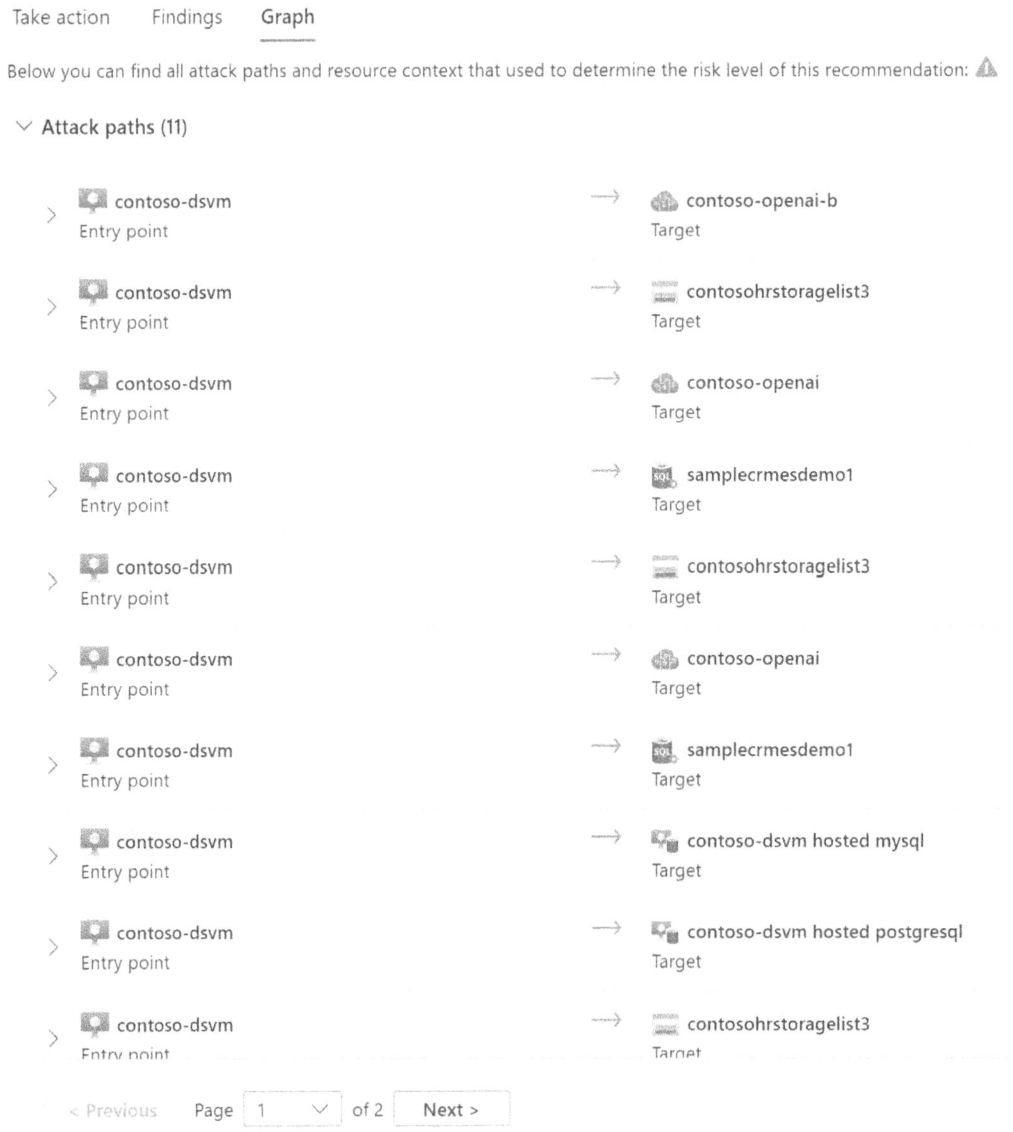

Take action      Findings      **Graph**

Below you can find all attack paths and resource context that used to determine the risk level of this recommendation: ⚠

∨ Attack paths (11)

| | | | |
|---|---|---|---|
| > | 🖥 contoso-dsvm<br>Entry point | ⟶ | ☁ contoso-openai-b<br>Target |
| > | 🖥 contoso-dsvm<br>Entry point | ⟶ | 🗄 contosohrstoragelist3<br>Target |
| > | 🖥 contoso-dsvm<br>Entry point | ⟶ | ☁ contoso-openai<br>Target |
| > | 🖥 contoso-dsvm<br>Entry point | ⟶ | 🗄 samplecrmesdemo1<br>Target |
| > | 🖥 contoso-dsvm<br>Entry point | ⟶ | 🗄 contosohrstoragelist3<br>Target |
| > | 🖥 contoso-dsvm<br>Entry point | ⟶ | ☁ contoso-openai<br>Target |
| > | 🖥 contoso-dsvm<br>Entry point | ⟶ | 🗄 samplecrmesdemo1<br>Target |
| > | 🖥 contoso-dsvm<br>Entry point | ⟶ | 🗄 contoso-dsvm hosted mysql<br>Target |
| > | 🖥 contoso-dsvm<br>Entry point | ⟶ | 🗄 contoso-dsvm hosted postgresql<br>Target |
| > | 🖥 contoso-dsvm<br>Entry point | ⟶ | 🗄 contosohrstoragelist3<br>Target |

< Previous      Page  1  ∨  of 2      Next >

*Figure 5.18: List of attack paths*

This is very powerful information because, here, you can see all attack paths (if you click on the > sign, you will see the entire attack path graph) that were identified just because this recommendation was not remediated.

# Data security posture

One built-in functionality in Defender CSPM is the data-aware security capability. This capability helps uncover misconfigurations across cloud resources that can lead to sensitive data exposure or data breaches. To discover the data, Defender for Cloud works behind the scenes, discovering a selected number of assets in your cloud data stores. After discovery is done, it will apply a process called **smart sampling**, to discover evidence of sensitive data issues. Agentless scanning architecture enables low-friction onboarding, and this is possible because of the sensitive data[1] discovery option (*Figure 5.19*) that is activated (set to **On**) when you enable the Defender CSPM plan.

Sensitive data discovery | Sensitive data discovery automatically discovers managed cloud data resources containing sensitive data at scale. This feature accesses your data, it is agentless, uses smart sampling scanning, and integrates with Microsoft Purview sensitive information types and labels. Learn more | Off / On

*Figure 5.19: List of attack paths (for better visualization, refer to https://packt.link/ gbp/9781836204879)*

A sample of this option is shown in the figure above, which is part of the Defender CSPM settings, as shown at the beginning of this chapter (*Figure 5.2*). The insights of this discovery will appear in risk-based recommendations, attack paths, and when using the Cloud Security Explorer functionality in Defender for Cloud.

> It is important to mention that the initial discovery for storage can take up to 24 hours. If you add a new AWS S3 bucket or GCP storage bucket post-initial discovery, these resources can take up to 48 hours to appear.

Below you have a list of potential attack paths that will surface insights related to sensitive data:

- Internet-exposed Azure Blob containers/AWS S3 buckets with sensitive data are publicly accessible

- Internet-exposed Azure/AWS managed database instance allows basic authentication method

- Internet-exposed VM/EC2 instance has high severity vulnerabilities AND hosted database

- Internet-exposed VM/EC2 instance has high severity vulnerabilities AND read permission to a data resource with sensitive data

- Internet-exposed hosted SQL database[2] has a common username AND allows code execution on the VM

- Private Blob container/AWS S3 bucket replicates data to publicly exposed Blob container/ AWS S3 bucket

> As you can see, these are multicloud scenarios, and that's because the data security posture supports Azure, AWS, and GCP.

While the default information types will cover most of the scenarios, it is also possible to perform some customization to better fit your organization's needs.

## Customization

Defender for Cloud uses the built-in sensitive information types[3], which are provided by Microsoft Purview (even if you don't have a Purview license). Some of the information types and labels are enabled by default. A subset of them is supported by sensitive data discovery.

You can view the ones that are selected and customize (disabling the ones that are enabled or enabling the ones that are not enabled yet) on the data discovery page. To access this page, open the Defender for Cloud dashboard, click **Environment settings** in the **Management** section, and click the **Data sensitivity** tile on the right side, as shown in *Figure 5.20*:

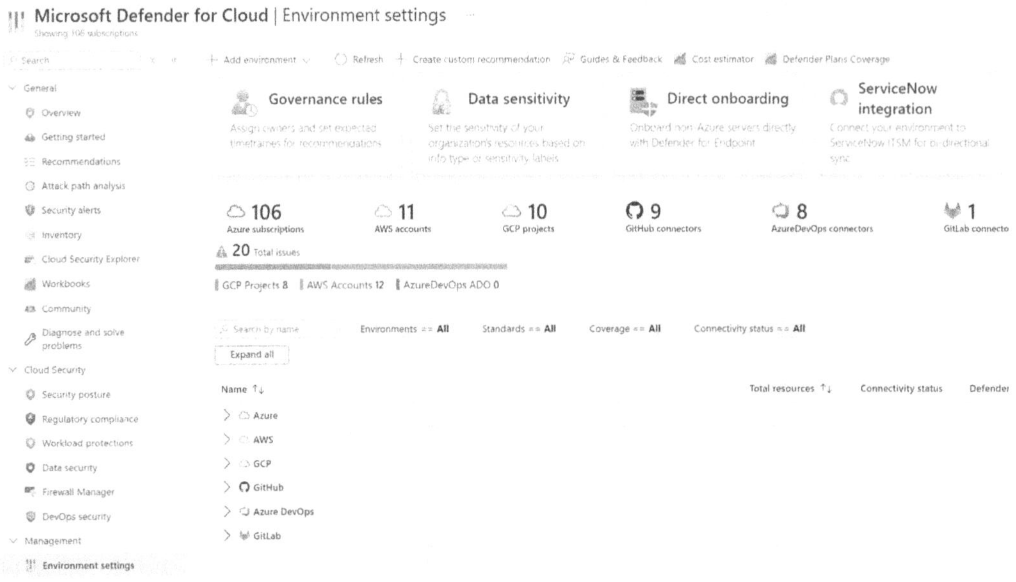

*Figure 5.20: Environment settings (for better visualization, refer to https://packt.link/ gbp/9781836204879)*

Once you click on this tile, the **Data sensitivity** page appears, as shown in *Figure 5.21*. Notice that, on this page, you see the sensitivity based on pre-defined info types (**Finance**, **PII**, **Credentials**, and **Other**).

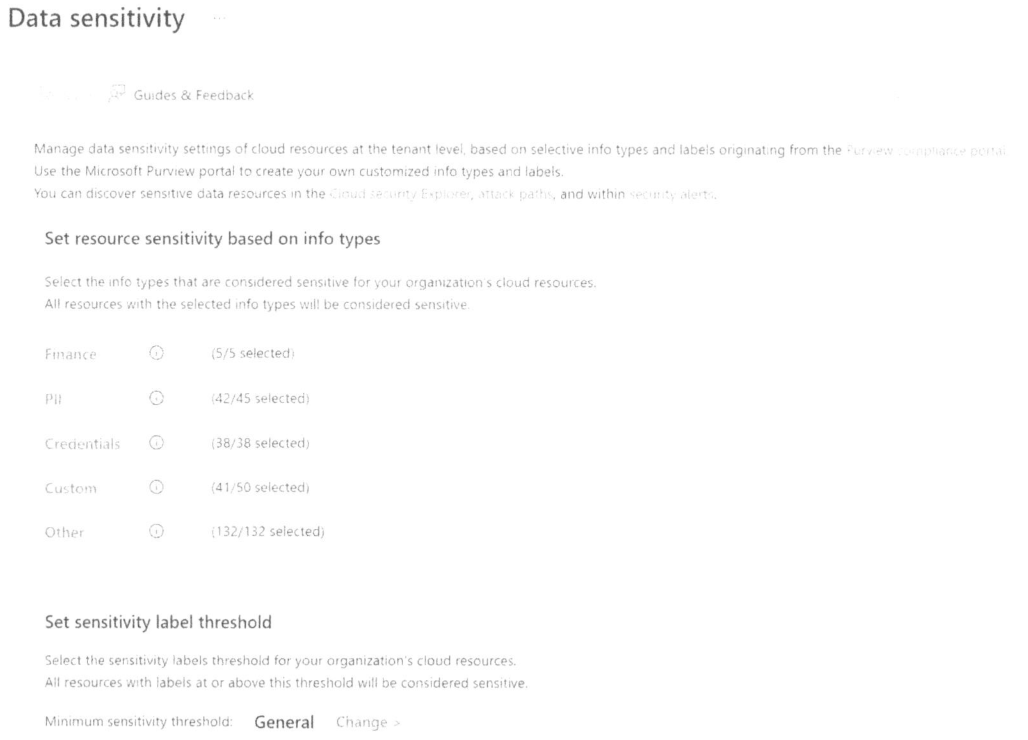

Data sensitivity

Guides & Feedback

Manage data sensitivity settings of cloud resources at the tenant level, based on selective info types and labels originating from the Purview compliance portal. Use the Microsoft Purview portal to create your own customized info types and labels.
You can discover sensitive data resources in the Cloud Security Explorer, attack paths, and within security alerts.

Set resource sensitivity based on info types

Select the info types that are considered sensitive for your organization's cloud resources.
All resources with the selected info types will be considered sensitive.

Finance          ⓘ        (5/5 selected)

PII              ⓘ        (42/45 selected)

Credentials      ⓘ        (38/38 selected)

Custom           ⓘ        (41/50 selected)

Other            ⓘ        (132/132 selected)

Set sensitivity label threshold

Select the sensitivity labels threshold for your organization's cloud resources.
All resources with labels at or above this threshold will be considered sensitive.

Minimum sensitivity threshold:     **General**    Change >

*Figure 5.21: Data sensitivity page (for better visualization, refer to https://packt.link/ gbp/9781836204879)*

Here, you can either go to each info type category (**Finance**, **PII**, **Credentials**, **Other**), or you can click **Custom** to enable other types that are relevant to your business. If you click the **Custom** option, the **Custom info types** blade appears, as shown in *Figure 5.22*:

## Custom info types ✕

Select custom info types

| 🔍 Search |
|---|

🔘 All

☑ Canada Passport Number copy

☑ DLP.PBI.KW

☑ DSR 23030

☐ MS.DSR.1000_Talents

☐ MS.DSR.DLP.EmailCredTuning

☑ MS.DSR.DLP.TRiPDFM

☐ MS.DSR.DV1

☐ MS.DSR.DV2

☐ MS.DSR.DV3

☐ MS.DSR.DV4

☐ MS.DSR.DV5

☐ MS.DSR.DV6

☑ MS.DSR.KW_KPH.AdditionalChecks.2ndElement

☑ MS.DSR.KW_KPH.AdditionalChecks.Anywhere

☑ MS.DSR.KW_KPH.AdditionalChecks.AnywhereQuotes

☑ MS.DSR.KW_KPH.AdditionalChecks.Proxmity

☑ MS.DSR.L1

☑ MS.DSR.O1

**Apply**  Cancel

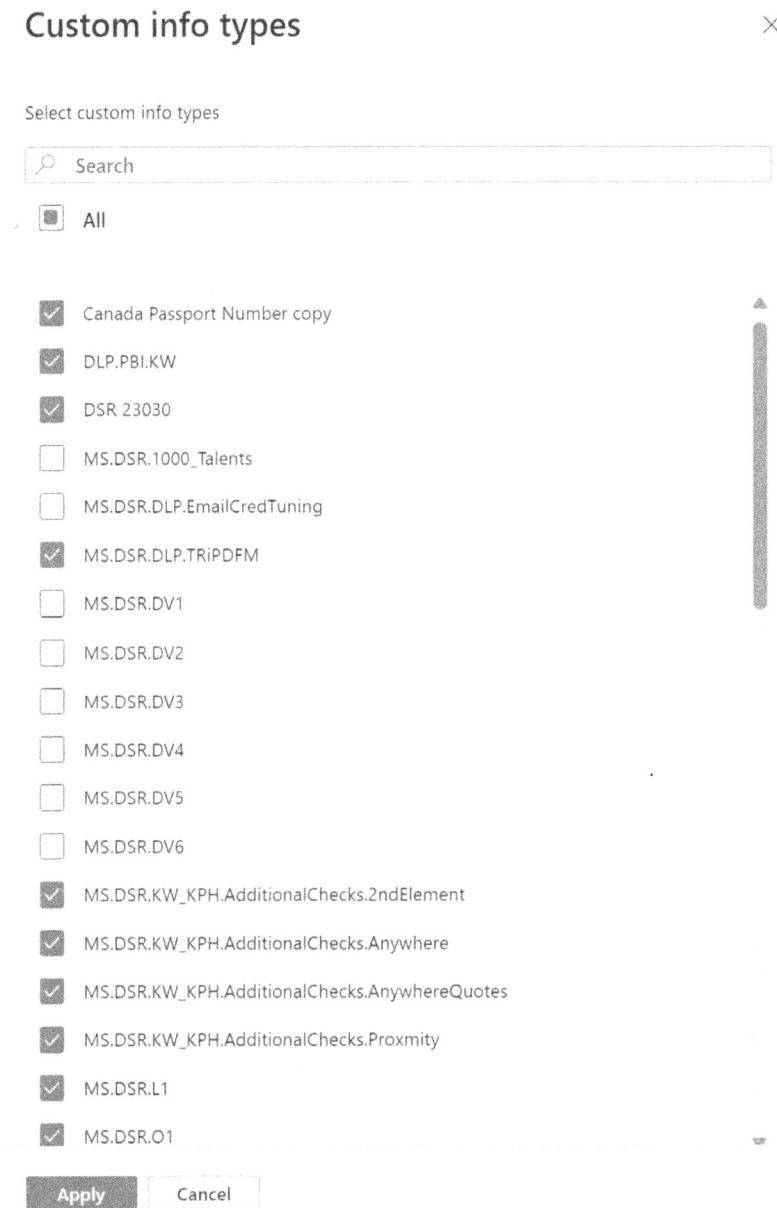

Figure 5.22: Custom info types blade

On this page, you can enable or disable the info types that are relevant to your business. After that, you can click the **Apply** button and then the **Save** button to commit the changes. Keep in mind that this is an optional process, and you should make these changes only if there is a business justification for doing so.

On the **Data sensitivity** page (the bottom part of the page), you can also set the sensitivity label threshold. All resources with labels at or above this threshold will be considered sensitive. If you click the **Change** option, the **Sensitivity label threshold** blade appears, as shown in *Figure 5.23*:

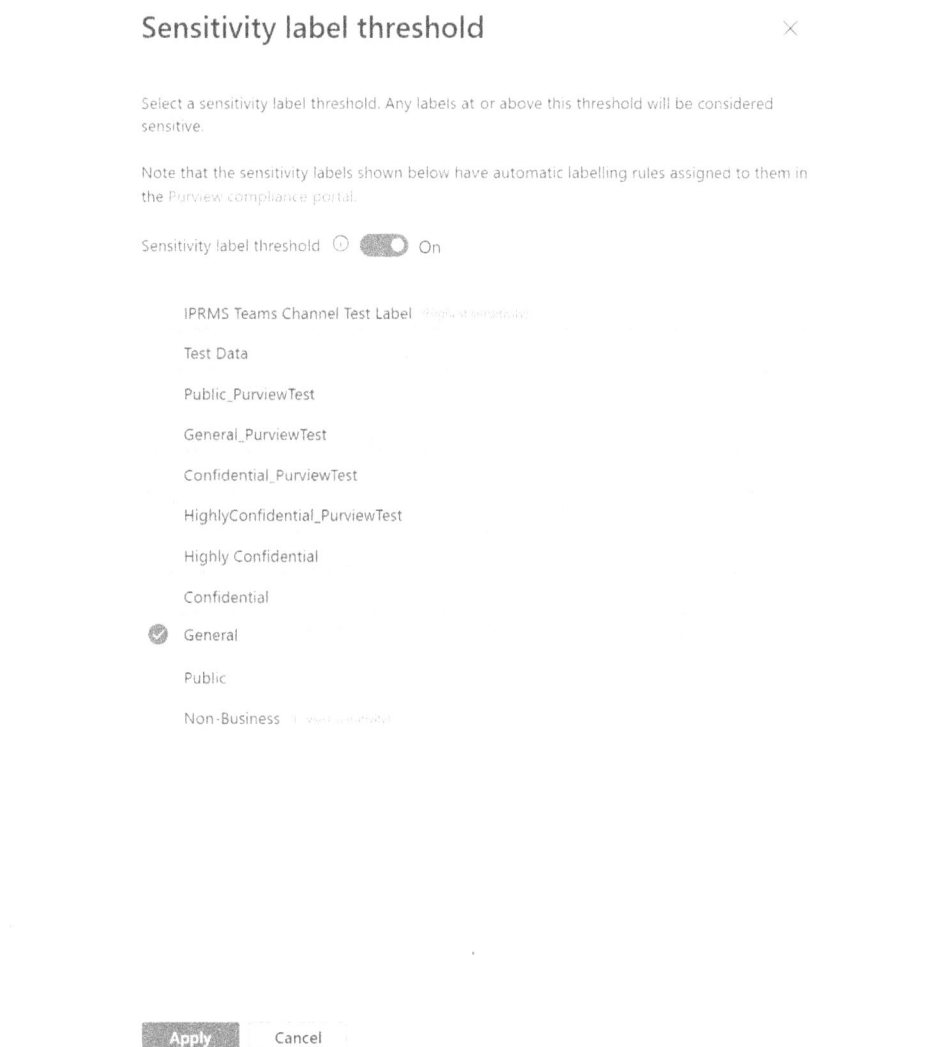

*Figure 5.23: Sensitivity label threshold blade*

By default, the threshold is already on, and the selected label represents the lowest setting that should be considered sensitive in your organization. For example, if you select **Confidential** as minimum, then **Highly Confidential** is also considered sensitive. However, **General**, **Public**, and **Non-Business** are not considered sensitive. Again, this is an optional setting. Since the default covers the most important scenarios, you should only change this setting if there is a business justification to do so.

While there are multiple locations on the Defender for Cloud dashboard that surface where data sensitivity is located, some organizations may have dedicated teams to deal with data security. For those teams, they can use the data security dashboard.

## Data security dashboard

One of the pillars of any security hygiene is visibility, and for the data security posture, this is no different. While the engine that performs the discovery of sensitive data brings you insights in different locations, you may feel a bit overwhelmed with the different places available in the Defender for Cloud dashboard where sensitive data is surfaced. To address this user experience, the Defender for Cloud team created a centralized location to see all data security-related insights—the data security dashboard.

To access this functionality, open the Defender for Cloud dashboard and click **Data security** in the **Cloud Security** section. *Figure 5.24* has an example of what a populated data security dashboard looks like:

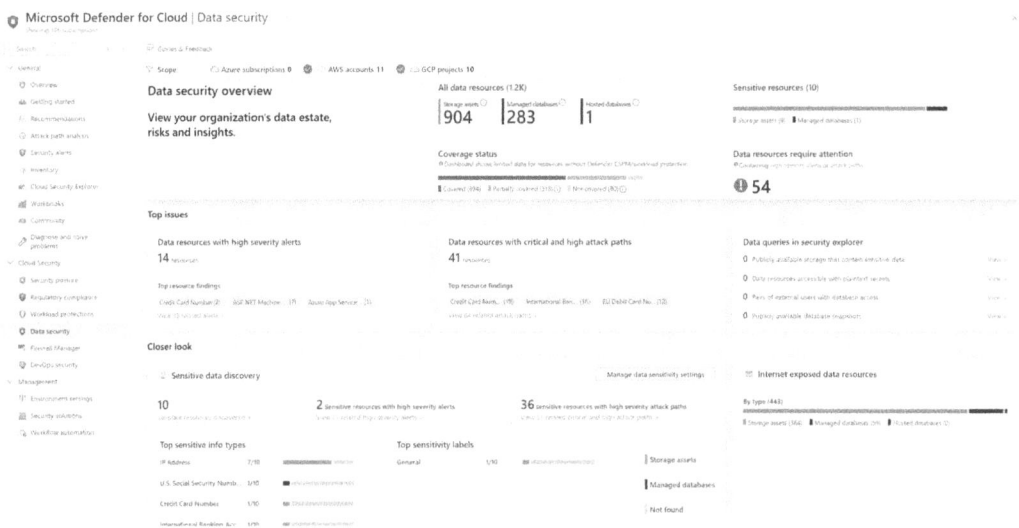

*Figure 5.24: Data security dashboard (for better visualization, refer to https://packt.link/gbp/9781836204879)*

The first part (top) of the dashboard (*Figure 5.25*) shows the number of storage assets, managed databases, and hosted databases. It also shows the coverage status, which basically means that full coverage is when Defender CSPM, Defender for Storage, and Defender for Database are enabled. The reason for that is that while Defender CSPM is the primary plan that needs to be enabled for the data security posture, these workloads also have threat protection available, and without the threat protection plan enabled for these workloads, you will not have visibility about current attacks against these workloads. If the subscription has Defender CSPM enabled but the threat protection plans are not enabled for the data-related workloads (such as Storage and Databases), then it will fit in the **Partially covered** category. The **Not covered** category means that none of the plans (including Defender CSPM) are enabled.

All data resources (1.2K)

| Storage assets | Managed databases | Hosted databases |
|---|---|---|
| 904 | 283 | 1 |

Coverage status
ⓘ Dashboard shows limited data for resources without Defender CSPM/workload protection

■ Covered (694)    ▨ Partially covered (318) ⓘ    Not covered (80) ⓘ

Sensitive resources (10)

▮ Storage assets (9)   ▮ Managed databases (1)

Data resources require attention
ⓘ Containing high severity alerts or attack paths

❶ 54

*Figure 5.25: First part of the dashboard (for better visualization, refer to https://packt.link/gbp/9781836204879)*

On the right, you have a count for sensitive resources, which gives you a breakdown of storage assets and managed databases. While, so far, these insights were more related to the data security posture, on the lower right side of this pane, you will have **Data resources require attention**, and you may see a hyperlink to alerts. In the example shown in *Figure 5.24*, you have **high severity** alerts. This means that there are, or there were, attacks against one of these data-related workloads (storage or databases). This information is only available if you have the threat protection plan for the data-related workload enabled.

In the second part of the dashboard, you have what is called the **Top issues** section, as shown in the example in *Figure 5.26*:

Top issues

| Data resources with high severity alerts | Data resources with critical and high attack paths | Data queries in security explorer |
|---|---|---|
| 14 instances | 41 instances | 0 Publicly available storage that could with sensitive data |
| | | 0 Data resources accessible with plaintext secrets |
| Top resource findings | Top resource findings | 0 Point of entry to users with database access |
| Credit Card Number (2)   ASP.NET Machine... (3)   Azure App Service... (7) | Credit Card Num... (19)   International Bank... (16)   EU Debit Card No... (12) | 0 Publicly available database endpoints |

*Figure 5.26: Second part of the dashboard (for better visualization, refer to https://packt.link/gbp/9781836204879)*

The second part (middle) of the dashboard is divided into three sections:

- Data resources with high severity alerts
- Data resources with critical and high attack paths
- Data queries in security explorer

These options are self-explanatory, and which one will be useful for you depends on what you are trying to accomplish. For example, if you work for the posture management team, and you are trying to be more proactive in your approach, you may want to look in the **Data resources with critical and high attack paths** section to address those attack paths before a threat action is able to exploit those paths.

The last part (bottom) of the dashboard is the **Closer look** section, which provides very useful information about sensitive data discovery breakdown, and internet-exposed data resources, as shown in the example in *Figure 5.27*:

*Figure 5.27: Third part of the dashboard (for better visualization, refer to https://packt.link/ gbp/9781836204879)*

> You can watch this (`https://bit.ly/CNAPPBook18`) episode of Defender for Cloud in the field where I interviewed the PM responsible for creating this dashboard. She gives a good demonstration of how to use this dashboard.

# Summary

In this chapter, we discussed how to onboard Defender CSPM. You learned about the use of the attack path to disrupt potential attacks and how to use the risk-based recommendations to prioritize what is important for your environment. Lastly, you learned about the data security posture, how discovery takes place, how to customize data sensitivity, and how to use the data security dashboard.

In the next chapter, you will learn how to onboard AWS and GCP to be managed in Defender for Cloud.

## Notes

1. You can see the list of supported data discovery types at `https://bit.ly/CNAPPBook17`.

2. Databases are scanned on a weekly basis.

3. You can see the list of sensitive information types at `https://bit.ly/CNAPPBook16`.

## Additional resources

- Watch the author Yuri Diogenes interviewing the developer responsible for creating the recommendation prioritization capability in Defender for Cloud in this episode of *Defender for Cloud in the Field*: `https://bit.ly/CNAPPBook19`

- Watch the author Yuri Diogenes interviewing the PM responsible for the Data Security Posture strategy in Defender for Cloud in this episode of *Defender for Cloud in the Field*: `https://bit.ly/CNAPPBook20`

- Watch the author Yuri Diogenes demonstrating the Attack Path capability in his presentation at Ignite 2023: `https://bit.ly/CNAPPBook21`

## Join our community on Discord

Read this book alongside other users. Ask questions, provide solutions to other readers, and much more.

Scan the QR code or visit the link to join the community.

`https://packt.link/SecNet`

# 6

# Multicloud

Organizations continue to adopt multiple cloud providers to provision different workloads. In a study done by Flexera[1] (State of Cloud Report), 92% of the respondents were using multicloud. While this decentralized approach may address different business needs, it also comes with challenges.

The lack of visibility and control over these workloads are the primary challenges inherent in this type of deployment. As you learned so far, CNAPP can overcome those challenges since it allows a centralized view of multicloud workloads.

Microsoft Defender for Cloud brings you centralized visibility of the security state of your workloads in Azure, **Amazon Web Services** (**AWS**), **Google Cloud Platform** (**GCP**), and on-premises. Defender for Cloud provides multicloud insights that can be used to prioritize recommendations and disrupt attacks using the attack path, including lateral movement across cloud providers in a multicloud environment.

This chapter covers:

- Connecting with AWS
- Connecting with GCP

## Connecting with AWS

To gain visibility to AWS resources and improve the security posture of your AWS environment, you will need to first deploy the AWS connector in Defender for Cloud. This connector is available for free as part of the Foundational CSPM, which gives the same foundational capabilities, such as security recommendations and secure score.

If you need more advanced CNAPP capabilities such as attack path analysis, you will need to enable Defender CSPM.

To better understand how these components work in a multicloud environment, let's see in more detail how Defender for Cloud connects with AWS, starting with the diagram shown in *Figure 6.1*:

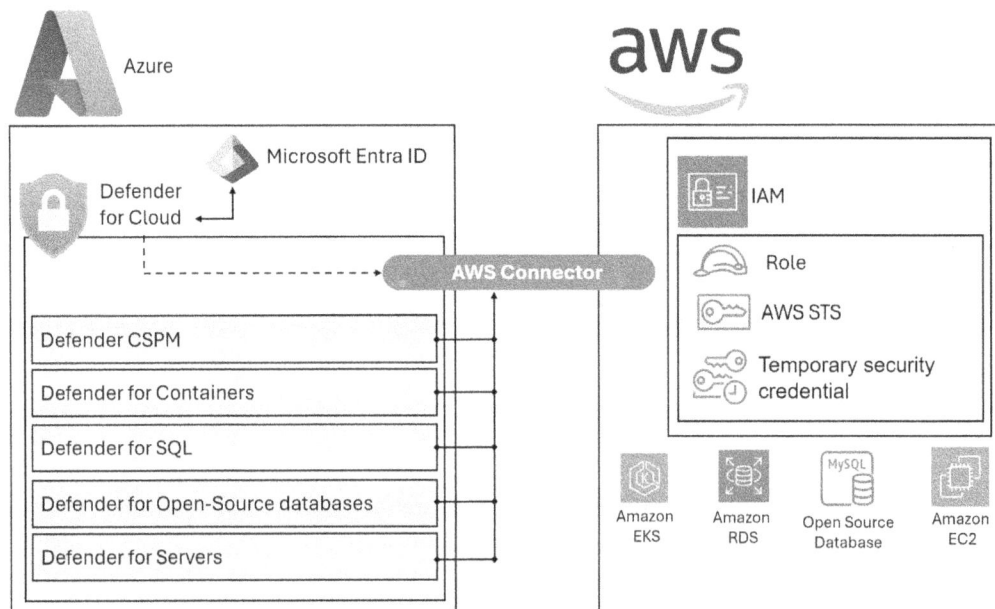

*Figure 6.1: Defender for Cloud connectivity with AWS*

The first part of this connectivity is to ensure that the AWS connector is properly configured. You will need Contributor permission in the Azure subscription, and Administrator permission on the AWS account.

During this configuration, the federated authentication between Defender for Cloud and AWS is created. The resources related to the authentication are created as a part of the CloudFormation template deployment, which uses an identity provider (OpenID Connect) and the **Identity and Access Management (IAM)** roles with a federated principal. Defender for Cloud acquires a Microsoft Entra token with a validity lifetime of 1 hour. This token is signed by the Microsoft Entra ID using the RS256 algorithm. The Entra token is exchanged with AWS short-living credentials. After the Microsoft Entra token is validated by the AWS identity provider, AWS STS exchanges the token with AWS short-living credentials, which are used by Defender for Cloud to scan the AWS account and discover the resources.

The conditions defined for the role level are used for validation within AWS and allow only the Microsoft Defender for Cloud CSPM application access to the specific role (and not any other Microsoft token).

Defender for Cloud will start populating the AWS recommendations[2] in the Defender for Cloud dashboard. However, if you need to protect the workloads in AWS, you will need to enable other Defender for Cloud plans, as shown in *Figure 6.1*. For example, if you need to protect Amazon EC2 instances, you will need to enable Defender for Servers at the connector level.

## Deploying the AWS connector

To deploy the AWS connector in Defender for Cloud, follow the steps below:

1. Open the Defender for Cloud dashboard and click **Environment settings**, in the **Management** section. The **Environment settings** page will appear, as shown in *Figure 6.2*:

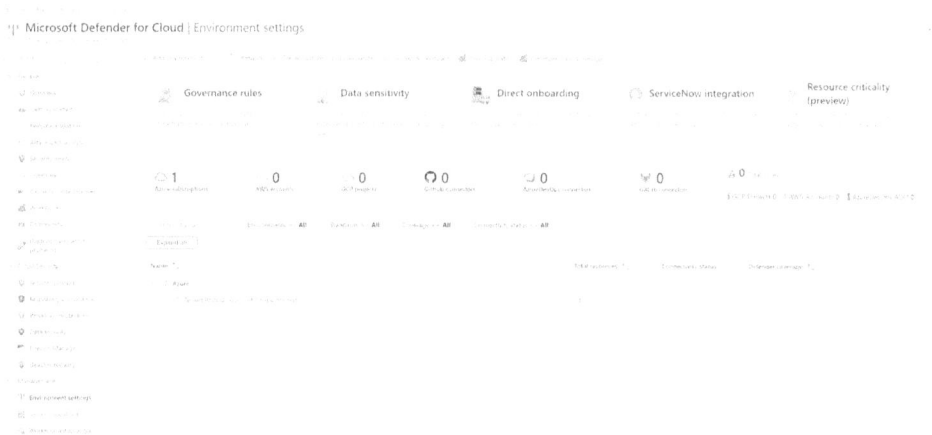

*Figure 6.2: Environment settings (for better visualization, refer to https://packt.link/gbp/9781836204879)*

2. On the top menu bar, click the **+ Add environment** button and click the **Amazon Web Services** option. The **Add AWS account** page will appear, as shown in *Figure 6.3*:

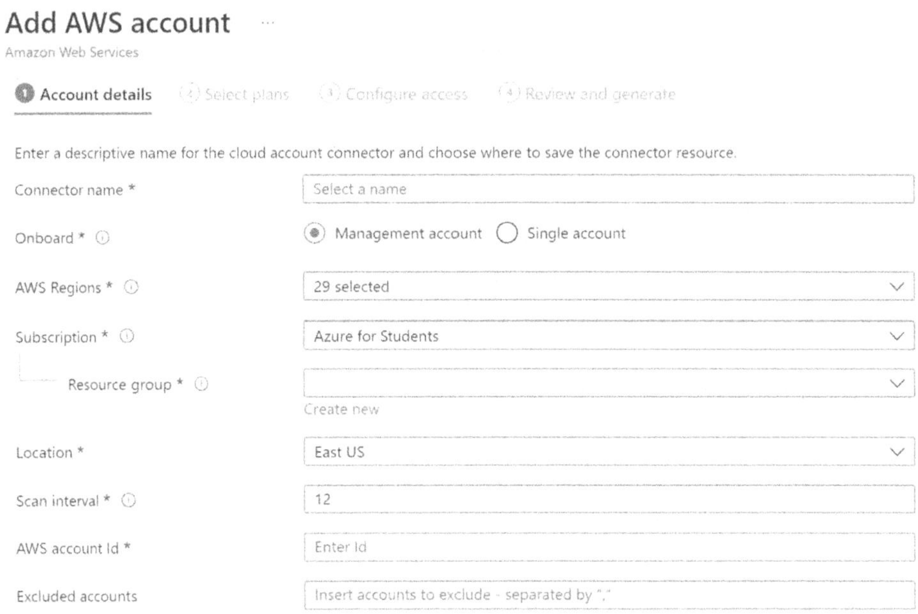

## Add AWS account

Amazon Web Services

1. **Account details**    2. Select plans    3. Configure access    4. Review and generate

Enter a descriptive name for the cloud account connector and choose where to save the connector resource.

| | |
|---|---|
| Connector name * | Select a name |
| Onboard * ⓘ | ⦿ Management account    ◯ Single account |
| AWS Regions * ⓘ | 29 selected     ⌄ |
| Subscription * ⓘ | Azure for Students     ⌄ |
| Resource group * ⓘ | ⌄ |
| | Create new |
| Location * | East US     ⌄ |
| Scan interval * ⓘ | 12 |
| AWS account Id * | Enter Id |
| Excluded accounts | Insert accounts to exclude - separated by "," |

*Figure 6.3: Add AWS account page (for better visualization, refer to https://packt. link/gbp/9781836204879)*

3. Type a name for this connector under the **Connector name** field.

4. Select the appropriate onboarding method (by default, **Management account**[3] is selected). If you are onboarding just one single AWS account, select **Single account**. If you need to onboard an AWS management account, select the **Management account** option. For this example, select **Single account**.

5. The **AWS Regions** field has all available regions selected by default, which means that Defender for Cloud will scan all those regions to try to find resources.

Next, you have the option to select the Azure subscription, which should be the same as where Defender for Cloud is enabled, and which resource group will be used by this connector. The **Resource group** selection is important because only users who have IAM permissions to that resource group (in which the security connector resides) can see recommendations in the Defender for Cloud dashboard for the AWS connector.

6. In the **Location** field, select the region in which the connector will be created.

7. In the **Scan interval** field, type the appropriate interval (between 1 hour and 24 hours) in which Defender for Cloud will scan your cloud resources.

8. In the **AWS account ID** field, type the identification number of your AWS account.

9. Since you changed the connector for a single account, the last field shown in *Figure 6.3* will not appear anymore. Click the **Next: Select plans >** button. The **Select plans** tab will appear, as shown in *Figure 6.4*:

*Figure 6.4: Selecting Defender for Cloud plans (for better visualization, refer to https://packt.link/gbp/9781836204879)*

Notice that you have two main sections, one on top for **Cloud Security Posture Management (CSPM)**, where you have **Foundational CSPM (Free)** and **Defender CSPM**. Leave this section as is, with both plans selected as **On**. The next section is **Cloud Workload Protection (CWP)**, which has the different threat protection plans for different workloads. For now (since the plans will be explained in later chapters), select **Off** for all these plans. Click the **Next: Configure access >** button.

The **Configure access** tab will appear, as shown in *Figure 6.5*:

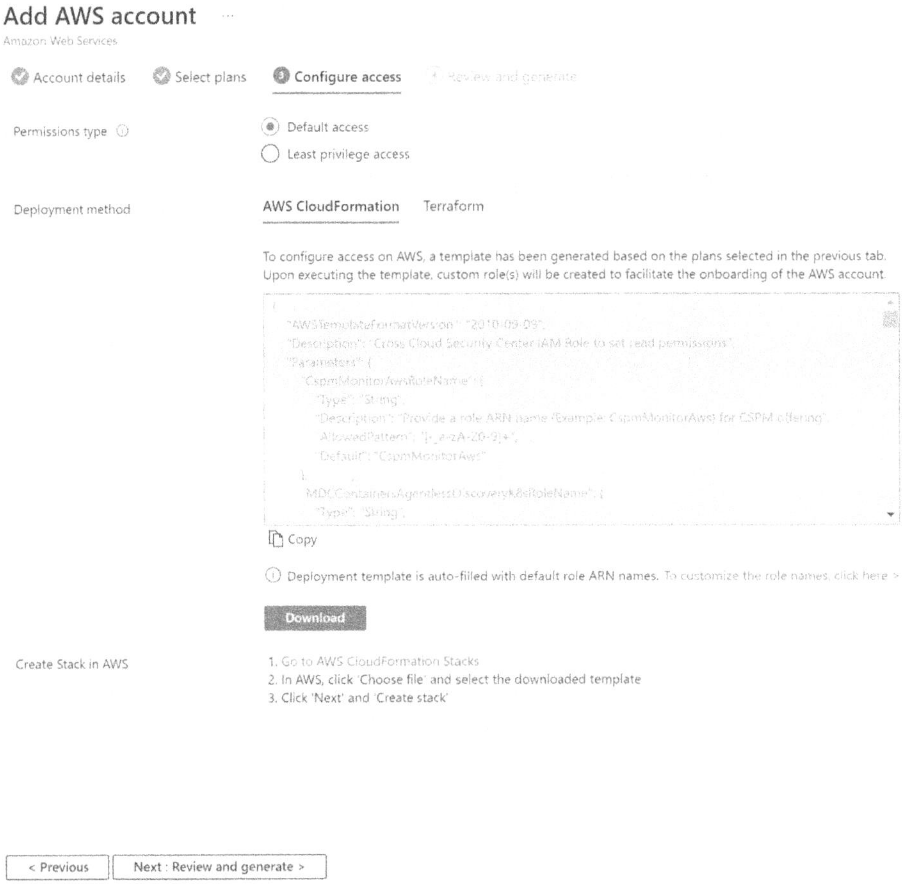

*Figure 6.5: Configure access page (for better visualization, refer to https://packt.link/gbp/9781836204879)*

If you select the **Least privilege access** option on the **Configure access** page, you will receive notifications on any new roles and permissions that are required to get full functionality for connector health.

10. On this page, click the **Download** button to download the template to your computer. Then, follow the three steps in the **Create Stack in AWS** section of the page. After finishing the second step of this list, you should see an AWS dashboard similar to *Figure 6.6*:

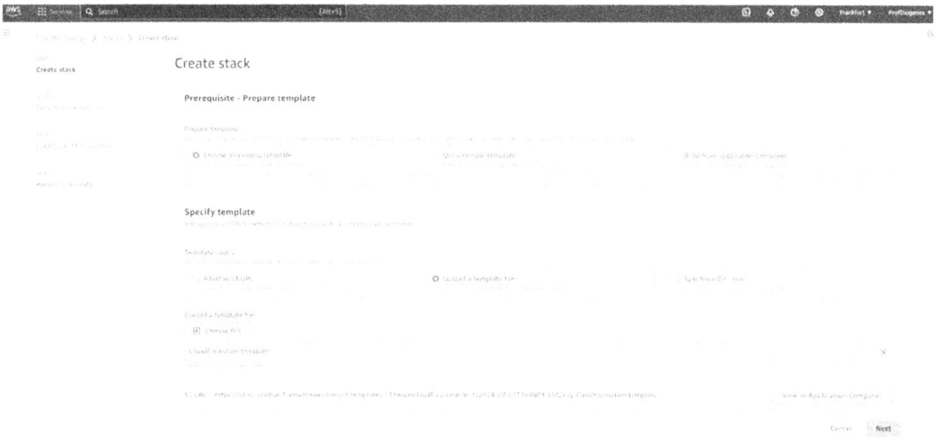

*Figure 6.6: AWS Create stack page (for better visualization, refer to https://packt.link/gbp/9781836204879)*

11. On the AWS **Create stack** page, click the **Next** button, which is the third step highlighted in the Defender for Cloud AWS connector (*Figure 6.5*). You will be requested to provide a name for your stack, as shown in *Figure 6.7*:

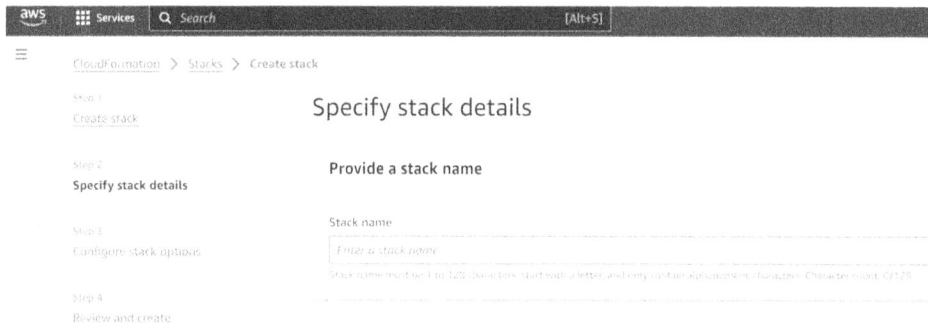

*Figure 6.7: Stack name (for better visualization, refer to https://packt.link/gbp/9781836204879)*

12. After typing the name, click the **Next** button and you will see the **Configure stack options** page. Here, you can (optionally) customize stack failure options, stack policy, rollback configuration, and notifications. For this example, leave all options as is and click the **Next** button.

The **Review and create** page will appear, as shown in *Figure 6.8*:

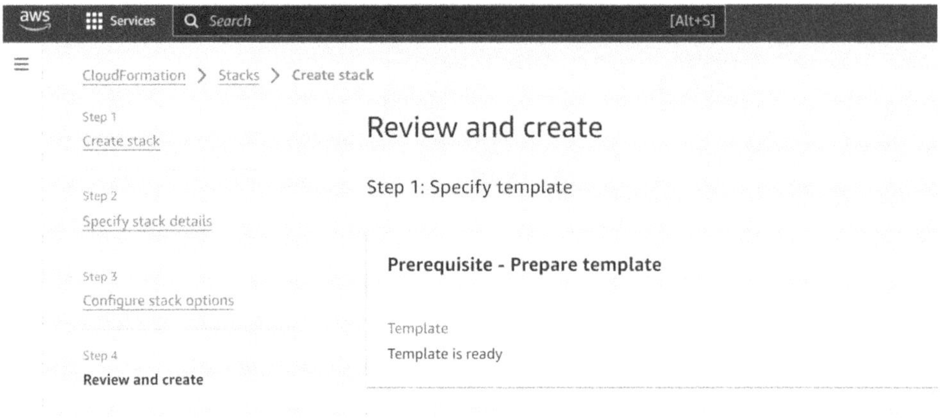

*Figure 6.8: Review and create page*

13. Select the checkbox next to **I acknowledge that AWS CloudFormation might create IAM resources with custom names** and click the **Submit** button. You will see the creation of the connector, where the initial status appears as **CREATE_IN_PROGRESS**, as shown in *Figure 6.9*:

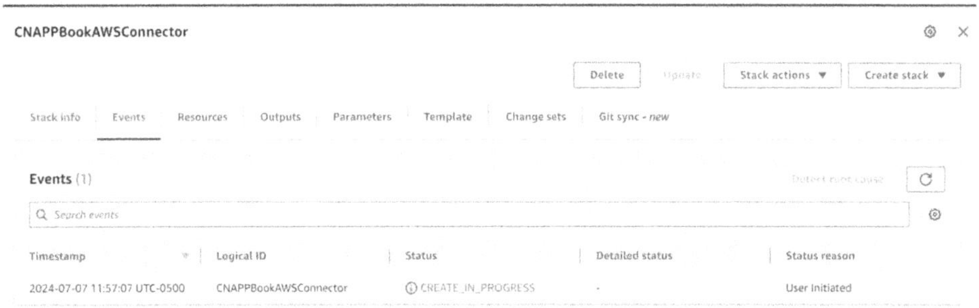

*Figure 6.9: Creating the AWS connector on the AWS side (for better visualization, refer to https://packt.link/gbp/9781836204879)*

14. When the status changes to **CREATE_COMPLETE**, go back to the Defender for Cloud page, where you were in *step 11*, click the **Next: Review and generate >** button, and then click the **Create** button. A new message will pop up notifying you that a new connector was created, as shown in *Figure 6.10*:

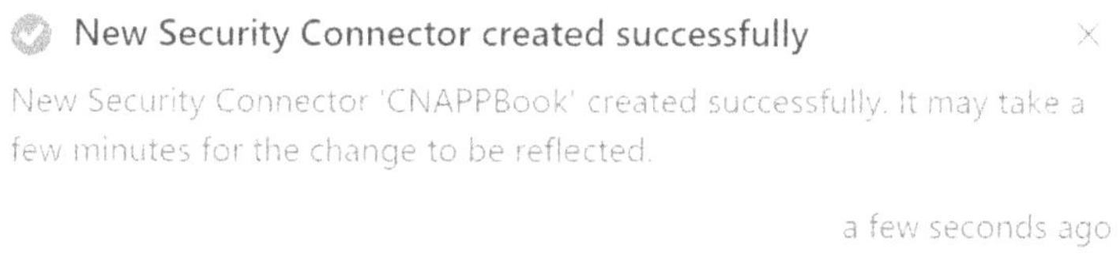

New Security Connector created successfully

New Security Connector 'CNAPPBook' created successfully. It may take a
few minutes for the change to be reflected.

a few seconds ago

Figure 6.10: Notification that a new connector was created

After a couple of minutes, refresh the **Environment settings** page in Defender for Cloud, and you
will see the new AWS account with the connectivity status showing as in progress, as shown in
*Figure 6.11*:

| Name ↑↓ | Total resources ↑↓ | Connectivity status | Defender coverage ↑↓ |
| --- | --- | --- | --- |
| ⌄ ◻ Azure | | | |
| ⟩ ◻ Tenant Root Group (1 of 1 subscriptions) | 2 | | |
| ⌄ AWS | | | |
| ◻ (CNAPPBook) | | ⊕ In progress | 1/4 plans |

Figure 6.11: AWS account connectivity status (for better visualization, refer to https://packt.
link/gbp/9781836204879)

Notice also that in the example of *Figure 6.11*, the **Defender coverage** column shows as **1 out of 4**,
and that's because, during the configuration, you selected only **Defender CSPM**, which, for this
example, is all you want.

## Reviewing initial assessment

You can view the initial assessment using the **Recommendations** dashboard. The recommendations for AWS are derived from **Microsoft Cloud Security Benchmark (MCSB)** and AWS Foundational Security Best Practices standard. To stay focused on only AWS recommendations, you
should ensure that the **Scope** bar has only **AWS accounts** selected, as shown in *Figure 6.12*:

⛶ Scope   ◯ ◻ Azure subscriptions 1   ◻ AWS accounts 1   ◯ ◻ GCP projects 0   ◯ ◯ GitHub connectors 0   ◯ ◻ AzureDevOps connectors 0   ◯ ◻ GitLab connectors 0

Figure 6.12: Changing the Scope to AWS account only (for better visualization, refer to https://
packt.link/gbp/9781836204879)

At this point, you should be able to see only the risk-based factor recommendations for your new AWS environment in Defender for Cloud, as shown in the example in *Figure 6.13*:

Figure 6.13: Risk-based factor recommendations for AWS (for better visualization, refer to https://packt.link/gbp/9781836204879)

Notice that some recommendations have a thunderbolt icon in the **Insights** column. When you see that, it means that there is an automatic fix for that recommendation. Let's open as an example the recommendation that says **VPC's default security group should restricts all traffic**, which has the information in the **Take action** tab (*Figure 6.14*):

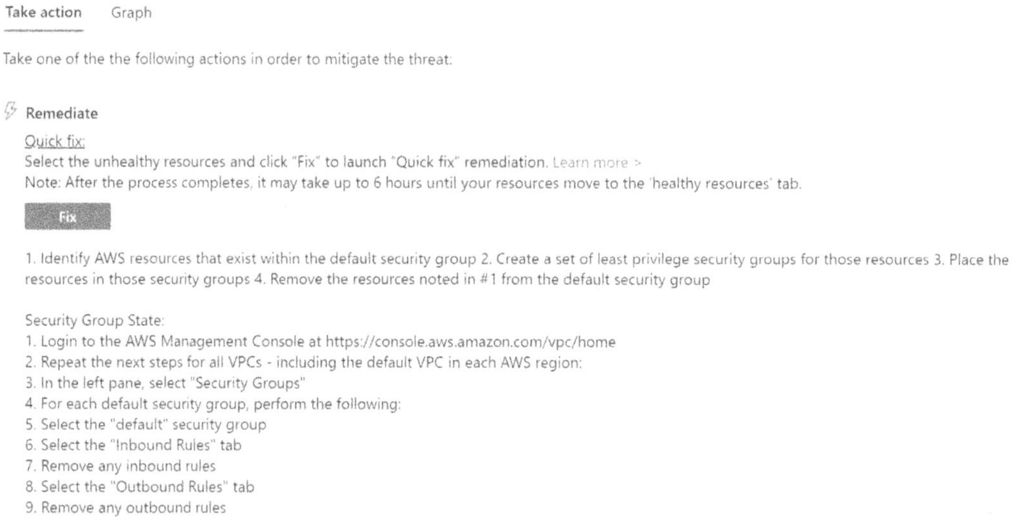

Figure 6.14: Fix button available (for better visualization, refer to https://packt.link/gbp/9781836204879)

The **Fix** button is available for this recommendation, which means that you can expedite the resolution of this recommendation directly from this dashboard. One important reminder that is stated is about the time to refresh; it says, **After the process completes, it may take up to 6 hours until your resources move to the healthy resources tab.** This is important data to keep in mind so that you have the right expectation regarding when the resource will show up as healthy in Defender for Cloud. It is worth mentioning that the refresh interval may vary according to the recommendation.

Defender CSPM performs read-only queries to the AWS resource APIs several times a day. While these read-only queries do not incur any charges (from an Azure and Defender for Cloud perspective), they will be registered in AWS CloudTrail in case you have enabled a trail for read events. These queries will be responsible for presenting the status of your recommendations in the Defender for Cloud dashboard. Keep in mind that logs in AWS CloudTrail can incur charges, the same as if you are using AWS GuardDuty. Read the article *How to better manage cost of API calls that Defender for Cloud makes to AWS* at `https://bit.ly/CNAPPBook25` to learn more about how to optimize this setting.

If you just need to query the resources that you have in AWS, then you can use the **Inventory** dashboard, which is available in the **General** section. There, you can change the **Environment** filter (*Figure 6.15*) to list only the AWS resources:

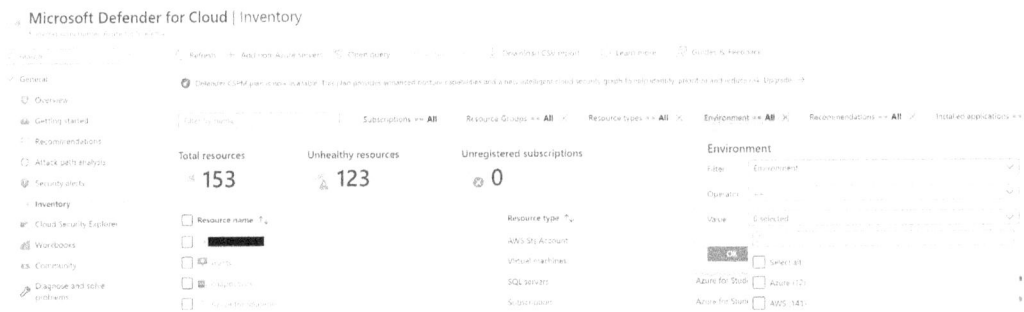

*Figure 6.15: Inventory (for better visualization, refer to https://packt.link/gbp/9781836204879)*

# Connecting with GCP

To gain visibility to GCP resources and improve the security posture of your GCP environment, you will need to first deploy the GCP connector in Defender for Cloud. This connector is available for free as part of the Foundational CSPM, which gives the same foundational capabilities, such as security recommendations and secure score. If you need more advanced CNAPP capabilities such as attack path analysis, you will need to enable Defender CSPM.

To better understand how these components work in a multicloud environment, let's see in more detail how Defender for Cloud connects with GCP, starting with the diagram shown in *Figure 6.16*:

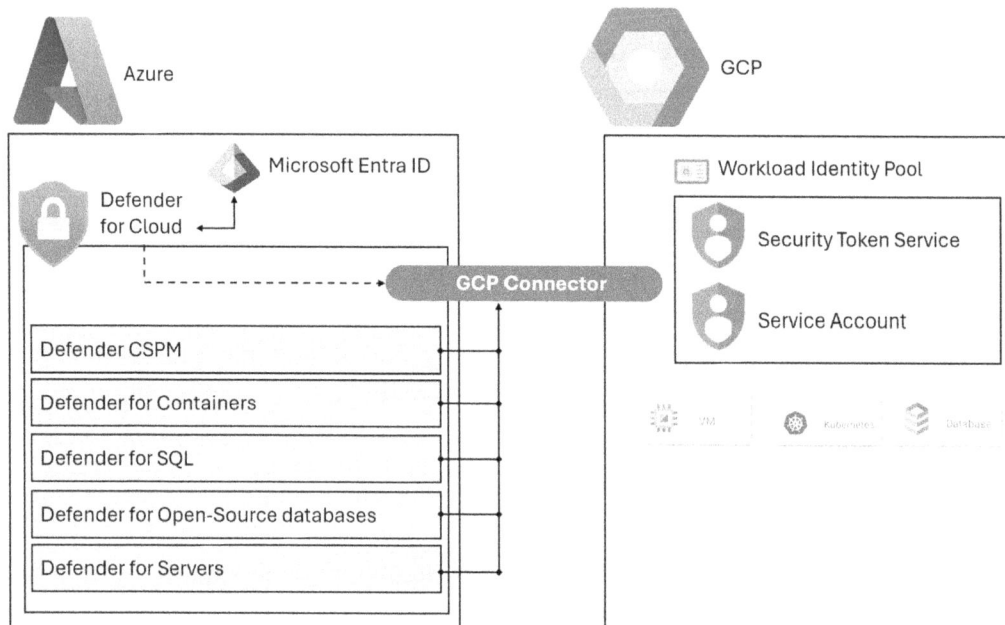

*Figure 6.16: GCP connector*

Defender for Cloud CSPM obtains a Microsoft Entra token signed by Microsoft Entra ID, which uses the RS256 algorithm and is valid for 1 hour. This token is then exchanged with Google's **Security Token Service (STS)** token. Google STS validates the token with the workload identity provider. Once validated, a Google STS token is returned to Defender for Cloud's CSPM service. Defender for Cloud CSPM uses the Google STS token to impersonate the service account and obtain service account credentials, which are then used to scan the project.

Just like the AWS connector, the GCP connector will also give you the option to enable threat protection plans to protect the supported workloads in GCP. To onboard the GCP connector, you will need to have the contributor level in the Azure subscription, and Owner permission on the GCP organization or project.

## Deploying the GCP connector

To deploy the GCP connector in Defender for Cloud, follow the steps below:

1.  Open the Defender for Cloud dashboard and click **Environment settings**, in the **Management** section. The **Environment settings** page will appear, as shown in *Figure 6.17*:

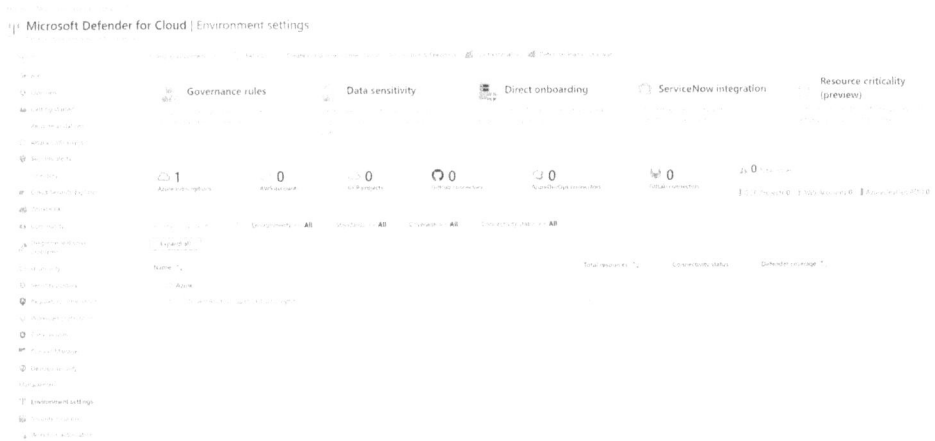

*Figure 6.17: Environment settings (for better visualization, refer to https://packt.link/gbp/9781836204879)*

2. On the top menu bar, click the **+ Add environment** button and click the **Google Cloud Platform** option. The **Add GCP project** page will appear, as shown in *Figure 6.18*:

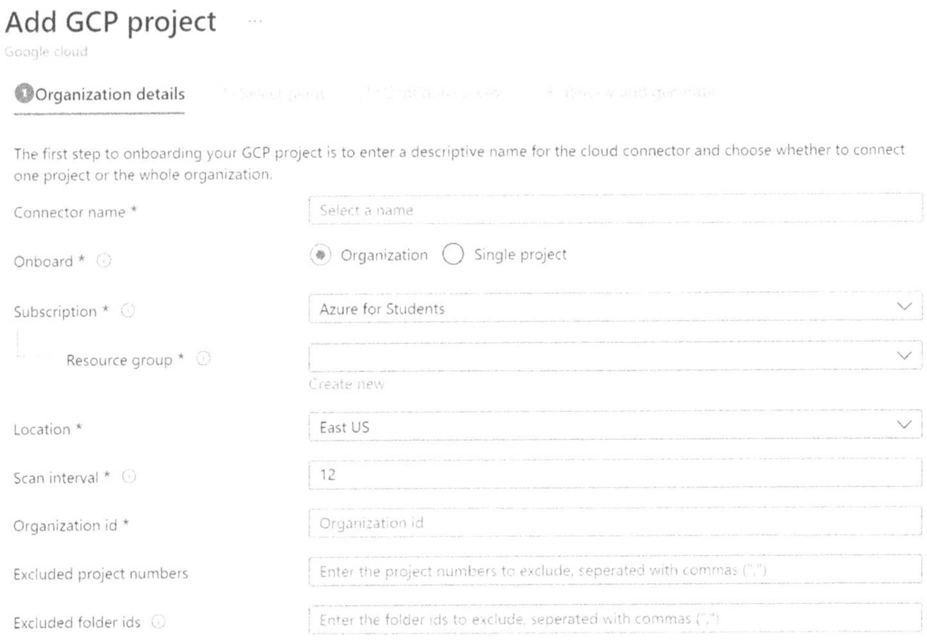

*Figure 6.18: Add GCP project page (for better visualization, refer to https://packt.link/gbp/9781836204879)*

3. Type a name for this connector in the **Connector name** field.

4. Select the appropriate onboarding method (by default, **Organization** is selected). For this example, select **Single project**.

5. Next, you have the option to select the Azure subscription, which should be the same as where Defender for Cloud is enabled, and which resource group will be used by this connector.

6. In the **Location** field, select the region in which the connector will be created.

7. In the **Scan interval** field, type the appropriate interval (between 1 hour and 24 hours) in which Defender for Cloud will scan your cloud resources. It is important to mention that some data collectors[4] have a fixed scan interval of 1 hour.

8. Since you selected **Single project**, you will see two different fields at the end that are not part of the screen shown in *Figure 6.18*. The fields are **GCP Project Number**, where you need to type the number of your GCP project, and **GCP Project ID**, where you need to type the ID for this project. Both pieces of information are available in the main GCP dashboard.

9. Click the **Next: Select plans >** button, and the **Select plans** tab will appear. Make sure the selection is similar to *Figure 6.19* (only Defender CSPM):

Figure 6.19: Selecting the plans (for better visualization, refer to https://packt.link/gbp/9781836204879)

10. After disabling the CWP plans and leaving only the CSPM plans enabled, click the **Next: Configure access >** button. The **Configure access** tab will appear, as shown in *Figure 6.20*:

*Figure 6.20: Configure access (for better visualization, refer to https://packt.link/ gbp/9781836204879)*

Follow the instructions on this page to copy the script and execute it in GCP Cloud Shell. The script is generated based on the plans that you selected to onboard (*Figure 6.19*). The script creates all required resources on your GCP environment so that Defender for Cloud can function properly.

During the execution of this script in your GCP project, you may receive the message below in the GCP Cloud Shell:

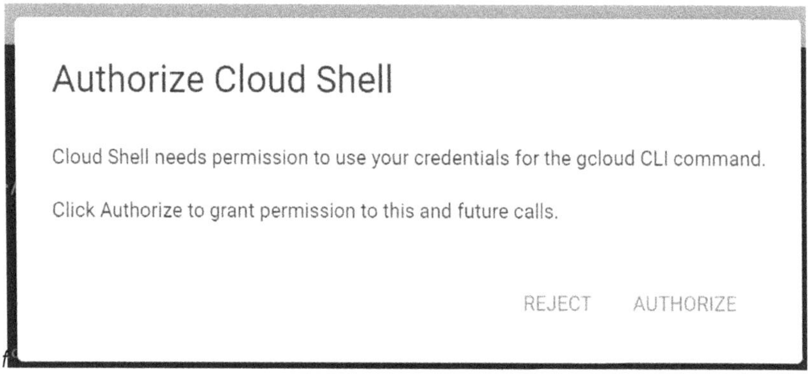

## Authorize Cloud Shell

Cloud Shell needs permission to use your credentials for the gcloud CLI command.

Click Authorize to grant permission to this and future calls.

REJECT     AUTHORIZE

*Figure 6.21: GCP Cloud Shell authorization message*

11. The script may take some minutes to finalize. When it finishes running, it will be back to the normal Cloud Shell prompt. At this point, go back to Defender for Cloud and click the **Next: Review and generate >** button.

12. In the **Review and generate** tab, click the **Create** button. You should see a similar notification, as shown in the example in *Figure 6.22*:

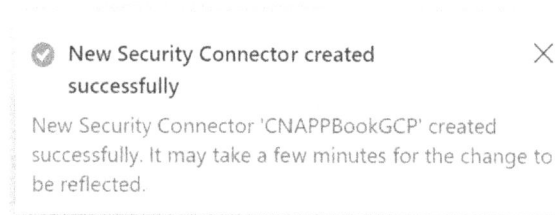

New Security Connector created successfully                              ✕

New Security Connector 'CNAPPBookGCP' created successfully. It may take a few minutes for the change to be reflected.

*Figure 6.22: Notification that a new connector was created*

13. After some minutes, you will see the addition of a new connector under **Environment settings,** and the connectivity status showing as **In progress**, as shown in *Figure 6.23*:

| Name ↑↓ | Total resources ↑↓ | Connectivity status | Defender coverage ↑↓ |
|---|---|---|---|
| ⌄ ☁ Azure | | | |
| ⟩ ⦾ Tenant Root Group (1 of 1 subscriptions) | 3 | | |
| ⌄ ☁ AWS | | | |
| ▩ ▬▬▬▬ CNAPPBook | 141 | ⊘ Connected | 1/4 plans |
| ⌄ ☁ GCP | | | |
| ⊚ ▬▬▬▬ CNAPPBookGCP | | ⊕ In progress… | 1/4 plans |

*Figure 6.23: Connectivity status for the new GCP connector (for better visualization, refer to https://packt.link/gbp/9781836204879)*

### Reviewing initial assessment

Just like explained before with the AWS connector, you can use the risk-based **Recommendations** page to visualize the assessments that were done and prioritize the remediation. GCP recommendations are derived from **Microsoft Cloud Security Benchmark (MCSB)** and GCP Default. You can also use the **Inventory** page in Defender for Cloud to visualize all GCP resources that were discovered. All the steps are the same, the only difference is that now you will filter for the GCP environment only.

## Summary

In this chapter, we discussed how to connect with AWS. You learned about the prerequisites, the architecture of the solution, and how to configure the AWS connector. You learned how to leverage the risk-based recommendations to prioritize what is important to remediate in the AWS environment. You also learned how to connect with GCP, and how to configure the GCP connector.

In the next chapter, you will learn how to onboard AWS and GCP to be managed in Defender for Cloud.

## Notes

1. You can download the Flexera State of Cloud Report from `https://bit.ly/CNAPPBook22`.

2. You can review the list of AWS recommendations supported by Defender for Cloud at `https://bit.ly/CNAPPBook23`.

3. Management account is an AWS term; to learn more about management accounts, visit `https://bit.ly/CNAPPBook24`.

4. By the time this book was published, the GCP data collectors that were using a fixed scan time of 1 hour were `ComputeInstance`, `ArtifactRegistryRepositoryPolicy`, `ArtifactRegistryImage`, `ContainerCluster`, `ComputeInstanceGroup`, `ComputeZonalInstanceGroupInstance`, `ComputeRegionalInstanceGroupManager`, `ComputeZonalInstanceGroupManager`, and `ComputeGlobalInstanceTemplate`.

## Additional resources

- Watch the author Yuri Diogenes interviewing the PM responsible for the GCP connector, during this episode of *Defender for Cloud in the Field*: `https://bit.ly/CNAPPBook26`

- For additional insights about multicloud solutions with Defender for Cloud, read the Multicloud Design Considerations Guide: `https://bit.ly/CNAPPBook27`

# Leave a Review!

Thank you for purchasing this book from Packt Publishing—we hope you enjoy it! Your feedback is invaluable and helps us improve and grow. Once you've completed reading it, please take a moment to leave an Amazon review; it will only take a minute, but it makes a big diff erence for readers like you. Scan the QR or visit the link to receive a free ebook of your choice.

```
https://packt.link/NzOWQ
```

# 7

# DevOps Security Capabilities

Defender for Cloud first started to incorporate DevOps security capabilities at Ignite 2022 when Microsoft announced the public preview of Defender for DevOps. During that time, the team of PMs that I was managing at Microsoft was fully engaged to help build this plan. We had already been working with Feature PMs and developers since the Defender for DevOps Private Preview release. It was great to see all the investment in that area coming to fruition at Ignite 2022.

Fast-forward to Ignite 2023. Many things changed, including how we approached this area. During my breakout session at Ignite[1], I announced the inclusion of DevOps security capabilities as part of Defender CSPM without extra charge. In other words, that was the end of the Defender for DevOps plan (which originally was going to be a paid plan), and the new phase of Defender CSPM with native DevOps security capabilities. The massive applause in the room reflected the audience's acceptance of that change, everyone loved the direction we were going in.

Defender CSPM's inclusion of DevOps security capabilities reflects the CNAPP architecture and how security posture management needs to be taken into consideration from code to runtime.

This chapter covers:

- DevOps security capabilities in Defender CSPM
- Connecting with GitHub
- Connecting with Azure ADO
- Connecting with GitLab

# DevOps security capabilities in Defender CSPM

DevOps security capabilities are an essential part of CNAPP, and everything starts with posture management. Defender CSPM brings native DevOps security capabilities that allow cloud security administrators to have full visibility of DevOps inventory and the security posture of preproduction application code across multi-pipelines. DevOps security capabilities in Defender CSPM include findings from code, secrets, and open-source dependency vulnerability scans. In addition to that, it also has capabilities to secure **Infrastructure as Code (IaC)** templates and container images. This helps minimize cloud misconfigurations from leaving the code and being provisioned in the production environments.

> By the time this chapter was written, DevOps security capabilities in Defender CSPM supported integration with GitHub Enterprise Cloud, GitLab SaaS, and **Azure DevOps (ADO)**.

It is important to emphasize that the most advanced DevOps security capabilities are exclusively available in Defender CSPM; however, some functionalities are also available using the Foundational CSPM free tier. These functionalities will vary according to the DevOps platform. The table below lists all the capabilities across the different platforms:

| Platform | Capability | CSPM plan | Observations |
|---|---|---|---|
| GitHub | Connect with GitHub repositories | Foundational and Defender CSPM | NA |
| | Recommendations to fix code vulnerabilities | Foundational and Defender CSPM | Requires the enablement of the paid feature in GitHub called GitHub Advanced Security[2], and Microsoft Security DevOps action |
| | Recommendations to discover exposed secrets | Foundational and Defender CSPM | Requires the enablement of the paid feature in GitHub called GitHub Advanced Security |
| | Recommendations to fix open-source vulner-abilities | Foundational and Defender CSPM | Requires the enablement of the paid feature in GitHub called GitHub Advanced Security |
| | Recommendations to fix infrastructure as code misconfigurations | Foundational and Defender CSPM | Requires the enablement of the paid feature in GitHub called GitHub Advanced Security, and Microsoft Security DevOps action |
| | Recommendations to fix DevOps environment misconfigurations | Foundational and Defender CSPM | NA |
| | Code to cloud mapping for Containers | Defender CSPM | Requires Microsoft Security DevOps action |
| | Code to cloud mapping for Infrastructure as Code templates | Defender CSPM | Requires Microsoft Security DevOps action |
| | Attack path analysis | Defender CSPM | Integrate insights from GitHub findings with the attack path |
| | Cloud security explorer | Defender CSPM | Integrate insights from GitHub findings with Cloud Security Explorer |

| | | | |
|---|---|---|---|
| GitLab | Connect GitLab projects | Foundational and Defender CSPM | NA |
| | Recommendations to fix code vulnerabilities | Foundational and Defender CSPM | Requires GitLab Ultimate edition |
| | Recommendations to discover exposed secrets | Foundational and Defender CSPM | Requires GitLab Ultimate edition |
| | Recommendations to fix open-source vulnerabilities | Foundational and Defender CSPM | Requires GitLab Ultimate edition |
| | Recommendations to fix infrastructure as code misconfigurations | Foundational and Defender CSPM | Requires GitLab Ultimate edition |
| | Cloud security explorer | Defender CSPM | Integrate insights from GitLab findings with Cloud Security Explorer |
| Azure DevOps | Connect Azure DevOps repositories | Foundational and Defender CSPM | NA |
| | Recommendations to fix code vulnerabilities | Foundational and Defender CSPM | Requires GitHub Advanced Security for ADO for CodeQL findings, and Microsoft Security DevOps extension |
| | Recommendations to discover exposed secrets | Foundational and Defender CSPM | Requires GitHub Advanced Security for ADO |
| | Recommendations to fix open-source vulnerabilities | Foundational and Defender CSPM | Requires GitHub Advanced Security for ADO |
| | Recommendations to fix infrastructure as code misconfigurations | Foundational and Defender CSPM | Requires Microsoft Security DevOps extension |
| | Recommendations to fix DevOps environment misconfigurations | Foundational and Defender CSPM | NA |
| | Pull request annotations | Defender CSPM | NA |

| | Code to cloud mapping for Containers | Defender CSPM | Requires Microsoft Security DevOps extension |
|---|---|---|---|
| | Code to cloud mapping for Infrastructure as Code templates | Defender CSPM | Requires Microsoft Security DevOps extension |
| | Attack path analysis | Defender CSPM | Integrate insights from ADO findings with the attack path |
| | Cloud security explorer | Defender CSPM | Integrate insights from ADO findings with Cloud Security Explorer |

Now that you understand the requirements and constraints for each DevOps platform, let's connect with GitHub and ADO.

# Connecting with GitHub

To configure the GitHub connector in Defender for Cloud, you need **Organization Owner** privileges on GitHub and must have **Contributor** or **Security Admin** privileges in the Azure subscription where Defender CSPM is enabled. You will notice that during the configuration of the GitHub connector, you will need to install the Microsoft Security DevOps action in GitHub, and to be able to do that, you will need **GitHub Write** privileges in your GitHub organization.

## Deploying the GitHub connector

To deploy the GitHub connector in Defender for Cloud, follow the steps below:

1.  Open the Defender for Cloud dashboard and click **Environment settings**, under the **Management** section. The **Environment settings** page appears as shown in *Figure 7.1*:

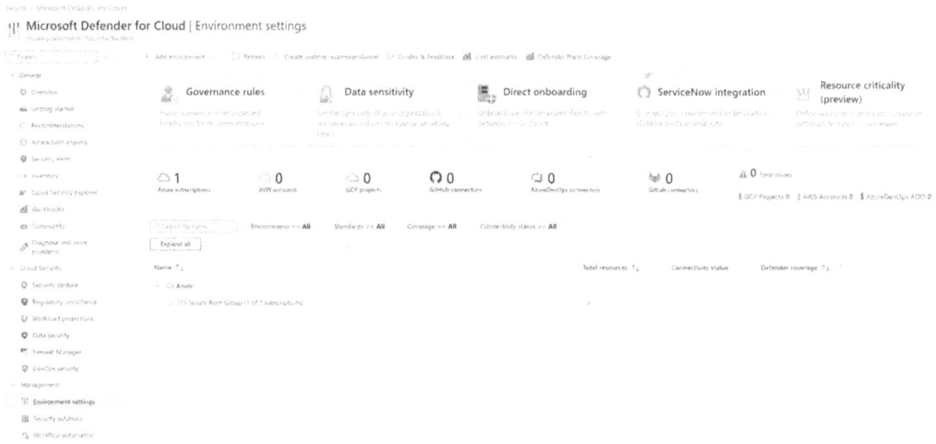

*Figure 7.1: Environment settings (for better visualization, refer to https://packt.link/gbp/9781836204879)*

2. On the top menu bar, click **+ Add environment** button and click the **GitHub** option. The **GitHub connection** page appears as shown in *Figure 7.2*:

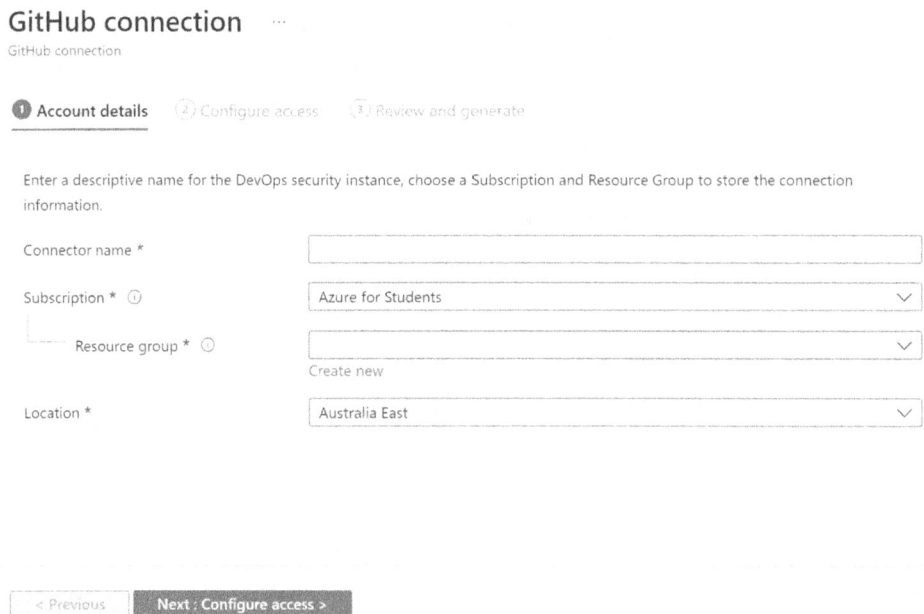

*Figure 7.2: Adding a GitHub connection (for better visualization, refer to https://packt.link/gbp/9781836204879)*

3. Type a name for this connector in the **Connector name** field.

4. Select the Azure **Subscription**, which should be the same where Defender for Cloud is enabled, and which **resource group** will be used by this connector.

5. In the **Location** field, select the region in which the connector will be created.

6. Click the **Next: Configure access >** button. The **Configure access** tab appears as shown in *Figure 7.3*:

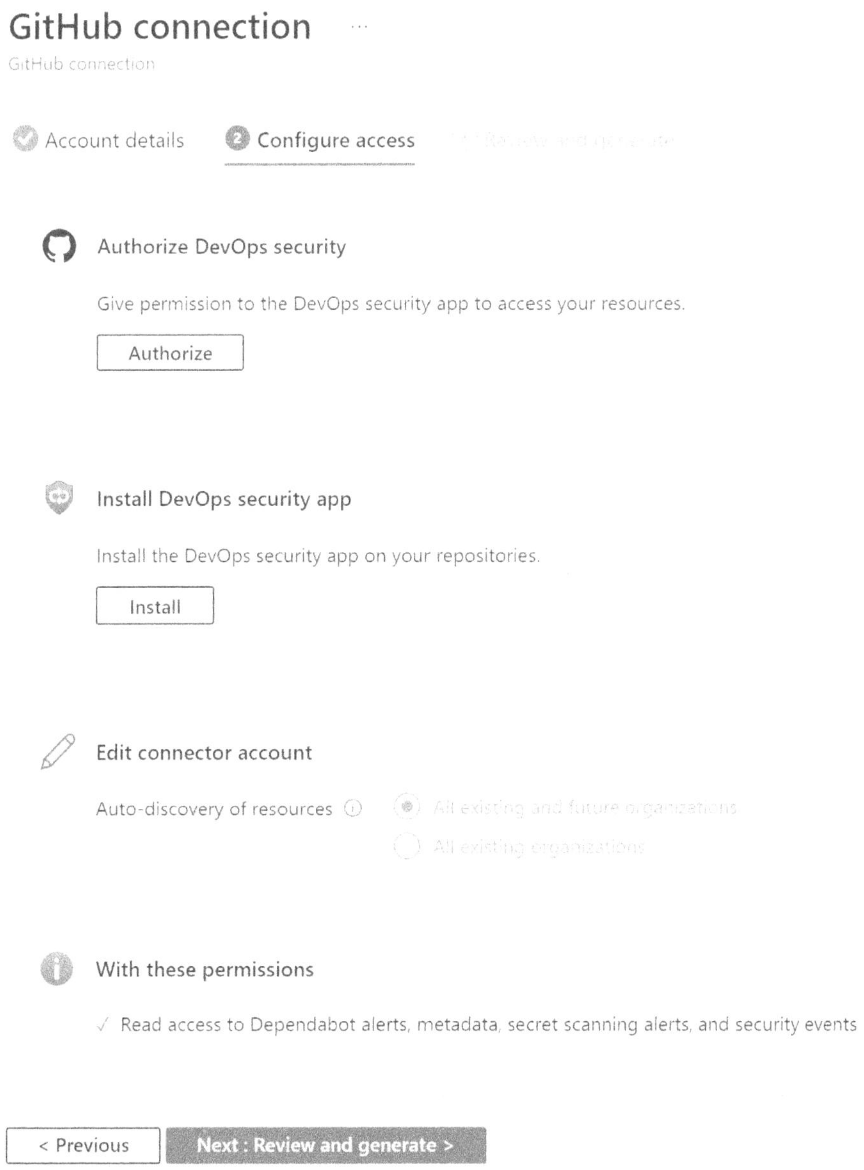

# GitHub connection

GitHub connection

✓ Account details    ② Configure access

Authorize DevOps security

Give permission to the DevOps security app to access your resources.

Authorize

Install DevOps security app

Install the DevOps security app on your repositories.

Install

Edit connector account

Auto-discovery of resources ⓘ   ● All existing and future organizations
                                 ○ All existing organizations

With these permissions

✓ Read access to Dependabot alerts, metadata, secret scanning alerts, and security events

< Previous    Next : Review and generate >

*Figure 7.3: Configure access for GitHub connector*

7.  Click the **Install** button to install the DevOps security app in your repository.

8.  If you have multiple repositories, you will be prompted to select the correct one, as shown in *Figure 7.4*. Select the correct one and click the **Continue** button:

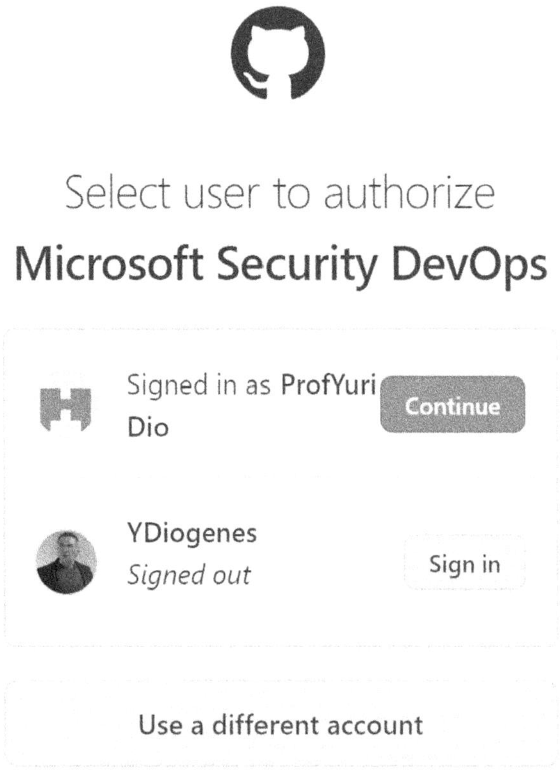

*Figure 7.4: Selecting the repository*

9. The **Install Microsoft Security DevOps** page appears as shown in *Figure 7.5*. There, you need to click on where you want to install.

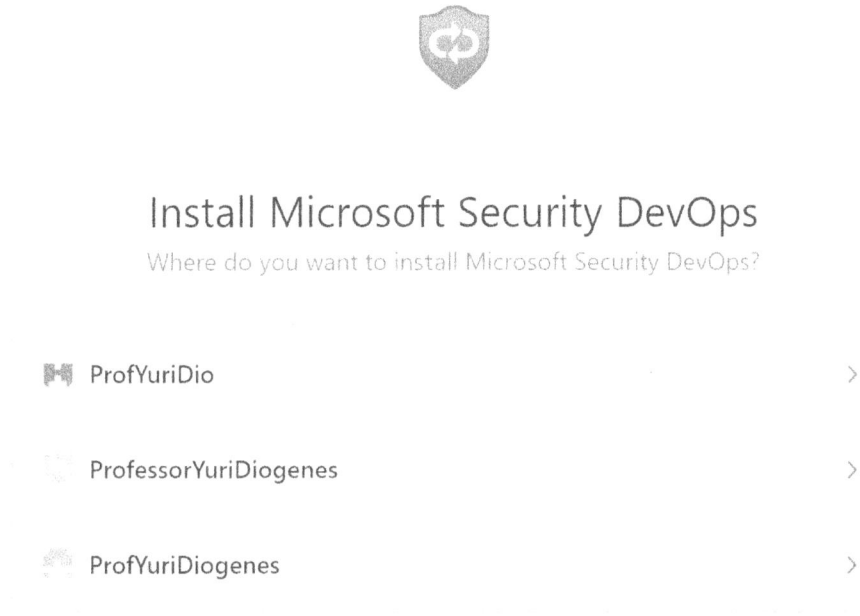

## Install Microsoft Security DevOps

Where do you want to install Microsoft Security DevOps?

| | |
|---|---|
| ProfYuriDio | > |
| ProfessorYuriDiogenes | > |
| ProfYuriDiogenes | > |

*Figure 7.5: Selecting where you want to install*

10. On the **Install Microsoft Security DevOps** page that appears (*Figure 7.6*), click the **Install** button.

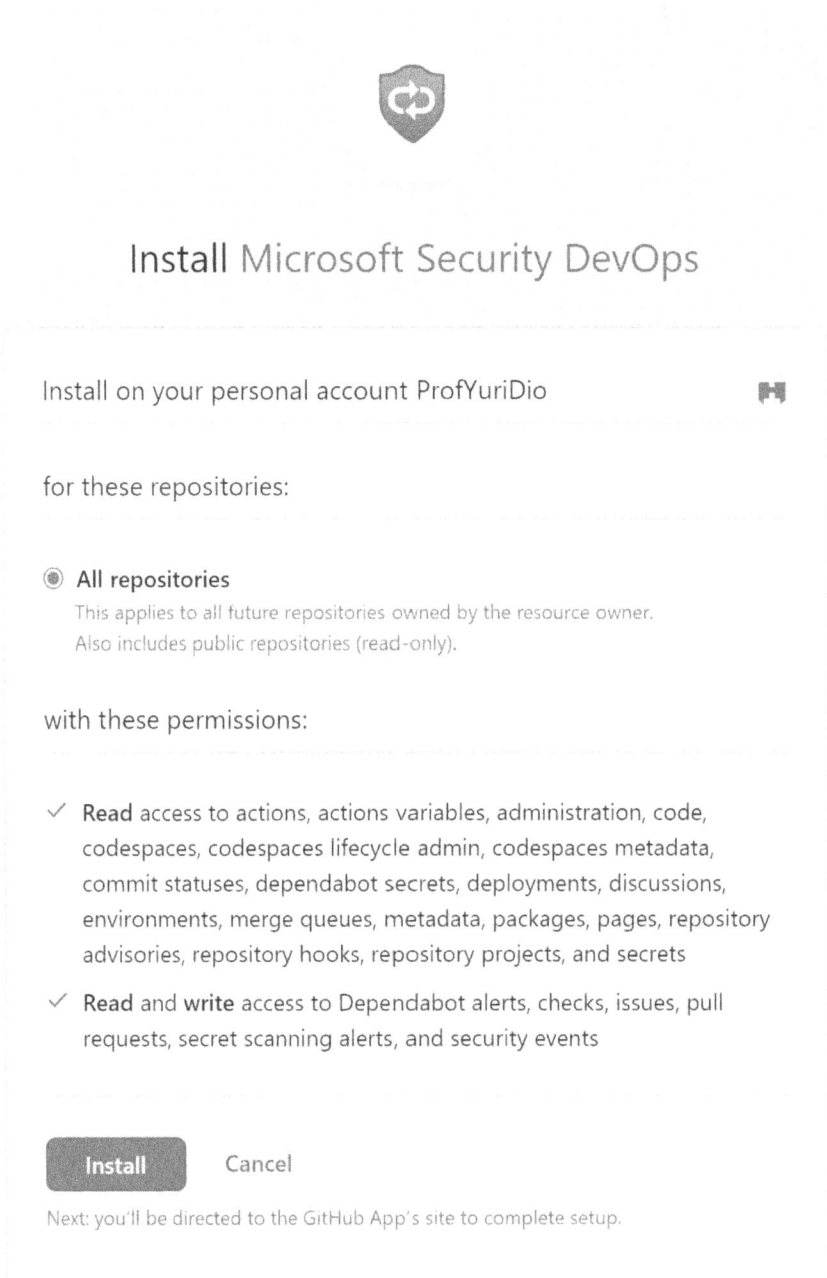

Install Microsoft Security DevOps

Install on your personal account ProfYuriDio

for these repositories:

◉ **All repositories**
This applies to all future repositories owned by the resource owner.
Also includes public repositories (read-only).

with these permissions:

✓ **Read** access to actions, actions variables, administration, code, codespaces, codespaces lifecycle admin, codespaces metadata, commit statuses, dependabot secrets, deployments, discussions, environments, merge queues, metadata, packages, pages, repository advisories, repository hooks, repository projects, and secrets

✓ **Read** and **write** access to Dependabot alerts, checks, issues, pull requests, secret scanning alerts, and security events

**Install**      Cancel

Next: you'll be directed to the GitHub App's site to complete setup.

*Figure 7.6: Acknowledging the privileges and installing*

11. Under the section **Edit connector account**, make sure that the option **All existing and future organizations** is selected (which it should already be by default). The goal is to ensure that **autodiscover** is enabled for all repositories in GitHub organizations and future ones that may be added.

12. Click the **Next: Review and generate>** button. The **Review and generate** tab appears as shown in *Figure 7.7*:

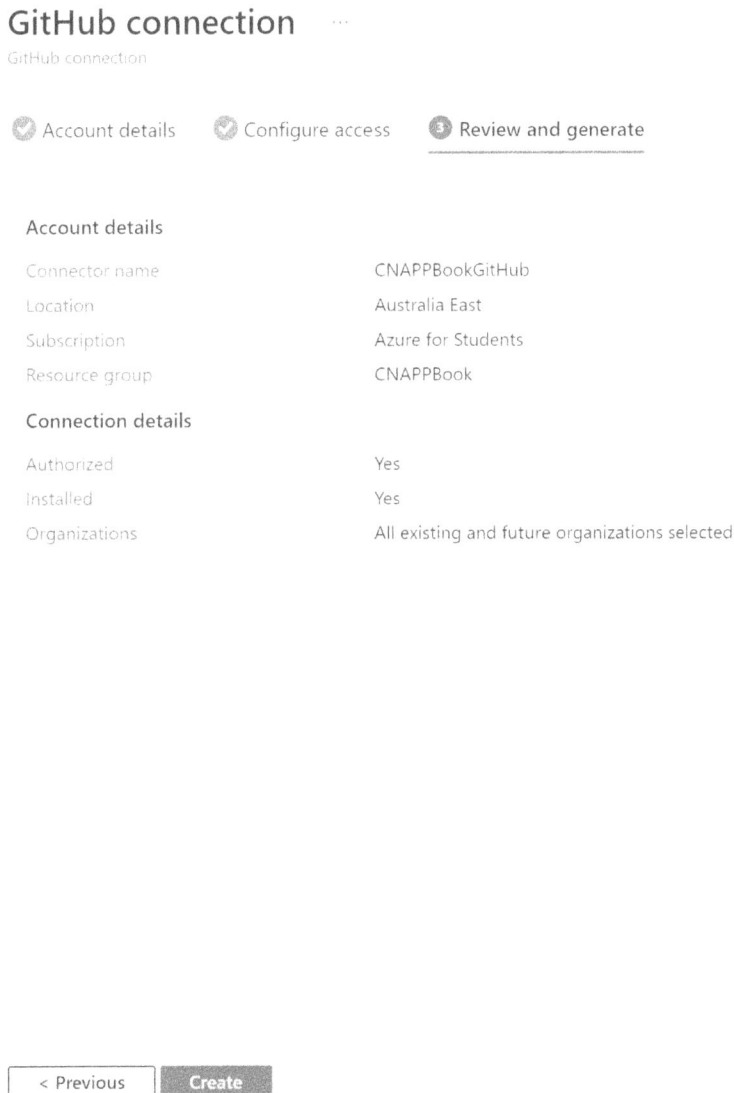

## GitHub connection
GitHub connection

✅ Account details    ✅ Configure access    ⑤ Review and generate

### Account details

| Connector name | CNAPPBookGitHub |
| Location | Australia East |
| Subscription | Azure for Students |
| Resource group | CNAPPBook |

### Connection details

| Authorized | Yes |
| Installed | Yes |
| Organizations | All existing and future organizations selected |

< Previous    **Create**

*Figure 7.7: Review and generate tab*

13. Click the **Create** button to finalize. You should see a notification informing you that the connector was created, as shown in *Figure 7.8*:

✅ Connector created successfully                                              ✕

New GitHub connector 'CNAPPBookGHC' created successfully. It may take
a few minutes for the change to be reflected.

a few seconds ago

*Figure 7.8: AWS Create stack page*

If you receive an error creating the connector, make sure to review the JSON section of the Azure Activity Log that corresponds to the creation of the connector (event **Update Security Connector**). If you see an entry like the one below, you may be using the wrong credentials to authenticate into your GitHub repository or trying to log in to the wrong repository:

```
"properties": {
        "statusMessage": "{\"status\":\"Failed\",\
"error\":{\"code\":\"ResourceOperationFailure\",\"message\":\"The
resource operation completed with terminal provisioning state
'Failed'.\",\"details\":[{\"code\":\"TokenExchangeFailed\",\
"message\":\"GitHub OAuth token exchange failed\"}]}}"
```

After a couple of minutes, refresh the **Environment settings** page in Defender for Cloud, and you will see the new GitHub connector with the connectivity status showing as **In progress**, as highlighted in *Figure 7.9*:

| Name ↑↓ | Total resources ↑↓ | Connectivity status | Defender coverage ↑↓ |
|---|---|---|---|
| > ☁ Azure | | | |
| > ☁ AWS | | | |
| > ☁ GCP | | | |
| ∨ ◯ GitHub | | | |
| ◯ CNAPPBookGitHub | | 🌐 In progress... | 1/1 plans |

*Figure 7.9: GitHub connectivity status (for better visualization, refer to https://packt.link/gbp/9781836204879)*

If you want to have hands-on experience creating this connector, you can follow the steps from the Defender for Cloud Public Lab. Go to aka.ms/MDCLabs and follow module 15.

# Reviewing initial assessment

The first synchronization may take up to eight hours, so if you go to the **Recommendations** page right after the onboarding process, you will not see recommendations for your GitHub repository. Once the synchronization is completed, you can use the same approach that you used to see AWS and GCP recommendations (showed in the previous chapter) by looking at the **Recommendations** dashboard and filtering only for GitHub, as shown in *Figure 7.10*:

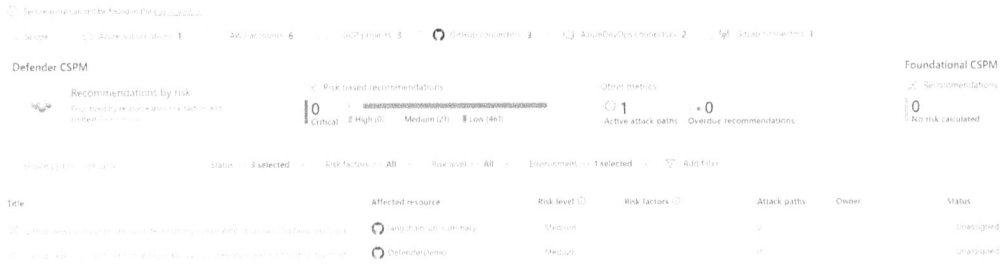

*Figure 7.10: Recommendations dashboard (for better visualization, refer to https://packt.link/gbp/9781836204879)*

This is the most appropriate way to review what is important to be remediated in your environment since it will prioritize the recommendations based on risk factors. However, if you need a quick overview of all your repositories, you can leverage the dedicated DevOps security page called **DevOps security**. This page is located in the Defender for Cloud dashboard under the **Cloud Security** section, as shown in *Figure 7.11*:

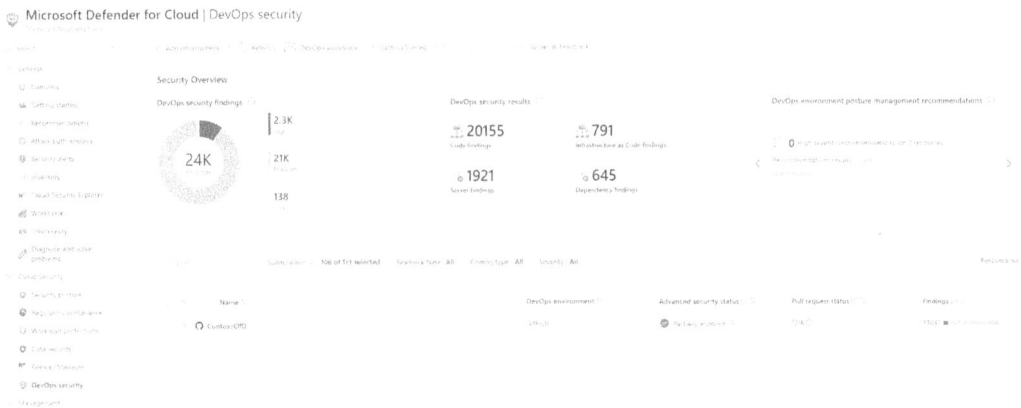

*Figure 7.11: DevOps security dashboard (for better visualization, refer to https://packt.link/gbp/9781836204879)*

On this dashboard, you will see a summary of the DevOps security findings across the connectors that you configured (pie chart on the left). On the right side of the dashboard, you will see the total number of DevOps security scan findings related to code, secrets, dependency, and Infrastructure as Code. In the lower part of the screen, you have the connector and the connector's status, including the status of the **GitHub Advanced Security** capability enablement.

# Remediating recommendations

GitHub-related recommendations will appear in Defender for Cloud just like any other recommendations; however, the remediation steps will involve taking actions in your GitHub repository. Let's use the recommendation below as an example:

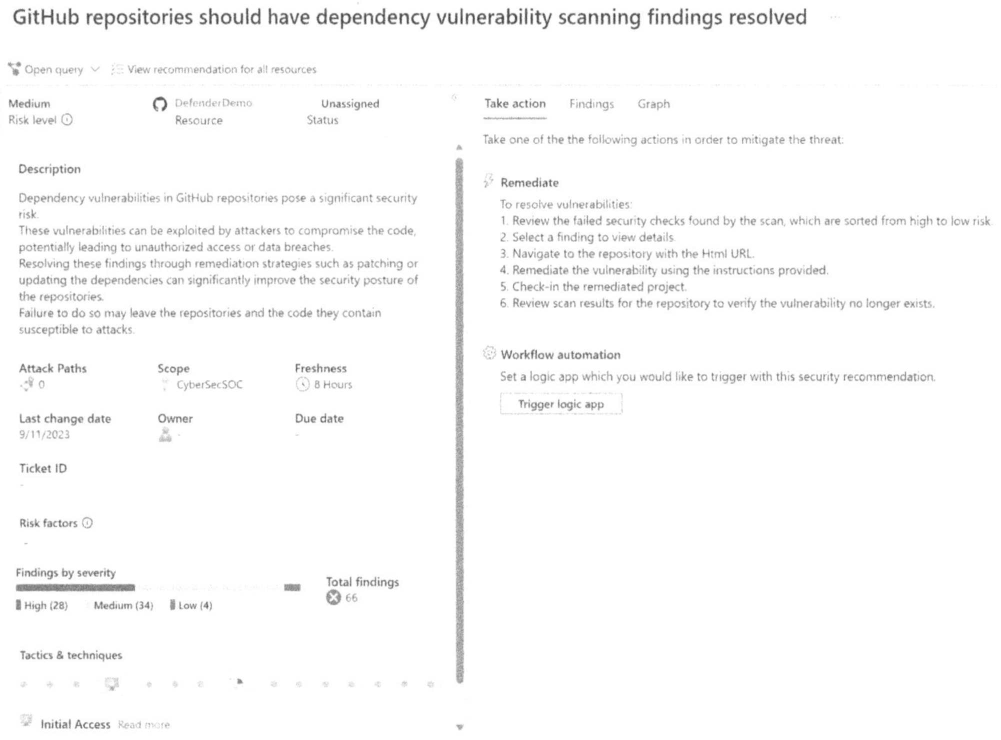

*Figure 7.12: GitHub recommendation (for better visualization, refer to https://packt.link/gbp/9781836204879)*

As you can see, the overall experience is the same, but when you read the steps in the **Take action** tab, it has the details about the remediation, and on step 3, it switches to GitHub. In the case of this recommendation: **GitHub repositories should have dependency vulnerability scanning findings resolved**, you still need to click the **Findings** tab to see what vulnerabilities were found. *Figure 7.13* below has an example of how this tab may look:

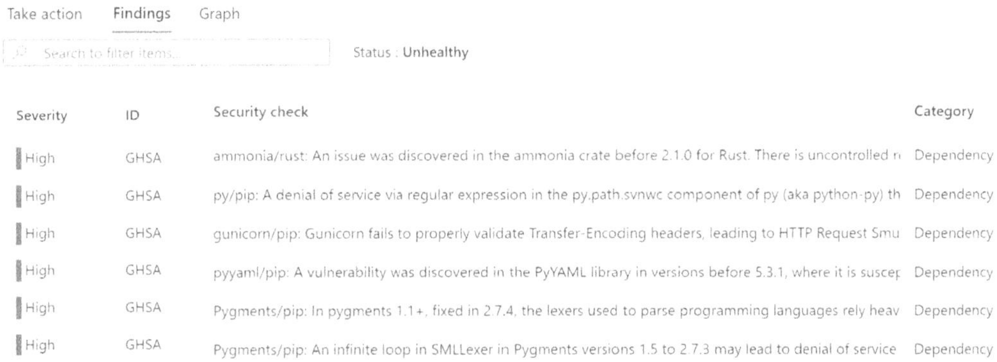

| Take action | Findings | Graph | | |
|---|---|---|---|---|
| Search to filter items... | | | Status : **Unhealthy** | |

| Severity | ID | Security check | Category |
|---|---|---|---|
| High | GHSA | ammonia/rust: An issue was discovered in the ammonia crate before 2.1.0 for Rust. There is uncontrolled r | Dependency |
| High | GHSA | py/pip: A denial of service via regular expression in the py.path.svnwc component of py (aka python-py) th | Dependency |
| High | GHSA | gunicorn/pip: Gunicorn fails to properly validate Transfer-Encoding headers, leading to HTTP Request Smu | Dependency |
| High | GHSA | pyyaml/pip: A vulnerability was discovered in the PyYAML library in versions before 5.3.1, where it is suscep | Dependency |
| High | GHSA | Pygments/pip: In pygments 1.1+, fixed in 2.7.4, the lexers used to parse programming languages rely heav | Dependency |
| High | GHSA | Pygments/pip: An infinite loop in SMLLexer in Pygments versions 1.5 to 2.7.3 may lead to denial of service | Dependency |

*Figure 7.13: Vulnerabilities found in the repository (for better visualization, refer to https:// packt.link/gbp/9781836204879)*

In the **Findings** tab, you can click on the security check that was done, and a blade appears on the right side with more details about the vulnerability, including hyperlinks at the bottom of the blade that will lead you directly to the GitHub repository:

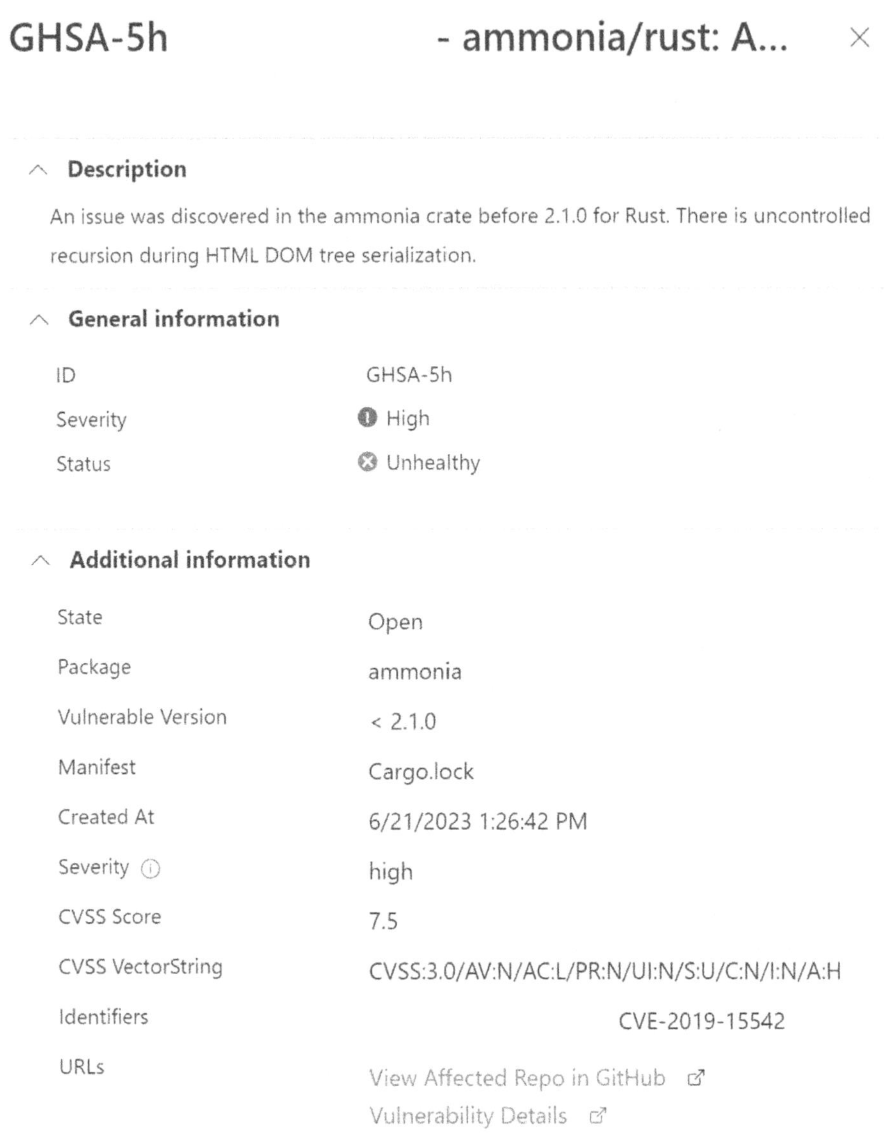

# GHSA-5h                        - ammonia/rust: A...        ✕

∧  **Description**

An issue was discovered in the ammonia crate before 2.1.0 for Rust. There is uncontrolled recursion during HTML DOM tree serialization.

∧  **General information**

| | |
|---|---|
| ID | GHSA-5h |
| Severity | ❶ High |
| Status | ✕ Unhealthy |

∧  **Additional information**

| | |
|---|---|
| State | Open |
| Package | ammonia |
| Vulnerable Version | < 2.1.0 |
| Manifest | Cargo.lock |
| Created At | 6/21/2023 1:26:42 PM |
| Severity ⓘ | high |
| CVSS Score | 7.5 |
| CVSS VectorString | CVSS:3.0/AV:N/AC:L/PR:N/UI:N/S:U/C:N/I:N/A:H |
| Identifiers | CVE-2019-15542 |
| URLs | View Affected Repo in GitHub  ↗ |
| | Vulnerability Details  ↗ |

*Figure 7.14: More details about the vulnerability*

In this case, there are two URLs at the bottom of this blade (this can vary according to the finding). If you click **View Affected Repo in GitHub**, it will open another tab in your browser to redirect to your GitHub repository.

It is important to keep in mind that you may have to authenticate your GitHub credentials to access this information. If you click the **Vulnerability Details** hyperlink, another tab will open and you will see the GitHub page that has more details about the CVE, as shown in *Figure 7.15*:

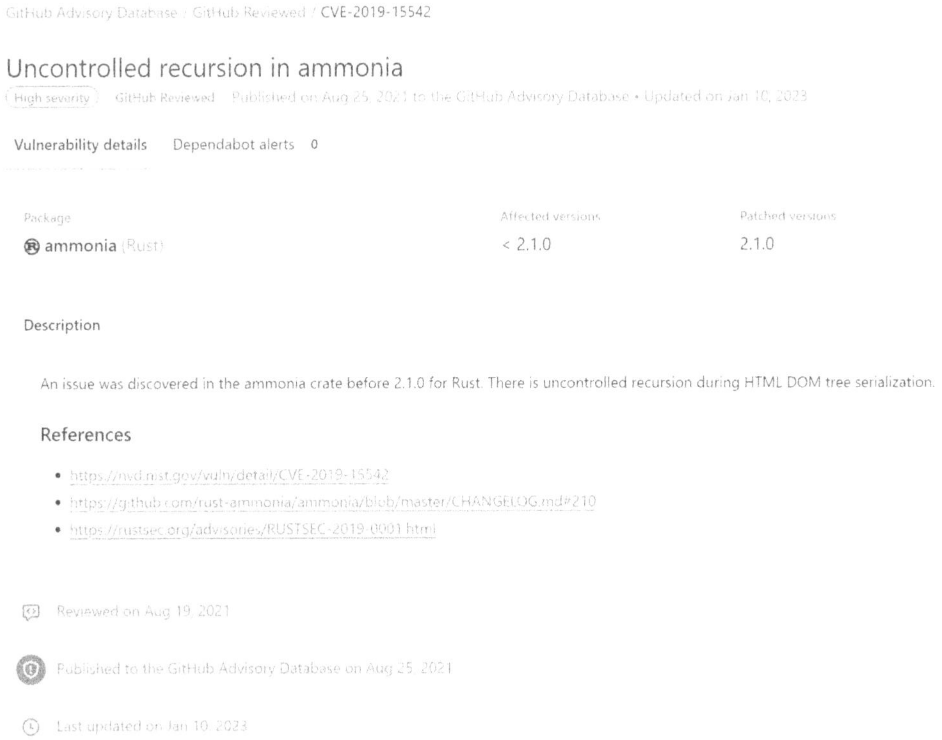

GitHub Advisory Database / GitHub Reviewed / CVE-2019-15542

## Uncontrolled recursion in ammonia
( High severity )  GitHub Reviewed   Published on Aug 25, 2021 to the GitHub Advisory Database • Updated on Jan 10, 2023

Vulnerability details      Dependabot alerts   0

| Package | Affected versions | Patched versions |
|---|---|---|
| ® ammonia (Rust) | < 2.1.0 | 2.1.0 |

Description

An issue was discovered in the ammonia crate before 2.1.0 for Rust. There is uncontrolled recursion during HTML DOM tree serialization.

References

- https://nvd.nist.gov/vuln/detail/CVE-2019-15542
- https://github.com/rust-ammonia/ammonia/blob/master/CHANGELOG.md#210
- https://rustsec.org/advisories/RUSTSEC-2019-0001.html

Reviewed on Aug 19, 2021

Published to the GitHub Advisory Database on Aug 25, 2021

Last updated on Jan 10, 2023

*Figure 7.15: More details about the CVE (for better visualization, refer to https://packt.link/ gbp/9781836204879)*

> You can see the full list of GitHub-related recommendations at `https://bit.ly/ CNAPPBook29`.

While this experience will not be new to most cloud security administrators who are used to dealing with security recommendations in Defender for Cloud, developers will also gain different insights with this integration.

*Figure 7.16* has an example of how security recommendations will be surfaced in GitHub:

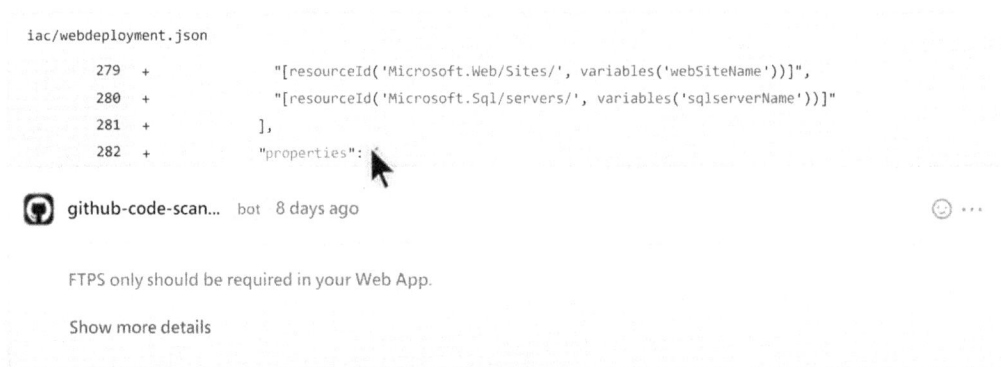

```
iac/webdeployment.json

        279  +                  "[resourceId('Microsoft.Web/Sites/', variables('webSiteName'))]",
        280  +                  "[resourceId('Microsoft.Sql/servers/', variables('sqlserverName'))]"
        281  +              ],
        282  +              "properties": {
```

github-code-scan...   bot   8 days ago                                             ☺ ...

FTPS only should be required in your Web App.

Show more details

*Figure 7.16: Developer experience in GitHub (for better visualization, refer to https://packt.*
*link/gbp/9781836204879)*

# Connecting with Azure DevOps

To configure the Azure DevOps connector in Defender for Cloud, you need the **Project Collection Administrator** privilege on the target organization on Azure DevOps and **Contributor** or **Security Admin** privileges in the Azure subscription where Defender CSPM is enabled. You will notice that during the configuration of the Azure DevOps connector, you will need to install the Microsoft Security DevOps extension in Azure DevOps, and to be able to do that, you will need **Azure DevOps Project Collection Administrator** privileges in your Azure DevOps organization.

## Deploying the Azure DevOps connector

To deploy the Azure DevOps connector in Defender for Cloud, follow the steps below:

1.  Open the Defender for Cloud dashboard and click **Environment settings**, under the **Management** section. The **Environment settings** page appears as shown in *Figure 7.17*:

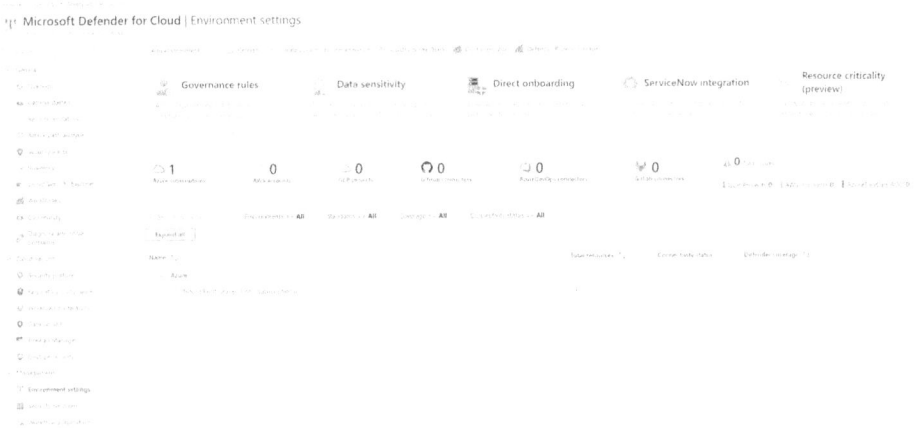

Figure 7.17: Environment settings (for better visualization, refer to https://packt.link/gbp/9781836204879)

2. On the top menu bar, click the **+ Add environment** button and click the **Azure DevOps** option. The **Azure DevOps connection** page appears as shown in *Figure 7.18*:

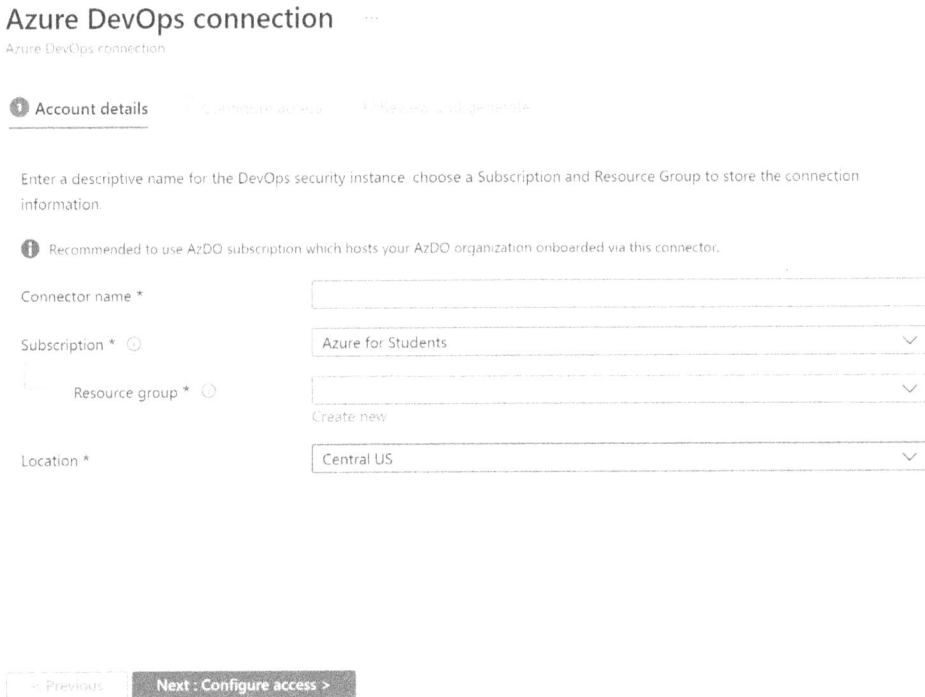

Figure 7.18: Azure DevOps connection page (for better visualization, refer to https://packt.link/gbp/9781836204879)

3. Type a name for this connector in the **Connector name** field.

4. Select the appropriate subscription and resource group for this connector.

5. In the **Location** field, select the appropriate region for this connector. Click the **Next: Configure access>** button and the **Configure access** tab appears, as shown in *Figure 7.19*:

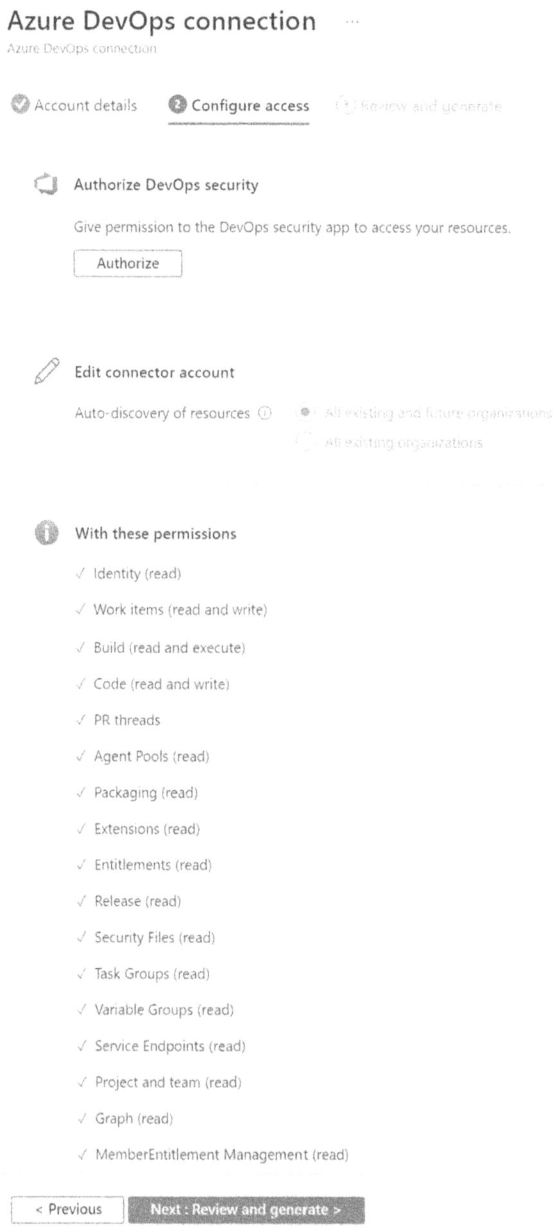

## Azure DevOps connection
Azure DevOps connection

✓ Account details    **2** Configure access    Review and generate

**Authorize DevOps security**

Give permission to the DevOps security app to access your resources.

Authorize

**Edit connector account**

Auto-discovery of resources ⓘ    All existing and future organizations

All existing organizations

**With these permissions**

✓ Identity (read)

✓ Work items (read and write)

✓ Build (read and execute)

✓ Code (read and write)

✓ PR threads

✓ Agent Pools (read)

✓ Packaging (read)

✓ Extensions (read)

✓ Entitlements (read)

✓ Release (read)

✓ Security Files (read)

✓ Task Groups (read)

✓ Variable Groups (read)

✓ Service Endpoints (read)

✓ Project and team (read)

✓ Graph (read)

✓ MemberEntitlement Management (read)

< Previous    **Next : Review and generate >**

*Figure 7.19: Configure access*

6.  Click the **Authorize** button and click the **Accept** button in the **Authorize DevOps security** pop-up message that appears. If you use the right credentials, you will see that the status changes to **Authorized**, as shown in *Figure 7.20*:

*Figure 7.20: Authorized status*

7.  Leave the option **All existing and future organizations** selected and click the **Next: Review and generate>** button. The **Review and generate** tab appears as shown in *Figure 7.21*. Click the **Create** button to finalize the configuration.

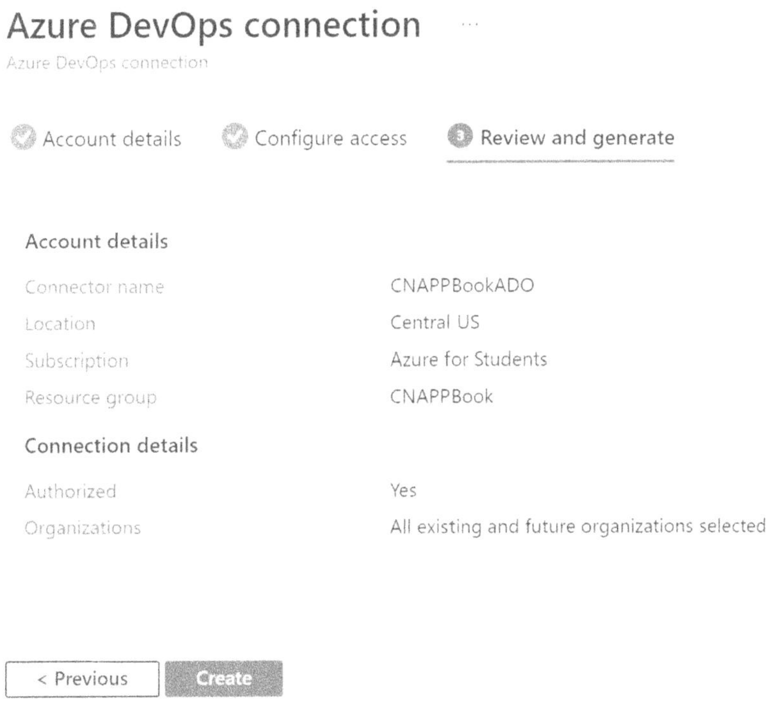

*Figure 7.21: Final stage to create the connector*

After a few minutes, you will see the addition of a new connector under the **Environment settings**, and the connectivity status showing as **In progress**, as shown in *Figure 7.22*:

*Figure 7.22: Connectivity status for the new ADO connector (for better visualization, refer to https://packt.link/gbp/9781836204879)*

# Reviewing initial assessment

The first synchronization may take up to eight hours, so if you go to the **Recommendations** page right after the onboarding process, you will not see recommendations for your Azure DevOps environment. Once the synchronization is completed, you can use the same approach that you used to see the GitHub recommendations earlier in this chapter, by looking at the **Recommendations** dashboard and filtering only for GitHub, as shown in *Figure 7.23*:

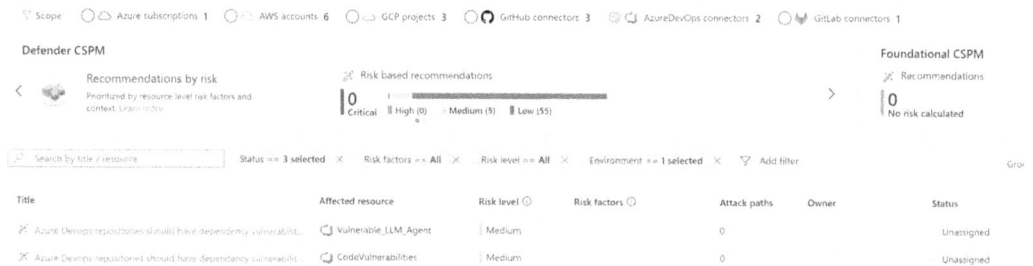

*Figure 7.23: ADO recommendations (for better visualization, refer to https://packt.link/ gbp/9781836204879)*

If you open the recommendation, you will have a similar experience to the one that you've seen with GitHub recommendations. Let's use the recommendation **Azure Devops repositories should have dependency vulnerability scanning findings resolved** as an example, which will show the following page once you open it:

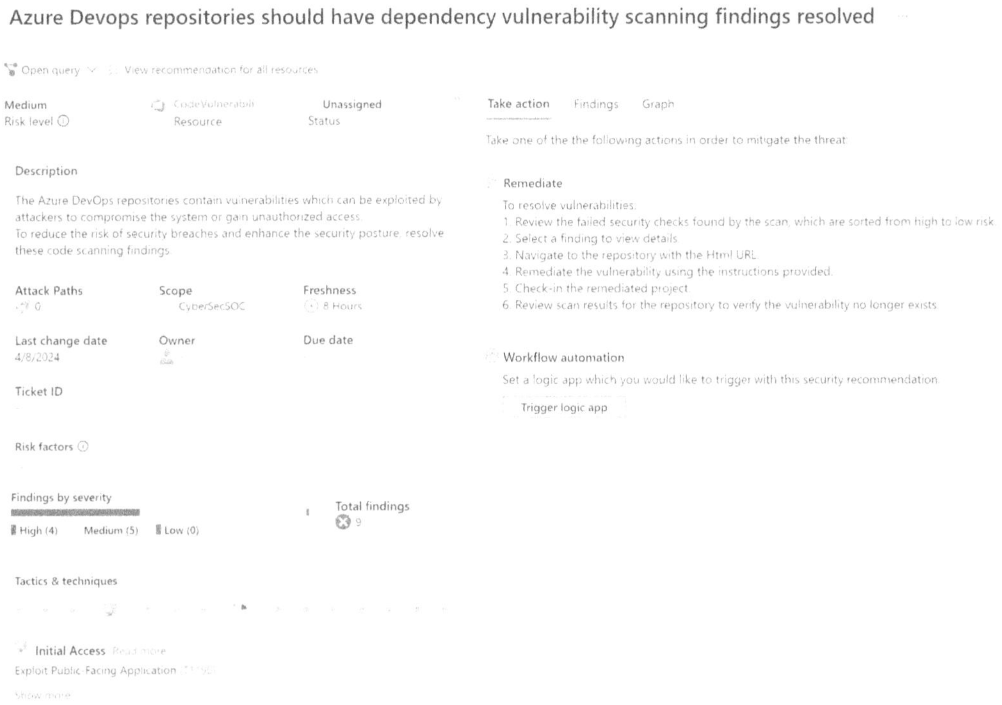

Azure Devops repositories should have dependency vulnerability scanning findings resolved

*Figure 7.24: ADO recommendation details (for better visualization, refer to https://packt.link/ gbp/9781836204879)*

Since this is a recommendation that deals with vulnerability scanning, you will need to click on the **Findings** tab to see more details, as was shown earlier in this chapter with the GitHub vulnerability findings. The experience is the same, including the option to click on the hyperlink and jump to the DevOps environment.

The only difference is that, in this case, you will jump to the ADO environment, as shown in *Figure 7.25*:

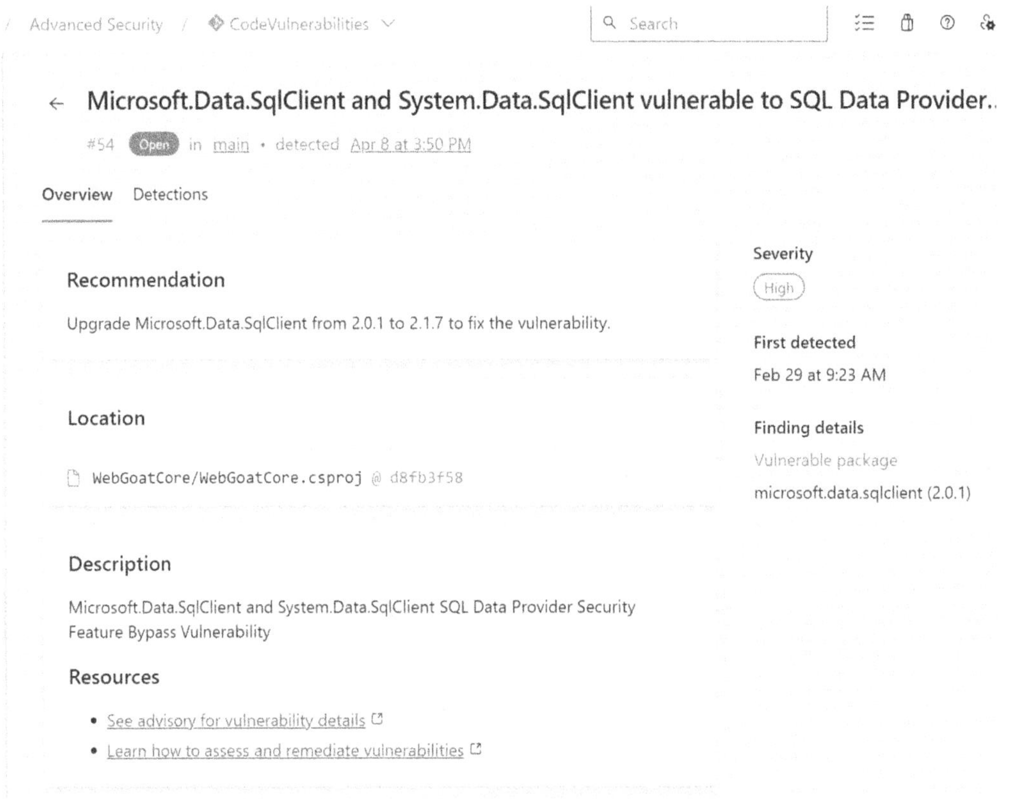

*Figure 7.25: Vulnerability description in ADO*

On the side of the developers, what they will see using ADO when they commit their code is a series of errors generated by the Microsoft Security DevOps extension, as shown in *Figure 7.26*:

Errors 3    Warnings 10

3. Checkov Error CKV_AZURE_190 - File: main.tf. Line: 5. Column 0.
MicrosoftSecurityDevOps

11. Checkov Error CKV2_AZURE_1 - File: main.tf. Line: 5. Column 0.
MicrosoftSecurityDevOps

BreakException: Guardian detected one or more breaking results.
MicrosoftSecurityDevOps

*Figure 7.26: Microsoft Security DevOps extension errors*

When opening the details of the scan, it is possible to see the tests that failed, as shown in *Figure 7.27*:

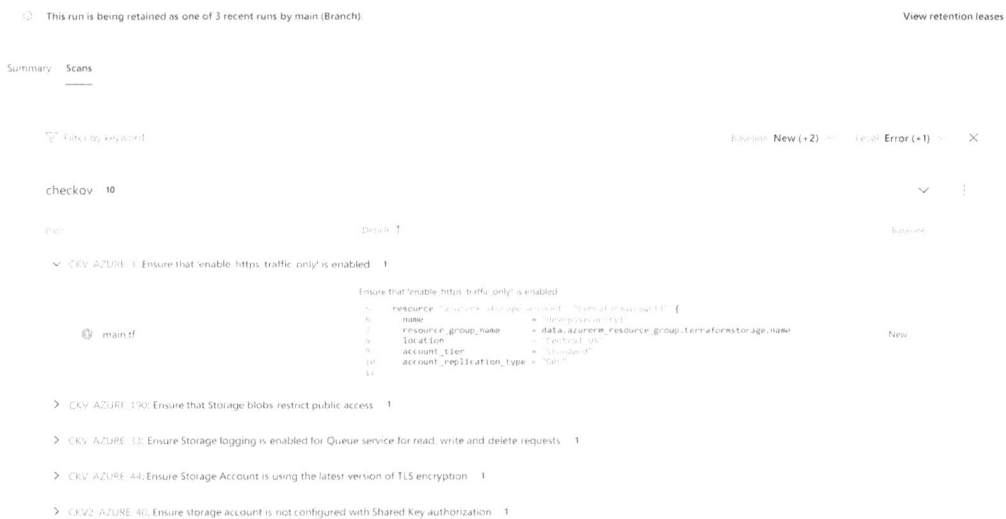

*Figure 7.27: Scan details (for better visualization, refer to https://packt.link/gbp/9781836204879)*

Code scanning insights will appear under the **Advanced Security** option in Azure DevOps with the results, as shown in *Figure 7.28*:

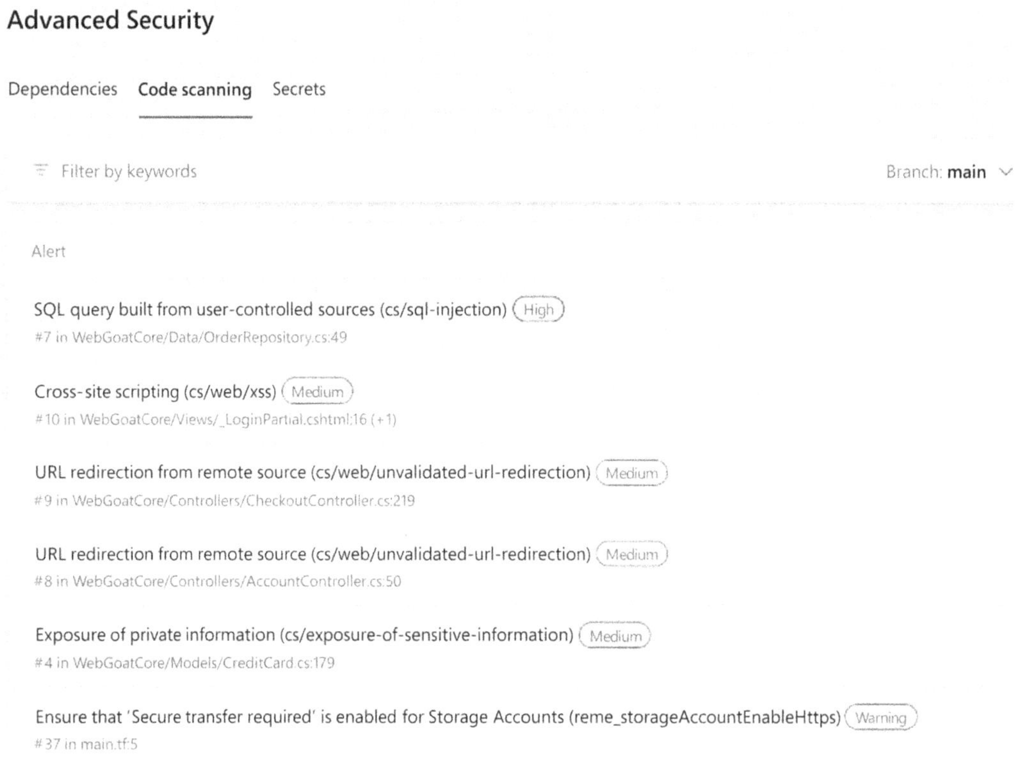

## Advanced Security

Dependencies    **Code scanning**    Secrets

---

≡  Filter by keywords                                                          Branch: **main**  ⌄

---

Alert

SQL query built from user-controlled sources (cs/sql-injection)  ( High )
#7 in WebGoatCore/Data/OrderRepository.cs:49

Cross-site scripting (cs/web/xss)  ( Medium )
#10 in WebGoatCore/Views/_LoginPartial.cshtml:16 (+1)

URL redirection from remote source (cs/web/unvalidated-url-redirection)  ( Medium )
#9 in WebGoatCore/Controllers/CheckoutController.cs:219

URL redirection from remote source (cs/web/unvalidated-url-redirection)  ( Medium )
#8 in WebGoatCore/Controllers/AccountController.cs:50

Exposure of private information (cs/exposure-of-sensitive-information)  ( Medium )
#4 in WebGoatCore/Models/CreditCard.cs:179

Ensure that 'Secure transfer required' is enabled for Storage Accounts (reme_storageAccountEnableHttps)  ( Warning )
#37 in main.tf:5

*Figure 7.28: Code scanning results*

# Pull request annotations

DevOps security capabilities in Defender CSPM highlight security findings as annotations in **Pull Requests (PRs)**. Cloud security administrators can activate PR annotations in Defender for Cloud to enable developers to address any identified issues. This approach helps prevent and resolve potential security vulnerabilities and misconfigurations before they reach the production stage. Instead of annotating all detected vulnerabilities in an entire file, this feature focuses on vulnerabilities introduced by the changes in the PRs. Developers can view these annotations in their source code management systems, while cloud security administrators can monitor any unresolved findings in Defender for Cloud. To enable PRs, you need to have write access (owner/contributor) to the Azure subscription.

To enable PR annotation for Azure DevOps, open the Defender for Cloud dashboard, and under the **Cloud Security** section, click the **DevOps security** option. Select the Azure DevOps environment that you want to configure (remember that this will only appear on your dashboard after the ADO connector is configured and fully synchronized) and click the **Manage resources** option in the top menu. The **Configuration** blade appears as shown in *Figure 7.29*:

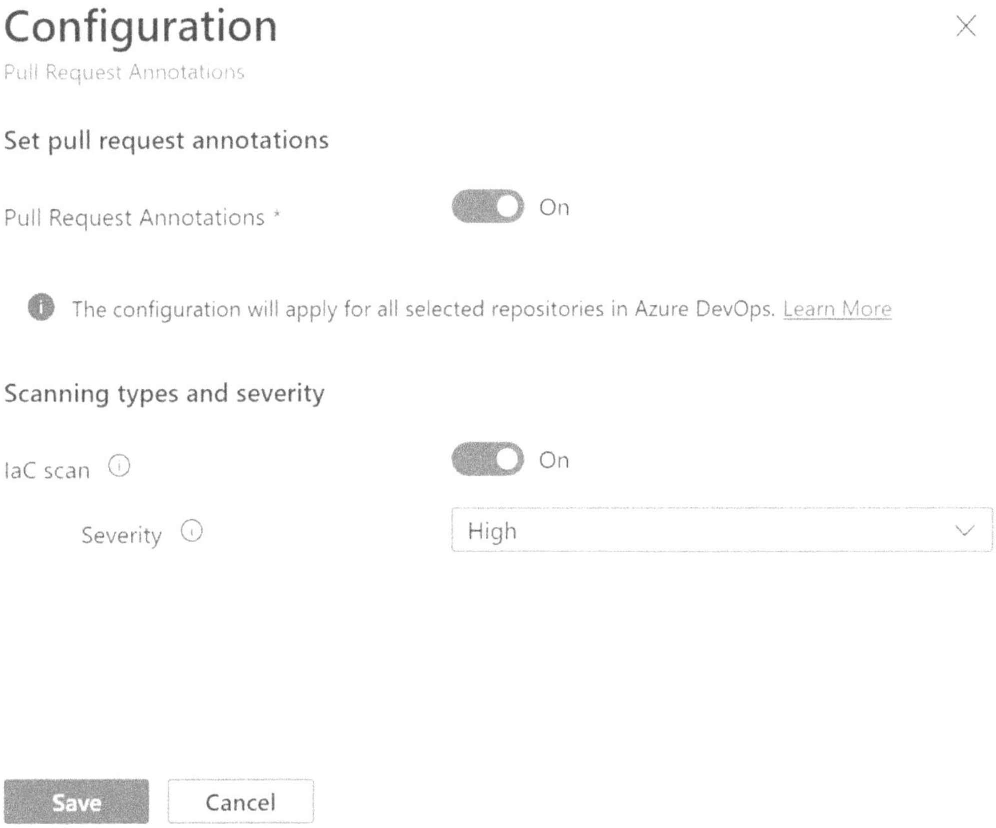

# Configuration ✕
Pull Request Annotations

## Set pull request annotations

Pull Request Annotations *                     🔘 On

    ⓘ  The configuration will apply for all selected repositories in Azure DevOps. Learn More

## Scanning types and severity

IaC scan  ⓘ                                    🔘 On

    Severity  ⓘ

| High | ⌄ |

| Save | Cancel |

*Figure 7.29: Pull request annotation*

Make sure the **Pull Request Annotations** toggle is set to **On**, and also make sure **the IaC scan** toggle is set to **on**, as shown in *Figure 7.29*. The **IaC scan** option enables PR annotations for IaC misconfigurations, and the severity dropdown allows you to select the severity level of annotations that will appear to the developers. Once you finish setting those options to **on**, click the **Save** button.

# Connecting with GitLab

To configure the GitLab connector in Defender for Cloud, you need to be the **Group Owner** on the GitLab group and have **Contributor** or **Security Admin** privileges in the Azure subscription where Defender CSPM is enabled. As mentioned before, this connector will only work if you are using the GitLab Ultimate edition.

## Deploying the GitLab connector

To deploy the GitLab connector in Defender for Cloud, follow the steps below:

1.  Open the Defender for Cloud dashboard and click **Environment settings**, under the **Management** section. The **Environment settings** page appears as shown in *Figure 7.30*:

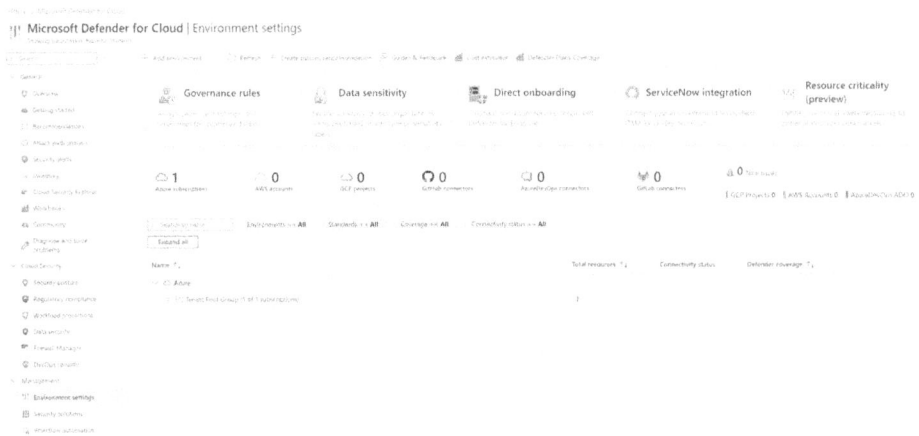

*Figure 7.30: Environment settings (for better visualization, refer to https://packt.link/ gbp/9781836204879)*

2. On the top menu bar, click the **+ Add environment** button and click the **GitLab** option. The **GitLab connection** page appears, as shown in *Figure 7.31*:

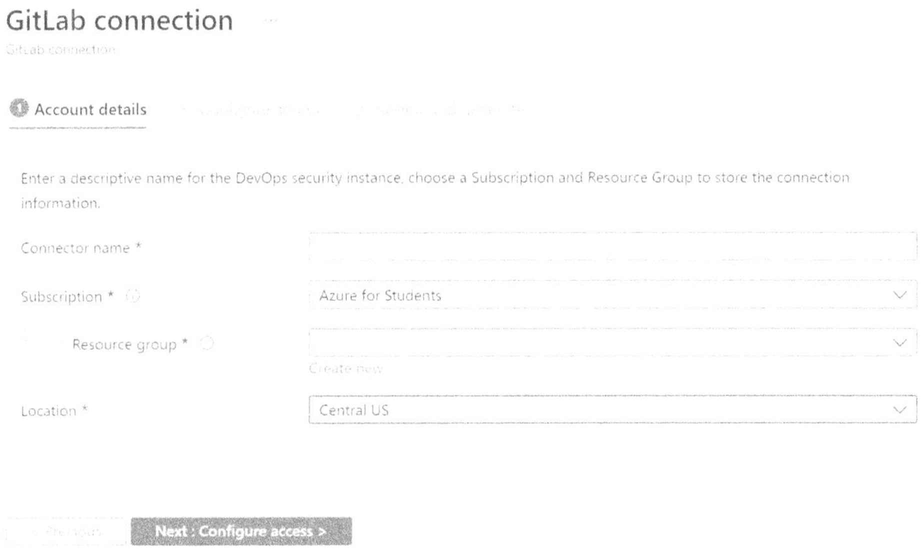

GitLab connection

GitLab connection

1 Account details

Enter a descriptive name for the DevOps security instance, choose a Subscription and Resource Group to store the connection information.

Connector name *

Subscription *     Azure for Students

Resource group *

Create new

Location *     Central US

Previous     **Next : Configure access >**

*Figure 7.31: GitLab connection (for better visualization, refer to https://packt.link/gbp/9781836204879)*

3. Type a name for this connector in the **Connector name** field.

4. Select the appropriate subscription and resource group for this connector.

5.  In the **Location** field, select the appropriate region for this connector. Click the **Next: Configure access>** button and the **Configure access** tab appears, as shown in *Figure 7.32*:

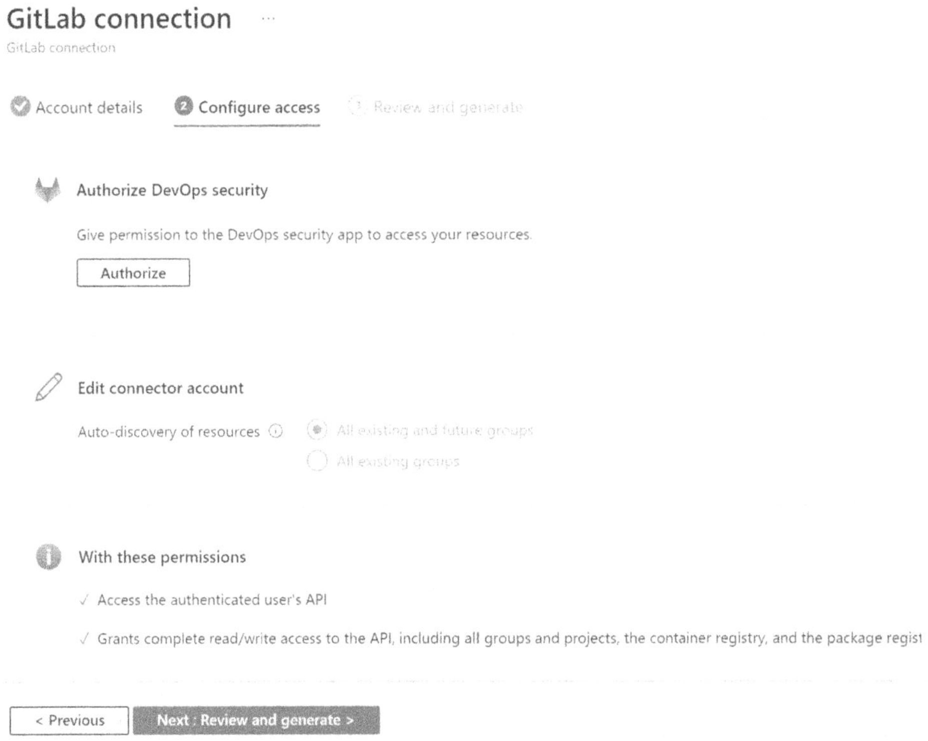

GitLab connection   ...
GitLab connection

✔ Account details       ② **Configure access**       ③ Review and generate

🦊  Authorize DevOps security

Give permission to the DevOps security app to access your resources.

[ Authorize ]

✏️  Edit connector account

Auto-discovery of resources ⓘ      ⦿ All existing and future groups
                                     ◯ All existing groups

ⓘ  With these permissions

✓ Access the authenticated user's API

✓ Grants complete read/write access to the API, including all groups and projects, the container registry, and the package regist

[ < Previous ]  [ **Next : Review and generate >** ]

*Figure 7.32: Configure access tab in the GitLab connector (for better visualization, refer to https://packt.link/gbp/9781836204879)*

6.  Click the **Authorize** button, in the pop-up window, read the list of permission requests, and then select the **Accept** button.

7.  Leave the option **All existing and future organizations** selected and click the **Next: Review and generate>** button. In the **Review and generate** tab, click the **Create** button to finalize the configuration.

After the full synchronization, GitLab recommendations will start populating in Defender for Cloud and are also accessible via the **Recommendations** page, as shown in the example below:

*Figure 7.33: GitLab recommendation (for better visualization, refer to https://packt.link/gbp/9781836204879)*

# Summary

In this chapter, we discussed the DevOps security capabilities available in Defender CSPM. We learned about the prerequisites and how to connect Defender for Cloud with GitHub, Azure DevOps, and GitLab. We also learned about how the recommendations from GitHub and ADO appear in Defender for Cloud, and how developers will experience these recommendations on their platform.

In the next chapter, we will learn how to manage security recommendations with the governance capability in Defender CSPM.

# Notes

1. You can view the Ignite presentation at `https://bit.ly/CNAPPBook21`.

2. You can obtain more information about GHAS at `https://docs.github.com/en/get-started/learning-about-github/about-github-advanced-security`.

# Additional resources

- Watch the author Yuri Diogenes interviewing a DevOps security PM during this episode of *Defender for Cloud in the Field*: `https://bit.ly/CNAPPBook30`

- Test these concepts on your own environment following the Defender for Cloud Public Labs and perform modules 14 and 15: `https://aka.ms/MDCLabs`

# Join our community on Discord

Read this book alongside other users. Ask questions, provide solutions to other readers, and much more.

Scan the QR code or visit the link to join the community.

https://packt.link/SecNet

# 8

# Governance and Continuous Improvement

Security posture management is a journey that never ends, and that's why many are describing the cybersecurity field as an infinite game. It is an ongoing process of continuous improvement where you learn from incidents and incorporate the lessons learned to improve your protection.

A key part of this continuous improvement cycle is to ensure that you have governance in place to hold stakeholders accountable for their assets. In a cloud environment, there are many workloads, and these workloads may be owned by different people from different departments. Due to this highly distributed scenario, it becomes challenging for cloud security administrators to handle security recommendations on their own; they need to share these responsibilities with workload owners. This is especially true when it comes to complex cloud deployments where multiple departments are involved. Therefore, it's essential to have good governance in place where people need to work together to handle posture management issues and assist the security administrators within the organization. With that in mind, having good governance in place becomes imperative to posture improvement.

Defender CSPM includes governance capabilities including native integration with the IT service management tool ServiceNow. These capabilities facilitate the assignment of ownership for security recommendations and the establishment of **Service-Level Agreements (SLAs)** to ensure workload owners address security recommendations.

This chapter covers:

- Governance
- Integration with ServiceNow
- Continuous improvement

# Governance

The intent of the governance feature in Defender for Cloud is to help cloud security administrators by providing a toolset to address security recommendations and define a due date or SLA to remediate them.

To accomplish this, the governance feature enables you to create governance rules. You will need **Contributor**, **Security Admin**, or **Owner** permissions on the Azure subscription where Defender CSPM is enabled in order to create governance rules.

To create governance rules, follow the steps below:

1. Open the Defender for Cloud dashboard and click **Environment settings**, under the **Management** section. The **Environment settings** page appears, as shown in *Figure 8.1*:

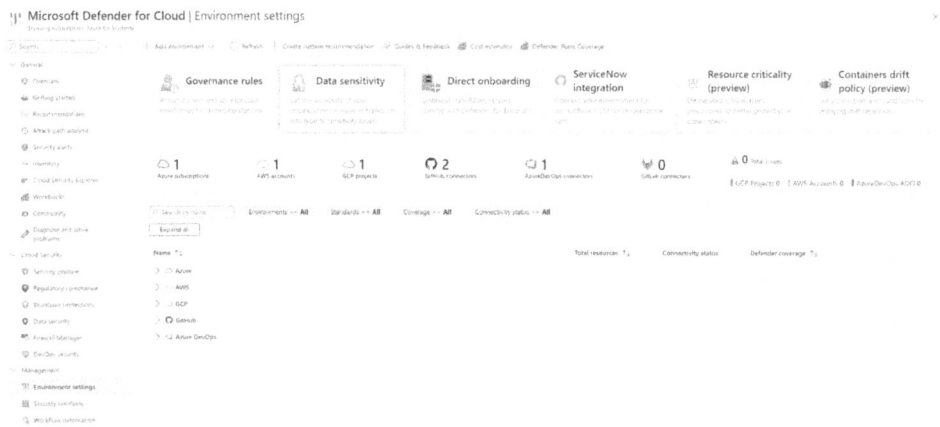

*Figure 8.1: Environment settings (for better visualization, refer to https://packt.link/gbp/9781836204879)*

2. On the top menu bar, click the **Governance rules** option. If this is the first time that you are creating a governance rule, the following page appears:

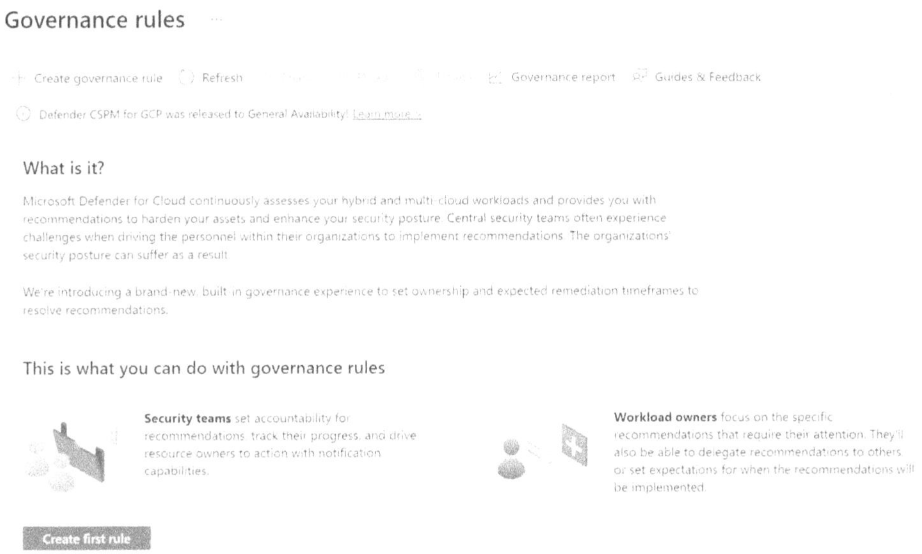

## Governance rules

*Figure 8.2: Governance rules page (for better visualization, refer to https://packt.link/gbp/9781836204879)*

3. Click the **Create first rule** button, and the **Create governance rule** blade appears, as shown in *Figure 8.3*:

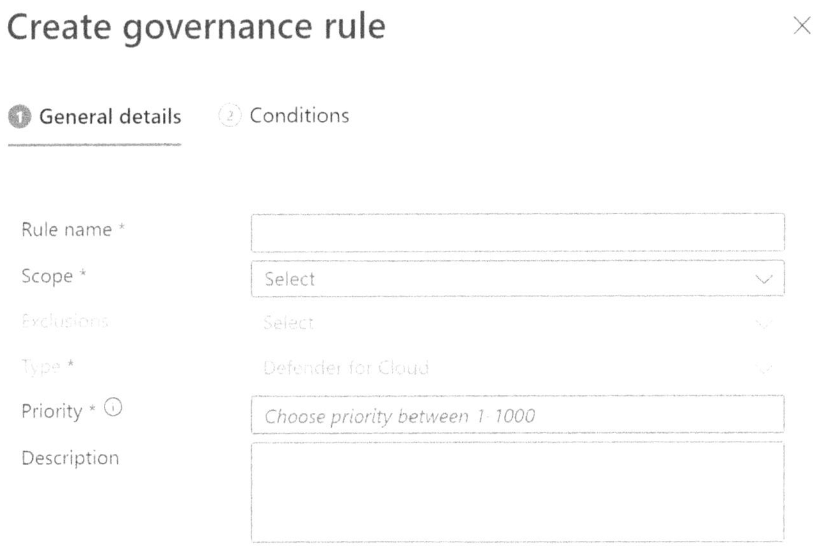

*Figure 8.3: General details of creating a governance rule*

4. For **Rule name**, type a name for this rule.

5. In the **Scope** field, click the dropdown and select the scope, which, in this case, will be the Azure subscription where you want to configure the governance. You could also select an AWS account or GCP project.

6. If appropriate, select the exclusion to define what doesn't apply to this rule. For example, if you select a management group and want to exclude specific subscriptions, you can use this field to select which subscriptions will not be applicable to this rule.

7. Leave Defender for Cloud for **Type**.

8. In the **Priority** field, type the priority for the rule. Rules are run in priority order from highest (1) to lowest (1000). Since many governance rules can apply to the same recommendation, the rule with the lowest priority value takes precedence.

9. After typing the priority, click the **Next** button (at the bottom of the blade). The **Conditions** tab appears, as shown in *Figure 8.4*:

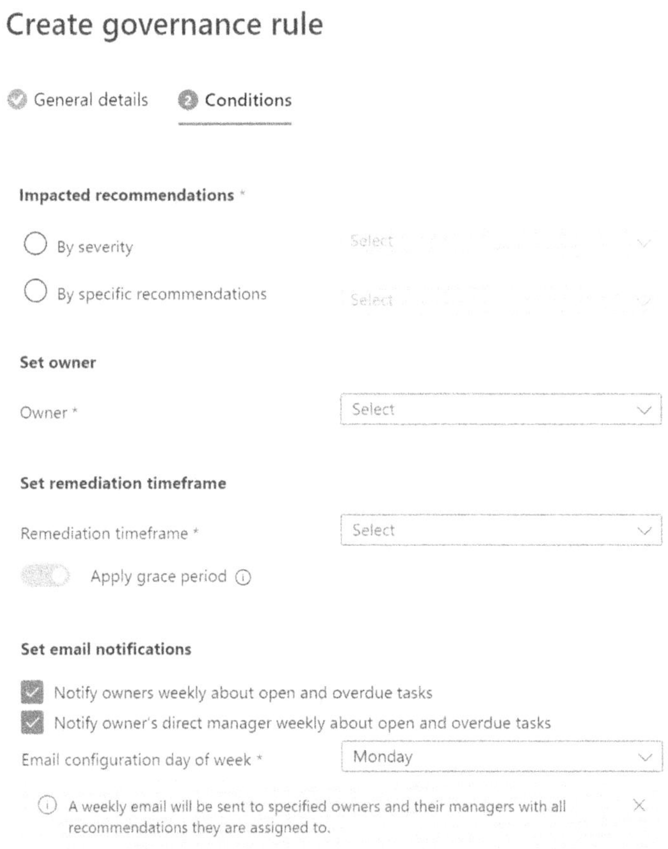

## Create governance rule

  ✅ General details    ② Conditions

**Impacted recommendations** *

  ◯ By severity        Select

  ◯ By specific recommendations      Select

**Set owner**

Owner *      Select

**Set remediation timeframe**

Remediation timeframe *      Select

  ◯ Apply grace period ⓘ

**Set email notifications**

☑ Notify owners weekly about open and overdue tasks
☑ Notify owner's direct manager weekly about open and overdue tasks

Email configuration day of week *      Monday

  ⓘ A weekly email will be sent to specified owners and their managers with all    ✕ recommendations they are assigned to.

*Figure 8.4: Conditions tab (for better visualization, refer to https://packt.link/ gbp/9781836204879)*

10. The first condition that you must select is the impacted recommendation. For that, you can select for it to be by severity or by specific recommendations. For this example, you can select the **By specific recommendations** option, and in the drop-down list, select **GitHub repositories should have code scanning enabled**.

11. The next field is **Owner**, which allows you to assign ownership for this role. You can select by email address or by resource tag. For this example, select **By email address** and type the email of the workload owner who will be responsible for this recommendation.

12. Next, you will select the due date. In the **Remediation timeframe** field, click the dropdown and select **14 days**. Notice that the available timeframe options are **7**, **14**, **30**, or **90 days**. This is the day after which the recommendation is found by the rule.

13. If the **Apply grace period** option is selected, the resources given a due date will not affect the secure score.

14. The section **Set email notifications** has two checkboxes selected by default to notify owners about open and overdue tasks and notify direct managers. You should set these options according to your company's security policy. Keep in mind that this option relies on the manager attribute in Entra ID.

15. Lastly, you can select the day of the week that the weekly email will be sent, using the option **Email configuration day of week**.

16. Click the **Create** button to finish. After a moment, the **Governance rules** page should be refreshed and you will see the rule, as shown in the screenshot below:

*Figure 8.5: Governance rule recently created (for better visualization, refer to https://packt.link/gbp/9781836204879)*

In a brand-new environment, it is easy to find the rules that you created, but in a production environment, you may have hundreds of these rules, which can make them challenging to find. To address this, you can use the **Add filter** button (*Figure 8.6*) and select the appropriate filter. For example, if you want to see all the rules assigned to a specific owner, you can select the **Owner** option and select the name of the owner from the list.

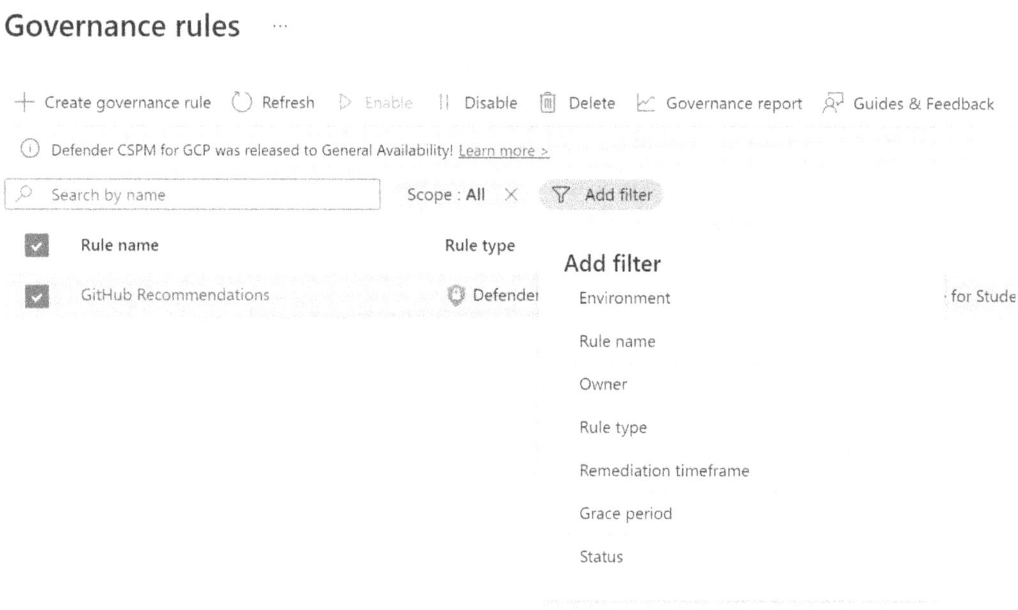

## Governance rules   ···

+ Create governance rule   ◯ Refresh   ▷ Enable   || Disable   🗑 Delete   ⌇ Governance report   ♟ Guides & Feedback

ⓘ Defender CSPM for GCP was released to General Availability! Learn more >

| 🔍 Search by name | Scope : **All** ✕   ▽ Add filter | | |
|---|---|---|---|
| ☑ **Rule name** | **Rule type** | **Add filter** | |
| ☑ GitHub Recommendations | 🔵 Defender | Environment | · for Stude |
| | | Rule name | |
| | | Owner | |
| | | Rule type | |
| | | Remediation timeframe | |
| | | Grace period | |
| | | Status | |

*Figure 8.6: Adding a filter (for better visualization, refer to https://packt.link/gbp/9781836204879)*

On this page, you can also access the built-in **Governance** workbook, which will give you an overview of the governance rules across one or more subscriptions (in the case of Azure) or other supported cloud providers (AWS and GCP). To see the workbook, click the **Governance report** button. An example of a populated report is shown in *Figure 8.7*:

Governance

Subscription

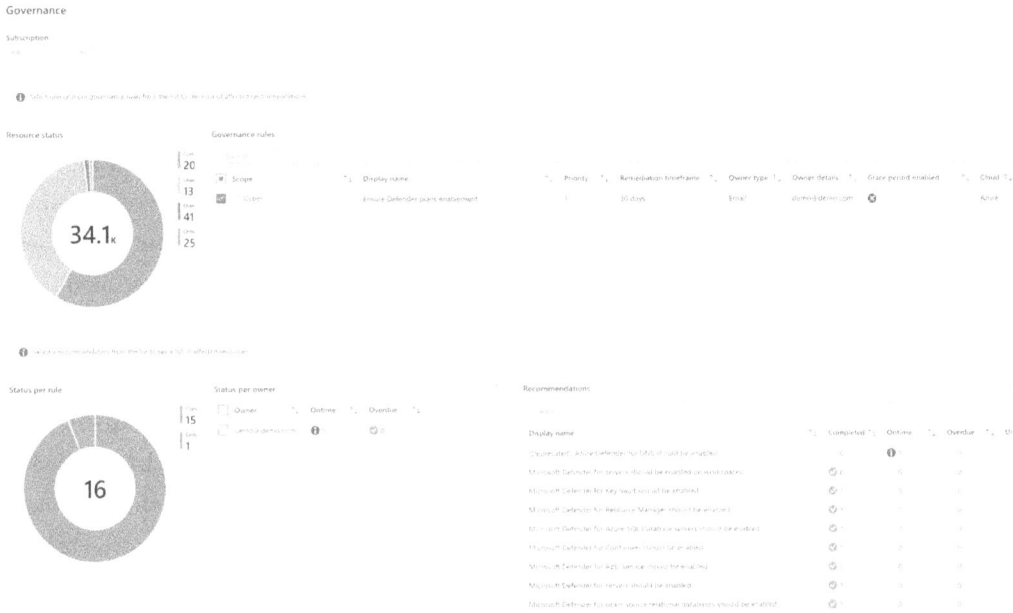

*Figure 8.7: Governance report (for better visualization, refer to https://packt.link/gbp/9781836204879)*

While the **Governance** report is a great place to go to have this one-stop view of the governance rules, you may want to see governance rules that have expired (also known as overdue) in more detail. To see overdue recommendations, go to the **Recommendations** page, and in the **Status** column, you will see what is shown in the example screenshot of *Figure 8.8*. Aside from **Overdue**, the status can also be on time, **completed**, or **unassigned**.

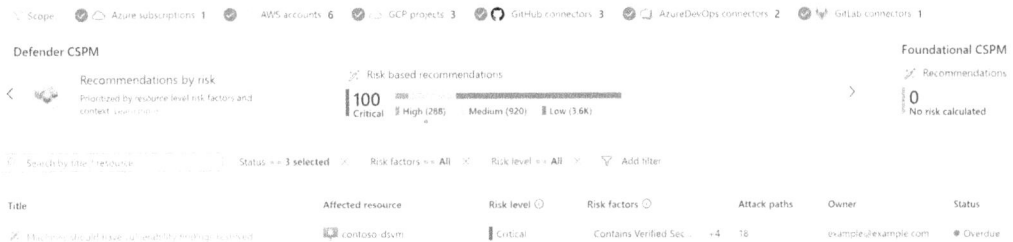

*Figure 8.8 Overdue recommendation (for better visualization, refer to https://packt.link/gbp/9781836204879)*

# Integration with ServiceNow

After the governance feature was released, the Defender for Cloud team identified another important native integration scenario for organizations that were using ServiceNow as their main **IT Service Management (ITSM)** tool. ServiceNow can be integrated with Defender for Cloud, enabling customers to prioritize the remediation of recommendations that impact their business and to leverage the existing process/tooling configuration in ServiceNow to route the requests to the correct organization structure. Defender for Cloud integrates with the ITSM module for incident management. Through this connection, customers can create and view ServiceNow tickets linked to recommendations directly from Defender for Cloud.

By default, the ticket between the platforms is synchronized, which means that the ticket status is verified to check if it is still in progress. If the ticket state changes to **Resolved**, **Canceled**, or **Closed** in ServiceNow, this change will be synchronized to Defender for Cloud and the assignment will be deleted. The synchronization also applies to the ticket owner; in other words, when the ticket owner changes in ServiceNow, the assignment owner also changes in Defender for Cloud to reflect that.

To configure this integration, you need to have an application registry in ServiceNow[1], and you need to have Security Admin, Contributor, or Owner permission in the Azure subscription where Defender CSPM is enabled.

> You can sign up for a trial version of ServiceNow using the Developer edition. Follow the guidelines in this article: `https://bit.ly/CNAPPBook32`.

## Configuring ServiceNow integration

After reviewing the prerequisites for this integration, follow the steps below to configure this capability in Defender for Cloud:

1. Open the **Defender for Cloud** dashboard and click **Environment settings**, under the **Management** section. The **Environment settings** page appears, as shown in *Figure 8.9*:

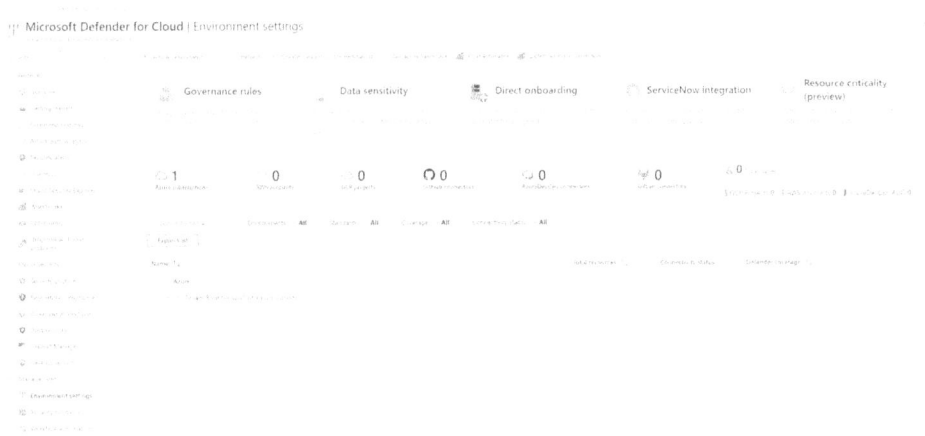

*Figure 8.9: Environment settings (for better visualization, refer to https://packt.link/gbp/9781836204879)*

2. On the top menu, click the **ServiceNow integration** option. The **ServiceNow integrations** page appears, as shown in *Figure 8.10*:

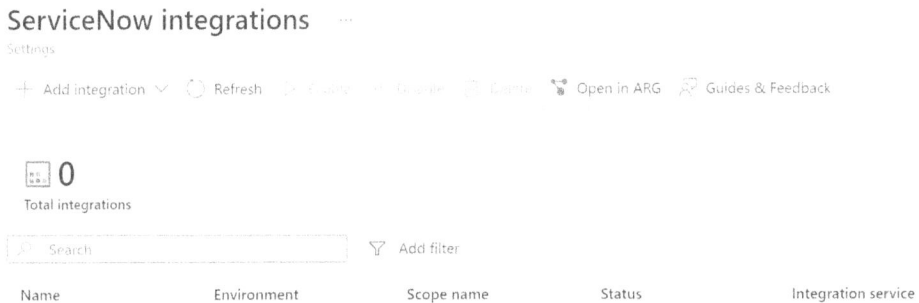

*Figure 8.10: ServiceNow integrations page (for better visualization, refer to https://packt.link/gbp/9781836204879)*

3.  Click the + **Add integration** button and click the **ServiceNow** option. The **Add integration** blade appears, as shown in *Figure 8.11*:

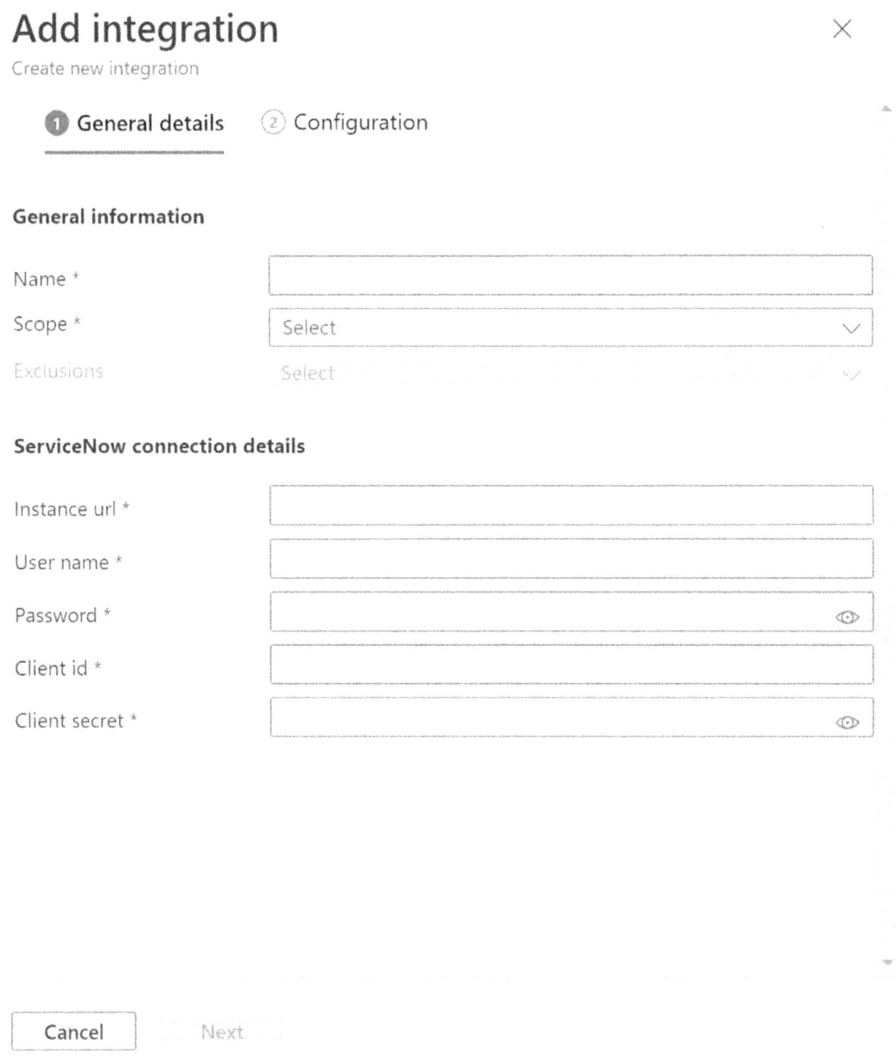

# Add integration                                                    ✕

Create new integration

①  **General details**      ②  Configuration

**General information**

Name *                  [                                                    ]

Scope *                 [ Select                                         ∨ ]

Exclusions              Select                                           ∨

**ServiceNow connection details**

Instance url *          [                                                    ]

User name *             [                                                    ]

Password *              [                                                  ◉ ]

Client id *             [                                                    ]

Client secret *         [                                                  ◉ ]

[ Cancel ]      Next

*Figure 8.11: Add integration page*

4.  In the **Name** field, type the name for this integration.

5.  Click the drop-down option in the **Scope** field and select the scope, which, in this case, will be the subscription. If applicable, the **Exclusions** option will become available.

6.  Type the instance URL, for example, `https://instanceid.service-now.com`, in the **Instance url** field.

7.  Type the username and password in the respective fields.

8.  Type the client identification in the **Client id** field.

9.  Type the secret in the **Client secret** field.

10. Click the **Next** button and the **Configuration** tab appears, as shown in *Figure 8.12*:

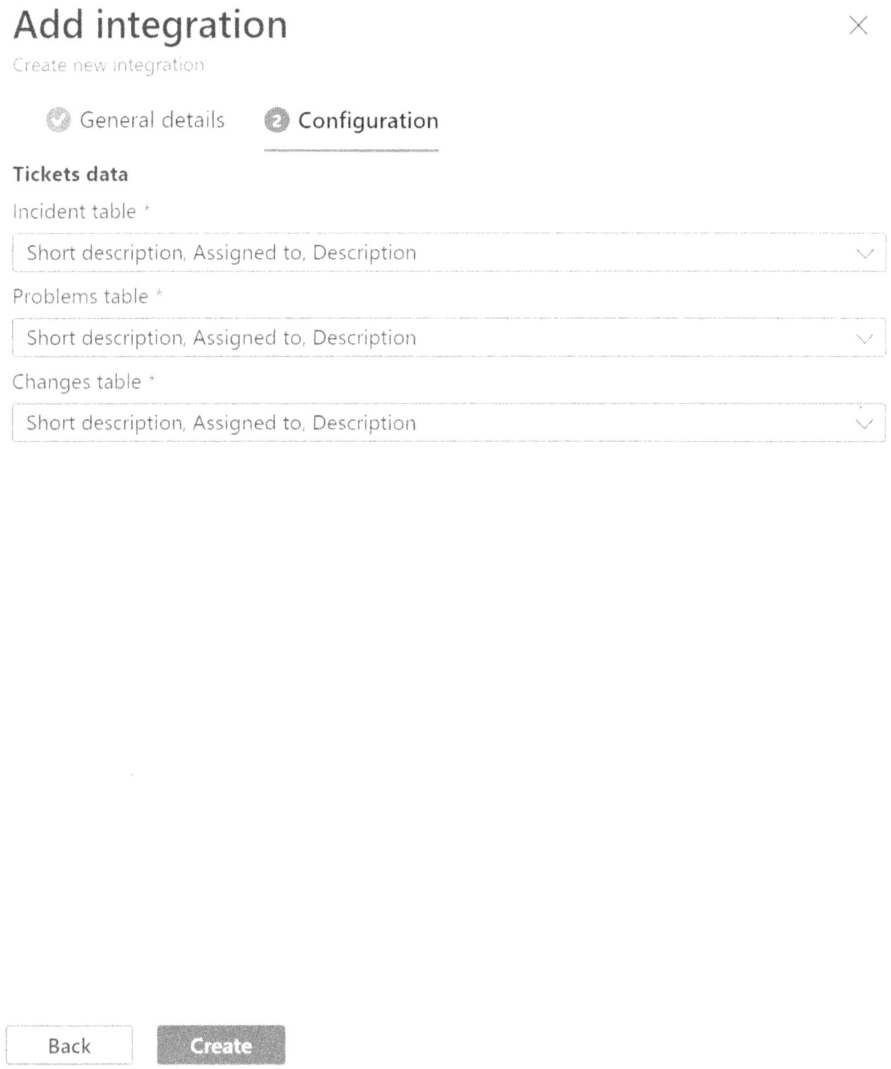

## Add integration                                                    ✕

Create new integration

✅ General details      ② Configuration

**Tickets data**

Incident table *

| Short description, Assigned to, Description                    ⌄ |

Problems table *

| Short description, Assigned to, Description                    ⌄ |

Changes table *

| Short description, Assigned to, Description                    ⌄ |

| Back |      **Create**

*Figure 8.12: Configuration tab*

11. Select the appropriate fields that you want to synchronize for the ticket in the incident, problems, and changes tables.

12. Click the **Create** button and a notification will appear confirming the creation. The new entry will appear on the page, as shown in *Figure 8.13*:

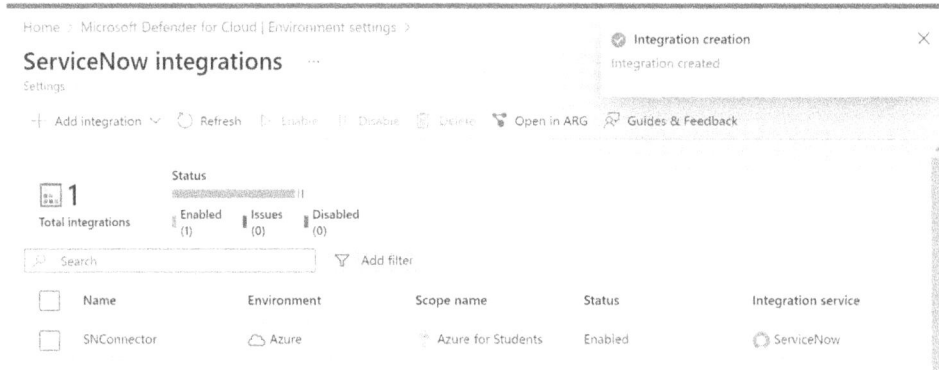

*Figure 8.13: Connection with ServiceNow created (for better visualization, refer to https://packt.link/gbp/9781836204879)*

Now that Defender for Cloud is integrated with ServiceNow, you can go back to the governance rule creation, and you will see that once you create a new rule, the **Type** dropdown now has two options, **Defender for Cloud** and **ServiceNow**, as shown in *Figure 8.14*:

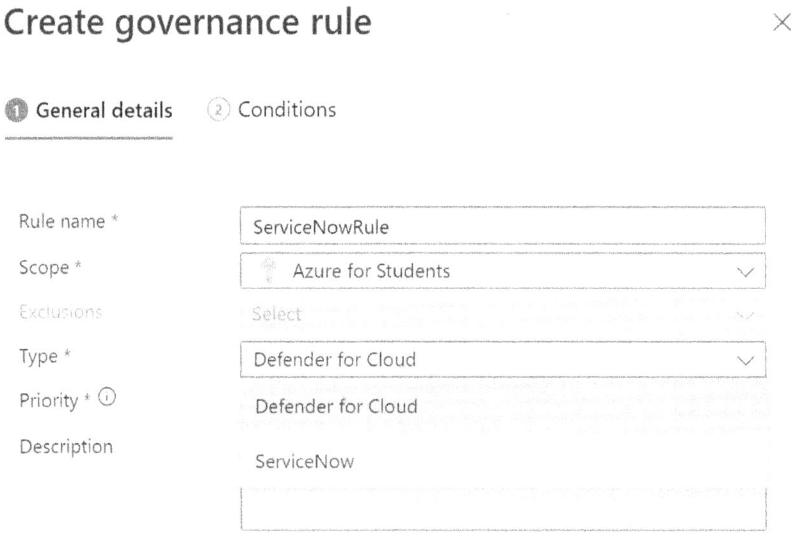

*Figure 8.14: New addition to the governance rule*

Upon selecting the type of rule as **ServiceNow**, you will see the addition of two new items in the blade, as shown in *Figure 8.15*:

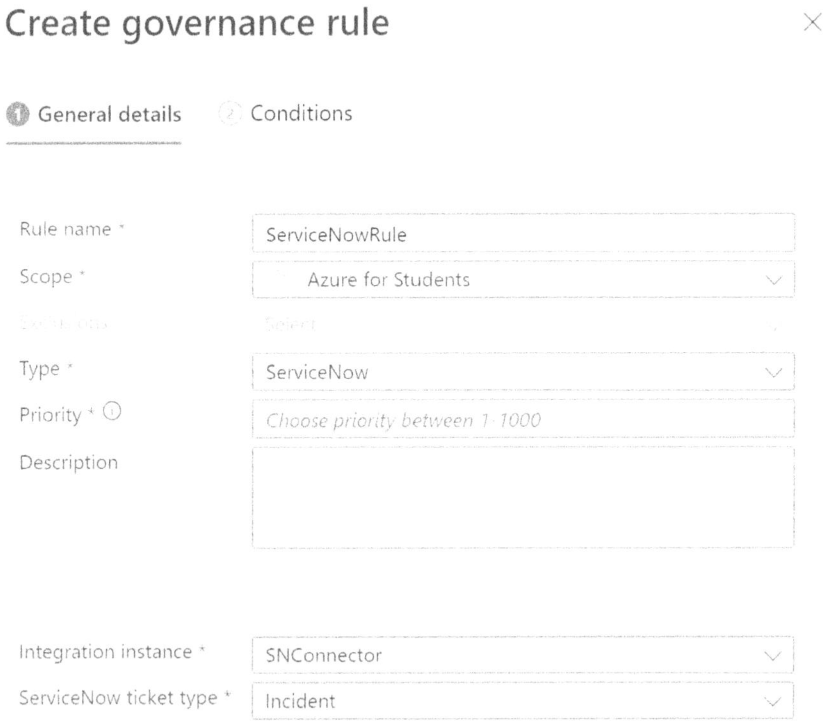

# Create governance rule

✕

● General details   ② Conditions

Rule name *

ServiceNowRule

Scope *

Azure for Students        ⌄

Select ⌄

Type *

ServiceNow        ⌄

Priority * ⓘ

*Choose priority between 1–1000*

Description

Integration instance *

SNConnector        ⌄

ServiceNow ticket type *

Incident        ⌄

*Figure 8.15: New addition to the governance rule*

The first item is the selection of the integration instance. You should click on the dropdown and select the instance that you created previously. The second item is the selection of the ServiceNow ticket type, which should be selected according to what you have in ServiceNow. The other steps to create the governance rule are the same as you learned in the previous section.

Once you finish creating the rule, you will notice the addition of a new rule on the **Governance rules** page, where the rule type is **ServiceNow**, as shown in *Figure 8.16*:

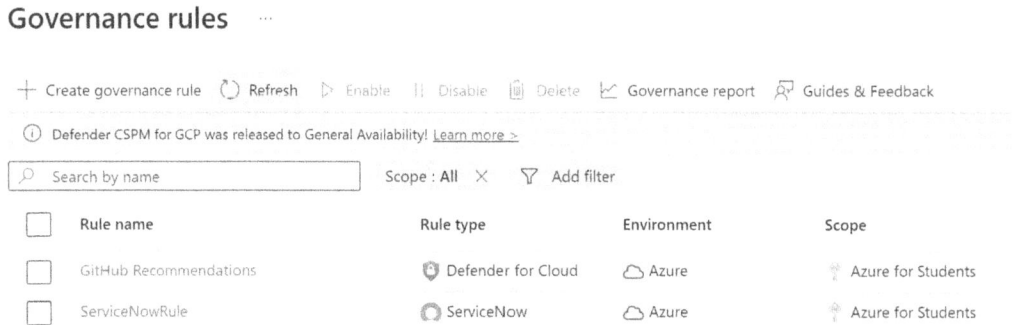

*Figure 8.16: New governance rule for ServiceNow (for better visualization, refer to https:// packt.link/gbp/9781836204879)*

# Delegate ownership

Ideally, you will reach a point where you will be working in a more proactive manner by creating governance rules upfront so that when new recommendations are triggered, actions can be automatically taken, like assigning ownership. However, when you are implementing governance for the first time in a production environment, where you have lots of security recommendations to address, your priorities will likely change.

In a production environment where you have just enabled Defender CSPM, you may be faced with hundreds of recommendations to address, so your first step will be to review the risk-based recommendations and see which ones are critical for your environment.

As you review these security recommendations, you may identify some that you should assign ownership to someone else to resolve, and that's where you can leverage the governance feature from the security recommendations dashboard. When you open the **Recommendations** page, and then open a security recommendation, you will notice on the right side the section **Delegate**, as shown in *Figure 8.17*:

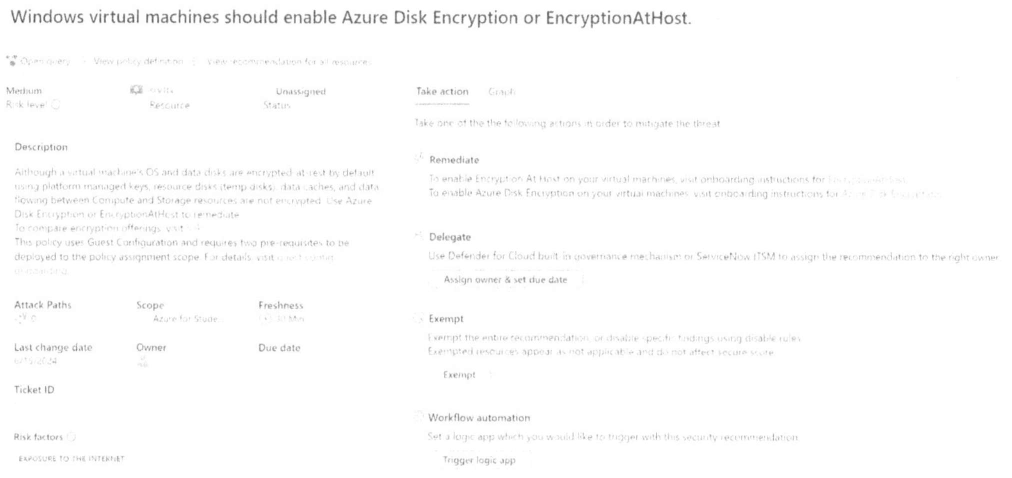

Windows virtual machines should enable Azure Disk Encryption or EncryptionAtHost.

Figure 8.17: Delegating ownership (for better visualization, refer to https://packt.link/
gbp/9781836204879)

From this page, click the **Assign owner & set due date** button; the **Create assignment** blade
appears, as shown in *Figure 8.18*:

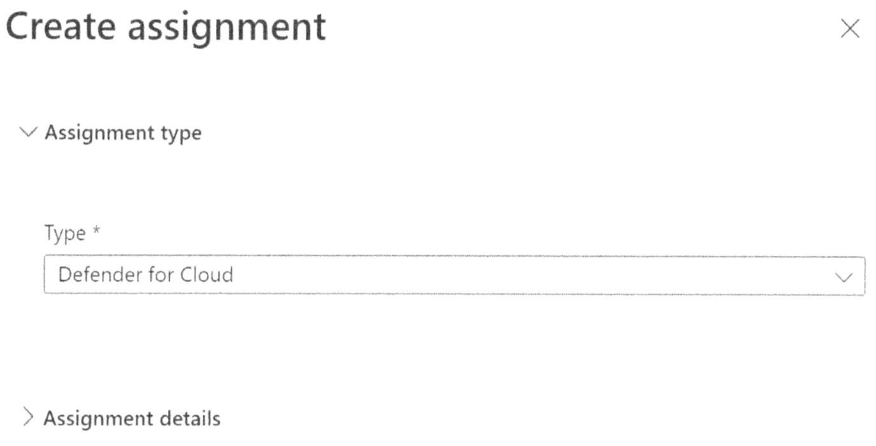

Figure 8.18: Create assignment

From this blade, you can select either Defender for Cloud or ServiceNow as your assignment type,
and based on your selection, the **Assignment details** section will change. The options are the
same ones that were presented previously in this chapter. The only difference is the approach
that you are taking to create the governance rule.

In other words, instead of creating governance rules proactively on the governance page, you are reviewing what is more critical to your environment and assigning ownership of the security recommendations.

*Figure 8.19* has an example of how this recommendation will appear in ServiceNow as a ticket once you finish creating the assignment (which will create a governance rule) in Defender for Cloud:

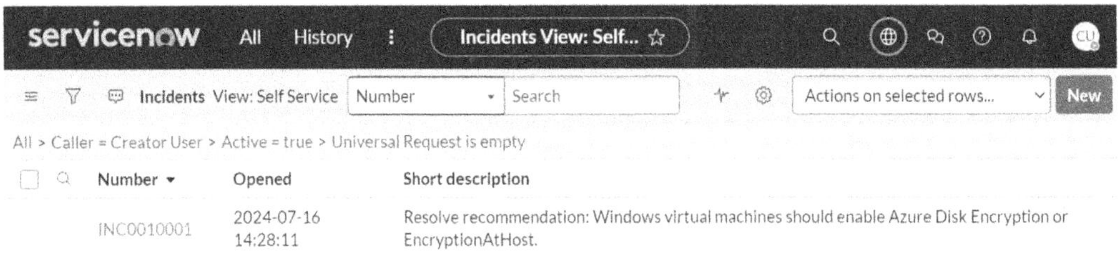

Figure 8.19: New incident in ServiceNow (for better visualization, refer to https://packt.link/gbp/9781836204879)

This ticket ID that appears in ServiceNow will also appear in the recommendation's details, in the **Ticket ID** field, as shown in *Figure 8.20*:

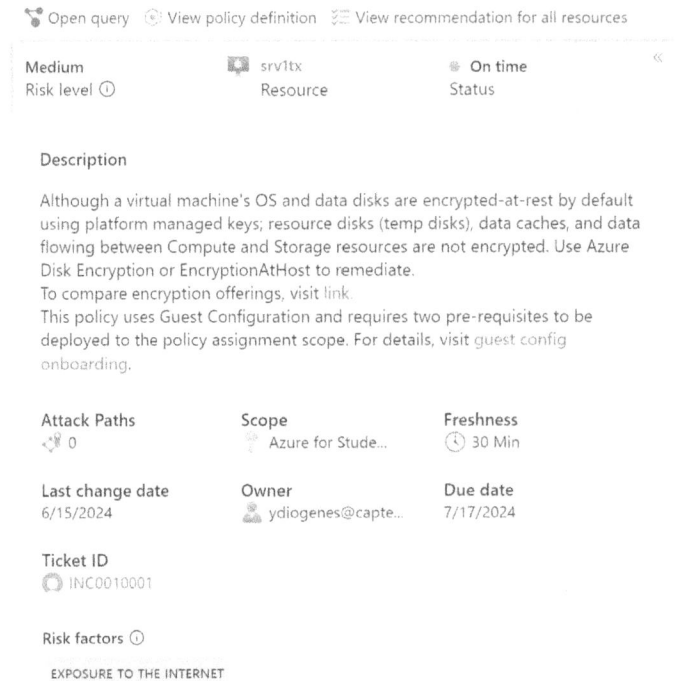

Figure 8.20: Synchronization of the Ticket ID field (for better visualization, refer to https://packt.link/gbp/9781836204879)

# Continuous improvement

While governance is imperative for continuous improvement, you also need to understand your environment and recognize that there might be situations where a security recommendation triggered by Defender for Cloud may not be applicable to your environment for several reasons.

Let's look at a scenario where a cloud security administrator receives the recommendation **Auditing on SQL server should be enabled** in Defender for Cloud. However, they know that to remediate this recommendation, they need to provision a new storage account to save the SQL auditing logs. At this point, the company doesn't have a budget for that, so the cloud security administrator will just have to assume the risk. To prevent this recommendation from continuing to show up, they can create an exemption. You need the **Owner** or **Security Admin** permission to create an exemption.

Exemptions rely on the **Microsoft Cloud Security Benchmark (MCSB)** initiative, which evaluates and retrieves resources' compliance state. However, the exemption option is not available for all recommendations because it is not supported by all recommendations. For a list of recommendations that do not support exemption, visit `https://bit.ly/CNAPPBook33`.

To access the exemption for a recommendation, open the recommendation itself, and in the right pane (the **Take action** tab), you will see the **Exempt** button, as shown in *Figure 8.21*:

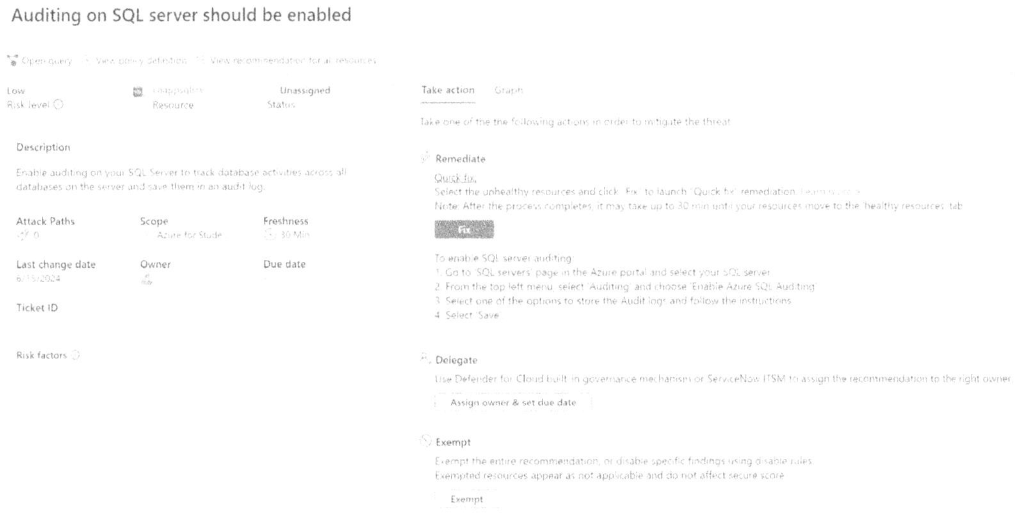

*Figure 8.21: Exemption (for better visualization, refer to https://packt.link/gbp/9781836204879)*

Click the **Exempt** button to open the **Exempt** blade with all the options that you can configure, as shown in *Figure 8.22*:

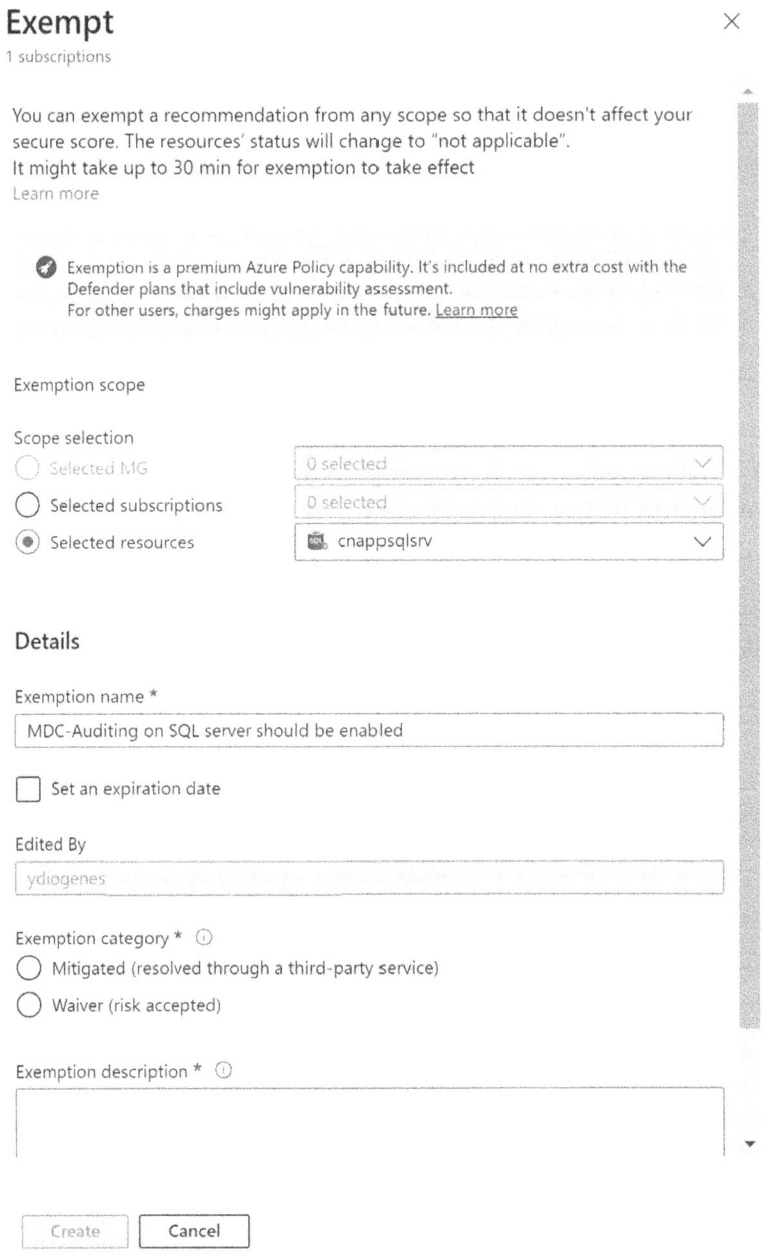

# Exempt                                                          ✕

1 subscriptions

You can exempt a recommendation from any scope so that it doesn't affect your secure score. The resources' status will change to "not applicable".
It might take up to 30 min for exemption to take effect
Learn more

> �’ Exemption is a premium Azure Policy capability. It's included at no extra cost with the Defender plans that include vulnerability assessment.
> For other users, charges might apply in the future. Learn more

Exemption scope

Scope selection

○ Selected MG                            | 0 selected                              ∨ |

○ Selected subscriptions           | 0 selected                              ∨ |

◉ Selected resources               | 🗄 cnappsqlsrv                          ∨ |


## Details

Exemption name *

| MDC-Auditing on SQL server should be enabled                                        |

☐ Set an expiration date

Edited By

| ydiogenes                                                                            |

Exemption category * ⓘ
○ Mitigated (resolved through a third-party service)
○ Waiver (risk accepted)

Exemption description * ⓘ

|                                                                                      |

[ Create ]   [ Cancel ]

*Figure 8.22: Creating an exemption*

On this page, you can only select the resource for this exemption because you activated the **Exemption** blade from the security recommendation itself, which is bound to the resource. In the **Details** section, you can add a name for the exemption and set an expiration date. It is recommended to set an expiration date because, most of the time, you want to revisit the decision of exempting a security recommendation. In this sample scenario, the exemption is created because of budget constraints, but this may not be the case six months from now, so setting the exemption to six months will trigger this recommendation again after the expiration date. You should also select the exemption category, which can be **Mitigated (resolved through a third-party service)** or **Waiver (risk accepted)**. In this example, it will be **Waiver**, since this security administrator is accepting the risk due to budget constraints. Make sure to type the rationale of this decision in the **Exemption description** field. Click the **Create** button to finish, and a notification will appear confirming that the exemption was created, as shown in *Figure 8.23*:

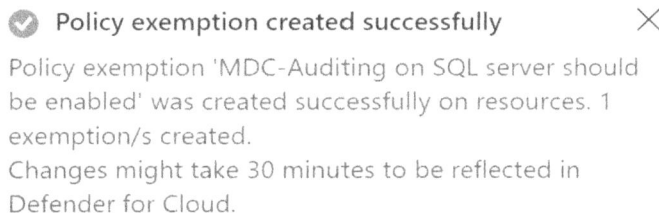

> ✅ **Policy exemption created successfully** ✕
>
> Policy exemption 'MDC-Auditing on SQL server should be enabled' was created successfully on resources. 1 exemption/s created.
> Changes might take 30 minutes to be reflected in Defender for Cloud.

*Figure 8.23: Policy exemption confirmation*

This type of fine-tuning will also impact your secure score, because recommendations that are exempted will not count toward the secure score. Having said that, do not use the exemption feature to mask problems in your environment and increase your secure score. Exemptions should be used as a last resort, must be evaluated case by case, and should have an expiration date.

With time, you may have multiple exemptions across different workloads. You can use the **Inventory** page in Defender for Cloud to visualize all resources that have exemptions. To do that, go to the Defender for Cloud dashboard, click **Inventory** under the **General** section, click the **Add filter** button, and select the **Contains exemptions** option, as shown in *Figure 8.24*:

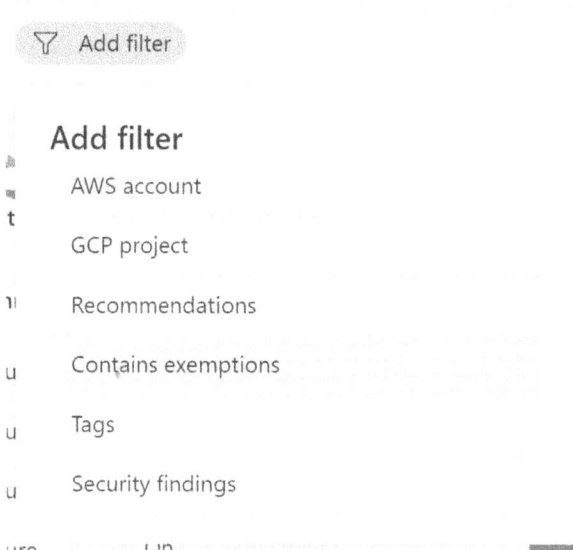

*Figure 8.24: Filtering for resources with exemptions*

You will need to select **Yes** (with exemptions) or **No** (without exemptions), and then click the **Apply** button. The result appears with all resources that have exemptions, as shown in *Figure 8.25*:

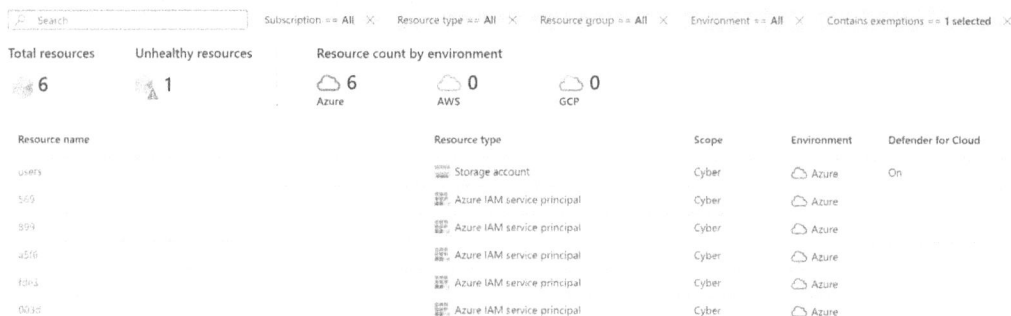

*Figure 8.25: Resources with exemptions (for better visualization, refer to https://packt.link/ gbp/9781836204879)*

# Final considerations

When you think about continuous improvement of your security posture, Defender for Cloud offers a set of capabilities that will assist you throughout this journey, which includes:

1.  Security recommendations based on the Microsoft Cloud Security Benchmark

2. Everything that starts with the initial ongoing assessment of your workloads and identification of the security state of these workloads

3. Secure score

4. Reflection of your current secure state, enabling you to use this number as a KPI for secure improvement over time

5. A risk-based approach to prioritize security recommendations

6. Prioritization of security recommendations based on a series of risk factors

7. Governance

8. Built-in governance capability to enable you to assign ownership to recommendations and establish a due date to remediate

9. Integration of governance capabilities with the ServiceNow system

10. Exemptions

11. Fine-tuning of security recommendations based on the reality of your environment

These capabilities are important, and they are not the type of capabilities that you use once and abandon; these are capabilities that are part of your posture management tools and will be used constantly.

# Summary

In this chapter, we discussed the use of the governance feature in Defender CSPM to assign ownership to security recommendations. You also learned how to integrate the governance feature with ServiceNow. You learned how to create exemptions to suppress recommendations that are not applicable to your environment and how to visualize all resources that have exemptions.

In the next chapter, you will learn how to perform proactive hunting with Cloud Security Explorer.

# Notes

1. You can obtain more information about how to create a ServiceNow API Client id and Client secret for the SCOM ServiceNow Incident Connector following this article from ServiceNow at `https://bit.ly/CNAPPBook31`.

# Additional resources

- Watch the author Yuri Diogenes interviewing the PM responsible for the governance feature during this episode of *Defender for Cloud in the Field*: `https://bit.ly/CNAPPBook34`
- Watch the author, Yuri Diogenes, interviewing the PM responsible for the ServiceNow integration feature during this episode of *Defender for Cloud in the Field*: `https://bit.ly/CNAPPBook35`

# Join our community on Discord

Read this book alongside other users. Ask questions, provide solutions to other readers, and much more.

Scan the QR code or visit the link to join the community.

`https://packt.link/SecNet`

# 9

# Proactive Hunting

As companies become more mature in their cloud security journey and the strengthening of their cloud security posture becomes second nature, it creates opportunities to be more proactive. A proactive approach to cloud security posture management is where companies should aim to operate, as this approach enables posture management teams to be more effective and anticipate potential attacks that could take place in the environment.

One of the key aspects of CNAPP is the process of scanning artifacts to create insights. Up until now, you've learned how these insights are used to enrich security recommendations and create attack paths. However, when an organization has achieved the level of cloud security maturity that allows it to invest in proactive hunting, those insights can also be leveraged to gain more information about potential areas that can be exploited.

Defender for Cloud has built-in capabilities that enable posture management teams to create queries against the data that was collected as part of the native CNAPP artifact scanning capability.

This chapter covers:

- Leveraging the insights collected by CNAPP
- Cloud Security Explorer
- Azure Resource Graph

## Leveraging the insights collected by CNAPP

In November 2021, Chen Zhaojun of Alibaba Cloud's security team[1] discovered a critical security flaw in the Apache Log4j 2 library, a popular logging tool used in many Java-based applications. Chen Zhaojun reported the vulnerability to the Apache Software Foundation, which started working on a fix for the problem.

In December, the vulnerability was publicly disclosed and, at this time, it was assigned the CVE identifier CVE-2021-44228, and this became the well-known Log4Shell vulnerability. Many organizations rushed to scan their environment to better understand if their systems were affected by this vulnerability.

Many organizations faced challenges in identifying this vulnerability when their platform didn't have a centralized manner to query all the data across multiple workloads and environments. Some organizations that were working with best-in-breed tools ended up paying a higher price for working in silos with different tools and without a centralized approach for data collection and investigation.

When a scenario like this takes place, it becomes critical to have a platform that can easily query data across all workloads, across multiple environments, and quickly obtain the results. This is one of the main differential factors of a CNAPP platform such as Defender for Cloud. It enables security posture management teams to easily query the artifact scanning results, also known as insights (from the CNAPP standpoint). *Figure 9.1* shows a representation of a multicloud environment when many workloads are interconnected and the security administrator searches from one single dashboard for vulnerabilities distributed across these environments.

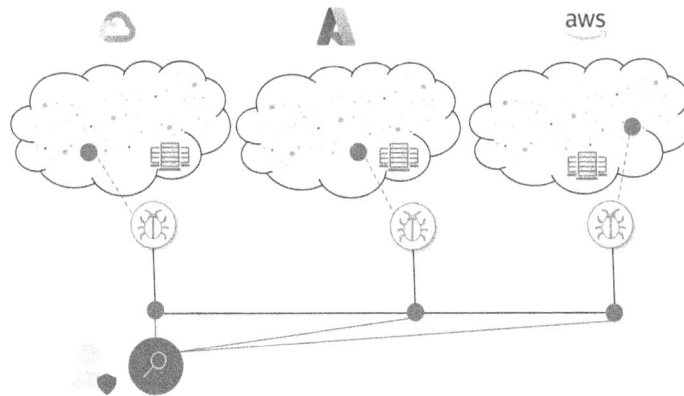

*Figure 9.1: Searching for vulnerabilities across multiple environments*

## Cloud Security Explorer

The extensions that are enabled with Defender CSPM allow the collection of insights using an agentless approach; in other words, no agent needs to be deployed to the workload to obtain the necessary data to generate the insight. These insights are used to generate the cloud security graph, which is a graph-based context engine that contains and correlates all this data and potential attacks.

The graph can be accessed using Cloud Security Explorer. To access Cloud Security Explorer, open the **Defender for Cloud** dashboard and click **Cloud Security Explorer** in the **General** section. The **Cloud Security Explorer** page appears as shown in *Figure 9.2*:

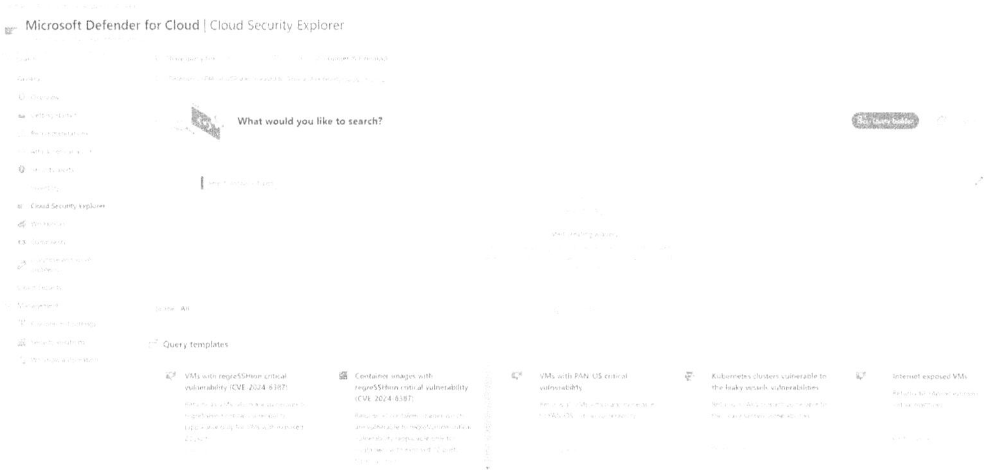

*Figure 9.2: Cloud Security Explorer (for better visualization, refer to https://packt.link/ gbp/9781836204879)*

Cloud Security Explorer is a blank canvas where you can start your query from scratch, or you can use the templates that are available in the lower part of the screen. You need the **Security Admin** role to properly perform queries, as this query will need read access across all objects that are part of the map.

For this initial example, start with a simple template that will check if there are VMs exposed to the internet and with high-severity vulnerabilities. To perform this query, click the **Open query** hyperlink in the template shown in *Figure 9.3*:

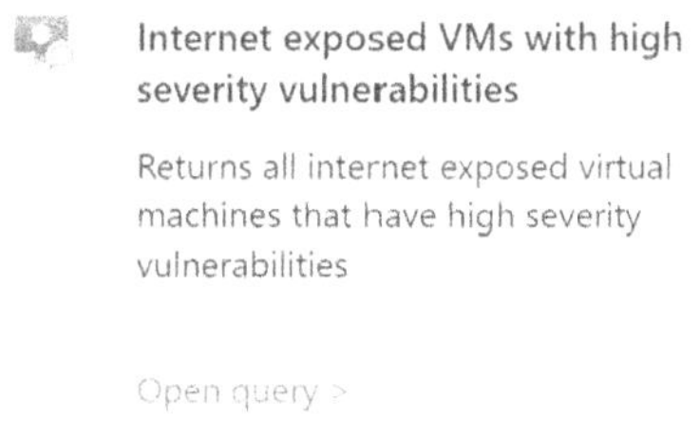

Internet exposed VMs with high severity vulnerabilities

Returns all internet exposed virtual machines that have high severity vulnerabilities

Open query >

*Figure 9.3: Cloud Security Explorer template (for better visualization, refer to https://packt. link/gbp/9781836204879)*

When you click on this template, Cloud Security Explorer will mount your query with the elements shown in *Figure 9.4*:

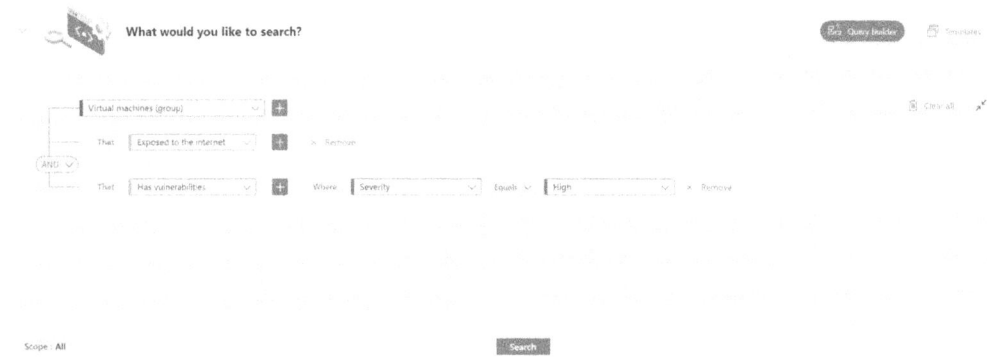

*Figure 9.4: Elements of the query (for better visualization, refer to https://packt.link/gbp/9781836204879)*

The first box (you will see a purple line in the left part of the box on your monitor's screen) represents the main entity that you will be using as the base of your query. Although the query in this example only has one entity, in more advanced queries, you can add more. The boxes underneath the main entity are the parameters and conditions. In the example in *Figure 9.4*, you have two conditions: **Exposed to the internet** AND (logical operator) **Has vulnerabilities**. In other words, for this query to produce results, both conditions must be true, since there is an AND as the logical operator. In this case, the second condition (**Has vulnerabilities**) also has another condition, which is the **Severity**, and a specific check (equal to **High**).

In the lower left part of the screenshot shown in *Figure 9.4*, you have a field called **Scope**, and it is set to **All**. This scope represents which environments are going to be queried. If you want to query in only certain environments, you can click the **Scope** option and select them, as shown in the example in *Figure 9.5*:

Select all
∨ ☑ Azure
    ☑ Azure for Students
∨ ☑ Github
    ☑ CNAPPBookGitHub
    ☑ CNAPPBookGHC
∨ ☑ GCP
    ☑ CNAPPBookGCP
∨ ☑ AzureDevOps
    ☑ CNAPPBookADO
∨ ☑ AWS
    ☑ CNAPPBook

**Apply**    Cancel

*Figure 9.5: Environment selection*

If you have multiple environments selected and you see the warning **Partial results expected**, it is because some of the selected environments (subscriptions in this case) do not have Defender CSPM enabled, as shown in *Figure 9.6*:

Scope : **2 selected**    ⚠ Partial results expected ⓘ

☰ **Results**

🔍 Search item

No Results

**Partial results expected**

**1/2 selected scopes are missing Defender CSPM plan**

Partial results will be displayed for these scopes. Enable plan to view information in the Security Explorer, and gain added posture capabilities for risk management.

🔗 To enable plan, visit "Environment settings" >

*Figure 9.6: Partial result warning*

After confirming the selection of the environments that are relevant to your query, click the **Search** button. The results will vary according to the environment and workloads you have, but they will always appear in the bottom part of the screen.

## Creating a custom query

Before starting a new query, you should understand what you want to find out. Knowing exactly what you are trying to discover will help you to be more effective in building the query; otherwise, you will waste too much time browsing around the options without doing anything productive. So, ask yourself and your teammates important questions, such as:

- What are we trying to find out?

    - Defining the main goal of the query can help you craft how this query is going to look in Cloud Security Explorer.

- Is this information that we are trying to find out related to a particular workload?

    - Defining which workload is most relevant to the query will help you define which entity (or entities) should be selected during the query in Cloud Security Explorer.

- Are there any specific conditions that we are looking for?

    - This is a critically important question to ask to narrow your search to only the specific conditions that are relevant to what you are trying to find with this query.

- Which environments are relevant for this query?

    - Although it is very appealing to just search across all environments, there will be times when you want to optimize your query to only specific environments. For example, if you know that the only affected environment is Azure, why are you going to add GCP and AWS as part of your query? You don't need to, and it doesn't add any value to you.

For this example, you want to identify all Azure Virtual machines that have keys and passwords and have high-severity recommendations unresolved. To get started, click the **Clear all** button, which will prompt you to confirm the operation, and you can click **Confirm**. Now you can start your custom query, and the first step is to select the main entity that you want to use in this query. Click the **Select resource types** drop-down box, and you will see the options available, as shown in *Figure 9.7*:

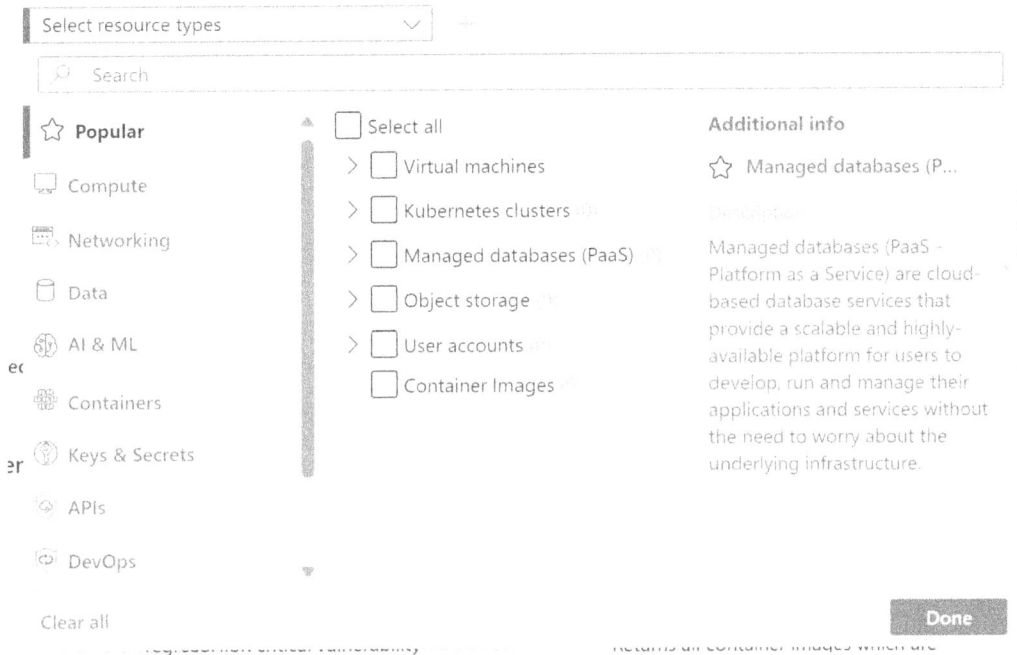

*Figure 9.7: Selecting the resource type*

On the left, you have the different resource categories, and as you select the category on the left, the right will show the resources available for that category. In the example shown in *Figure 9.7*, you have the **Compute** category, and the different resources available to query. Notice also that you have a greater than sign (>) beside each resource type, which means that there are different options to select. In the case of **Virtual machines**, the options are shown in *Figure 9.8*:

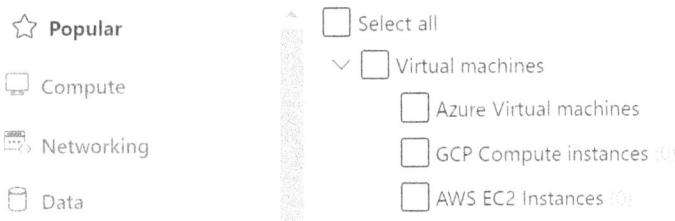

*Figure 9.8: Different types of VMs*

Select **Azure Virtual machines** for this example, and you will notice that the main entity gets selected. Now you need to click the plus sign (+) beside the entity to select the conditions, and then click the **Select condition** drop-down list, as shown in *Figure 9.9*:

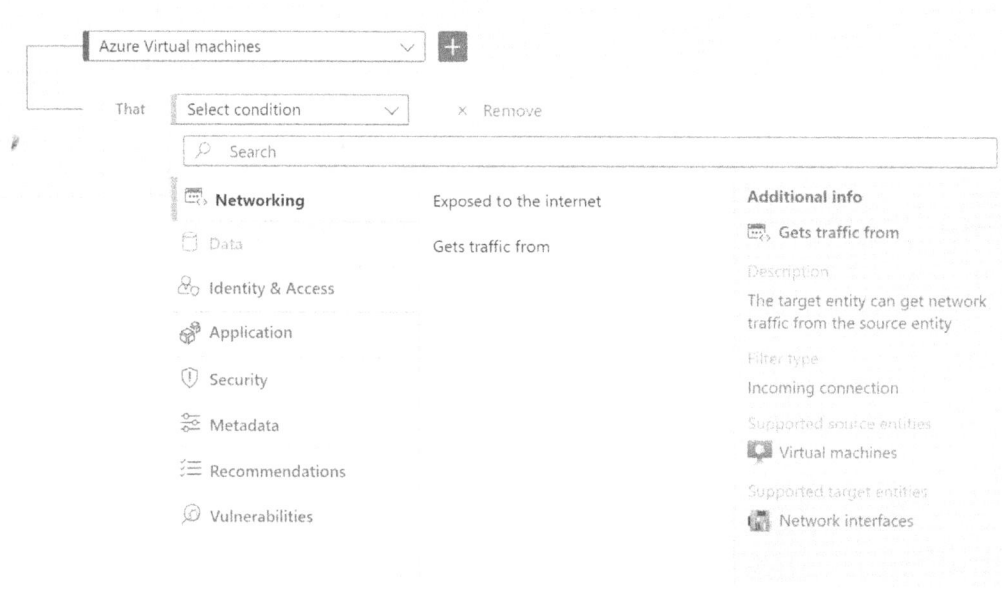

*Figure 9.9: Selecting condition in Azure Virtual machines (for better visualization, refer to https://packt.link/gbp/9781836204879)*

Since, for this query, the main condition is to find VMs with keys and passwords, you will select the **Identity & Access** category, and from there, select the **Contains** condition, as shown in *Figure 9.10*:

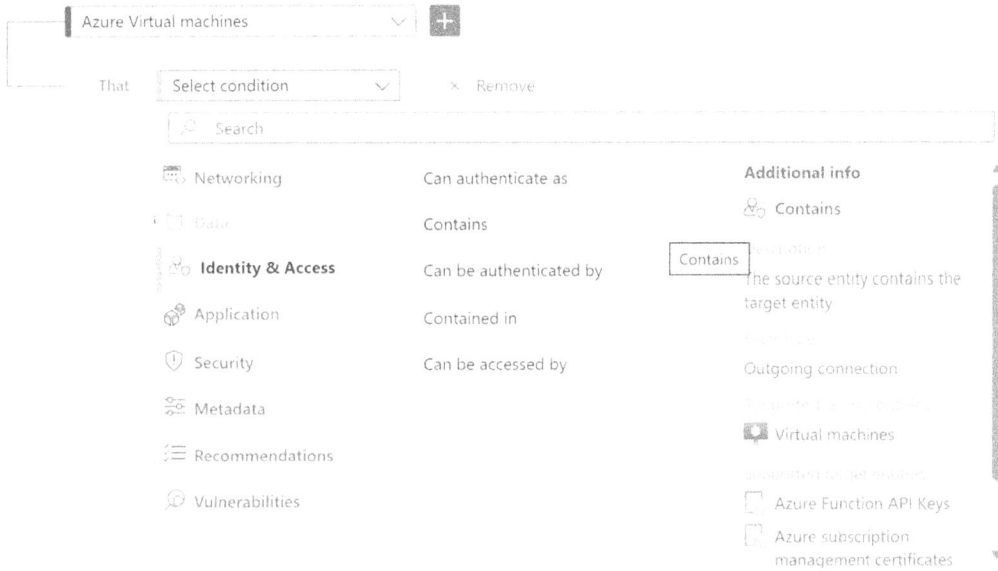

*Figure 9.10: Selecting the category (for better visualization, refer to https://packt.link/ gbp/9781836204879)*

Now click on the drop-down list for the condition. In the category, select **Keys & Secrets**, and on the right side, select the checkboxes for **Keys** and **Passwords**, as shown in *Figure 9.11*:

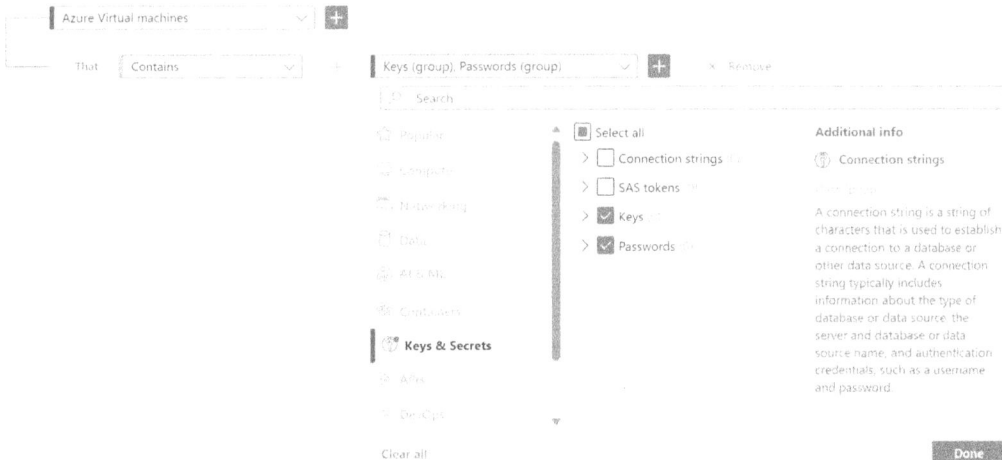

*Figure 9.11: Selecting keys and passwords (for better visualization, refer to https://packt.link/ gbp/9781836204879)*

After making the selection, click the **Done** button. The first part of your query is created; now you need the second part (high-severity security recommendations), which is another condition related to the main entity (**Virtual machines**). To add this second condition, click the plus sign (+) beside the main entity (**Azure Virtual machines**). You will see that an AND was added as a logical operator. Click the second condition drop-down list, click **Recommendations** on the left side, and click **By severity** on the right side, as shown in *Figure 9.12*:

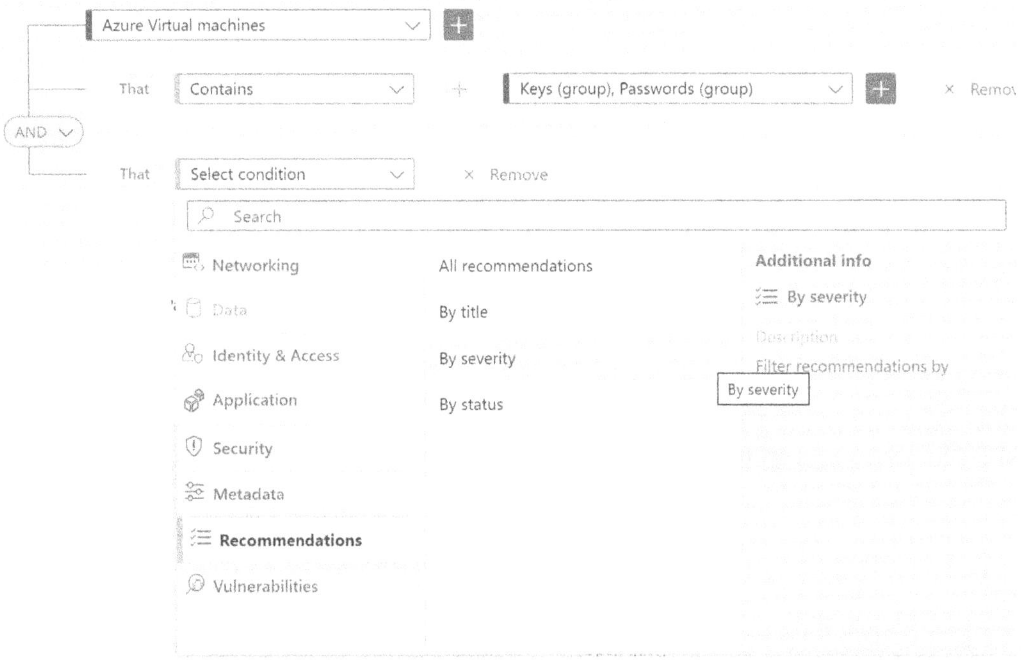

*Figure 9.12: Selecting recommendations by severity (for better visualization, refer to https://packt.link/gbp/9781836204879)*

Now you will see the addition of the **Has recommendations** condition; the **Severity** parameters are also added, and now you just need to click the **Equals** dropdown to select **High**, as shown in *Figure 9.13*:

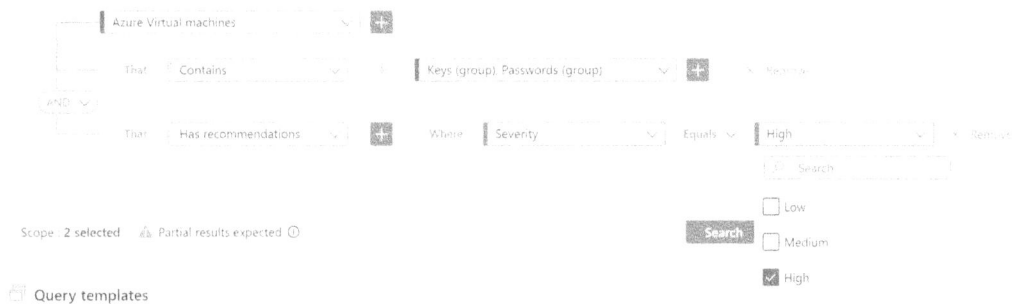

Figure 9.13: Selecting the final parameter for the query (for better visualization, refer to https://packt.link/gbp/9781836204879)

Once you finish, click the **Search** button to verify if you have machines that meet those conditions in your environment. The results can be also downloaded to a CSV file, by using the **Download CSV report** button and, if needed, you can share this query with another team member by using the **Share query link** button. Both buttons appear in the toolbar, as shown in *Figure 9.14*:

Figure 9.14: Additional options to share or download the query result (for better visualization, refer to https://packt.link/gbp/9781836204879)

The level of complexity of your query will vary according to the type of information you want to obtain, the level of granularity that you need, and how specific you are during the construction of your query. Cloud Security Explorer has the ability to accommodate high-complexity scenarios and still brings results very quickly.

# Azure Resource Graph

**Azure Resource Graph (ARG)** is particularly useful for scenarios that require deep insights and comprehensive management of resources. By leveraging its powerful querying capabilities, users can quickly gather information and generate detailed reports on resource configurations, compliance statuses, and operational metrics. ARG's query language is based on the **Kusto Query Language (KQL)**[2].

Since ARG allows custom queries using KQL and Defender for Cloud data is accessible via ARG, you can go even deeper if needed when you are doing proactive hunting. To improve the experience of switching from the Defender for Cloud dashboard and ARG, the Defender for Cloud team added a button called **Open query** in many parts of the dashboard, as shown in the example of the **Recommendations** page in *Figure 9.15*:

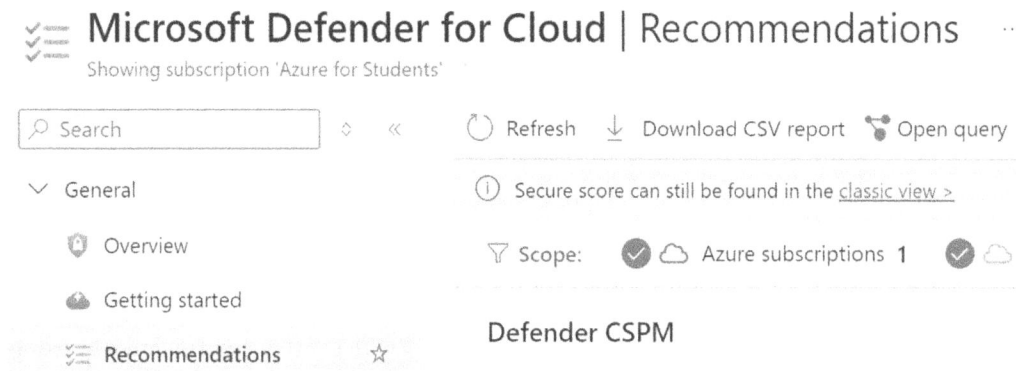

Figure 9.15: Open query button (for better visualization, refer to https://packt.link/ gbp/9781836204879)

When you click this button, you will be redirected to **Azure Resource Graph Explorer** within the context of the selected page (in this case), or recommendation (if you had one single recommendation open). *Figure 9.16* shows an example of how this looks for the **Recommendations** page.

Figure 9.16: Accessing Azure Resource Graph Explorer (for better visualization, refer to https:// packt.link/gbp/9781836204879)

From this page, you can click the **Run query** button and you will get the result. *Figure 9.17* has an example of the result for the **Recommendations** page:

Get started    Results    Charts    Messages

⤓ Download as CSV    📌 Pin to dashboard

| assess... ↑↓ | environment ↑↓ | statusChangeDate ↑↓ | riskLevel ↑↓ | riskFactors | attackPaths ↑↓ | statusCode ↑↓ | displayName ↑↓ | re |
|---|---|---|---|---|---|---|---|---|
| /subscriptio... | Azure | 2024-06-15T13:49:12... | 2 | \| Exposure to the Internet"\] | 0 | Unhealthy | Windows virtual mach... | /s |
| /subscriptio... | Azure | 2024-06-15T13:49:12... | 2 | \| Exposure to the Internet"\] | 0 | Unhealthy | Virtual machines and v... | /s |
| /subscriptio... | Azure | 2024-06-15T13:49:12... | 2 | \["Exposure to the Internet", "Weak Auth... | 0 | Unhealthy | SQL servers should ha... | /s |
| /subscriptio... | Azure | 2024-06-15T13:49:12... | 2 | \["Exposure to the Internet", "Weak Auth... | 0 | Unhealthy | Public network access ... | /s |
| /subscriptio... | Azure | 2024-06-15T13:49:12... | 2 | \["Exposure to the Internet", "Weak Auth... | 0 | Unhealthy | Private endpoint conn... | /s |

*Figure 9.17: ARG query result (for better visualization, refer to https://packt.link/gbp/9781836204879)*

In the first and second lines of the query shown in *Figure 9.16*, notice that you have the query being established to run under the **Microsoft.Security/assessment** context. This is because Defender for Cloud data is located in the **Microsoft.Security** resource provider, and the assessment is the location where the recommendations will appear[3].

In some situations, the **Open query** button may need additional context to open the right query (for example, when you are reviewing the **Machines should have vulnerability findings resolved** recommendation), and from that context, you click the **Open query** button; the following options (*Figure 9.18*) will appear:

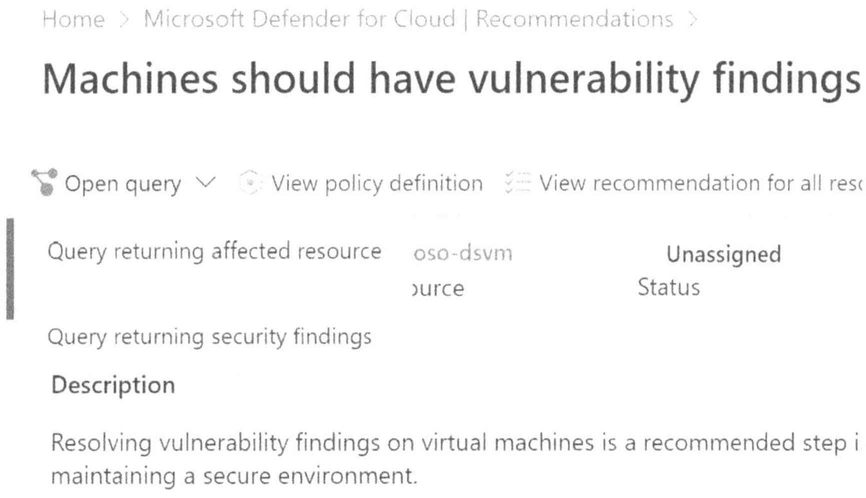

Home  >  Microsoft Defender for Cloud | Recommendations  >

# Machines should have vulnerability findings

🔗 Open query ∨    👁 View policy definition    ⊟ View recommendation for all reso

Query returning affected resource    oso-dsvm                    Unassigned
                                     ource                       Status

Query returning security findings

### Description

Resolving vulnerability findings on virtual machines is a recommended step i maintaining a secure environment.

*Figure 9.18: Additional context*

This is an important choice because you have two different queries available as part of this recommendation. Once you open ARG Explorer, you can also make modifications to the query to customize the way you prefer.

Another option is to click the **+ New query** button (see this button in *Figure 9.16*) in ARG and build a new query from scratch. For example, if you want to see all recommendations that are currently exempt, you can write the query by copying the KQL sample from https://bit.ly/CNAPPBook39, paste it in the ARG Explorer New query page, and click the **Run query** button. *Figure 9.19* has an example of the result of this query:

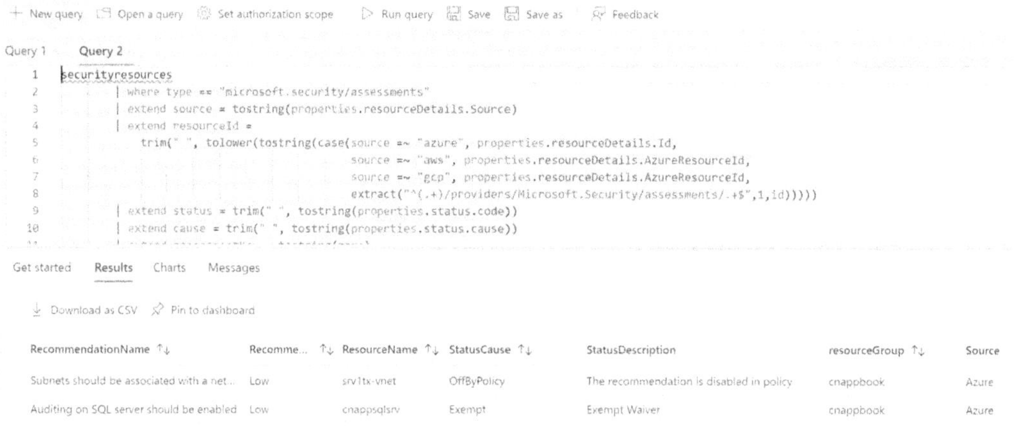

*Figure 9.19: Creating a custom query (for better visualization, refer to https://packt.link/gbp/9781836204879)*

You can use the **Save** button to save your query, and the **Save as** button to customize the name, description, and whether this is a private or shared query.

# Final considerations

Which Defender for Cloud capability you will utilize to perform proactive hunting is not that important; what is important is ensuring that your organization is driving continuous security posture improvement to elevate its cloud security maturity level. As I mentioned at the beginning, not all organizations are equipped and mature enough to start performing this type of task, because there are still the basics to be done, which are addressing critical recommendations and ensuring that there is no attack path open. The reason behind this rationale is one simple word: priority. Most companies have a limited number of resources (in this case, professionals), which means that they need to allocate their professionals to do what is most important for the organization at that time.

Once you have a good security posture in place and have addressed the most critical security recommendations, it is worth allocating professionals (although you may start with only one) to do proactive hunting.

In the beginning, proactive hunting may be very task-oriented, where the demand to find security flaws comes directly from your posture management team. However, with time, it is also common to incorporate threat intelligence to see if, based on this intel, your environment is at risk. One location where you can obtain this type of intel is `https://security.microsoft.com/intel-profiles` (you need to sign in with your Microsoft account). When you access this page (see the sample in *Figure 9.20*), you will see very useful information about threat actors and vulnerabilities.

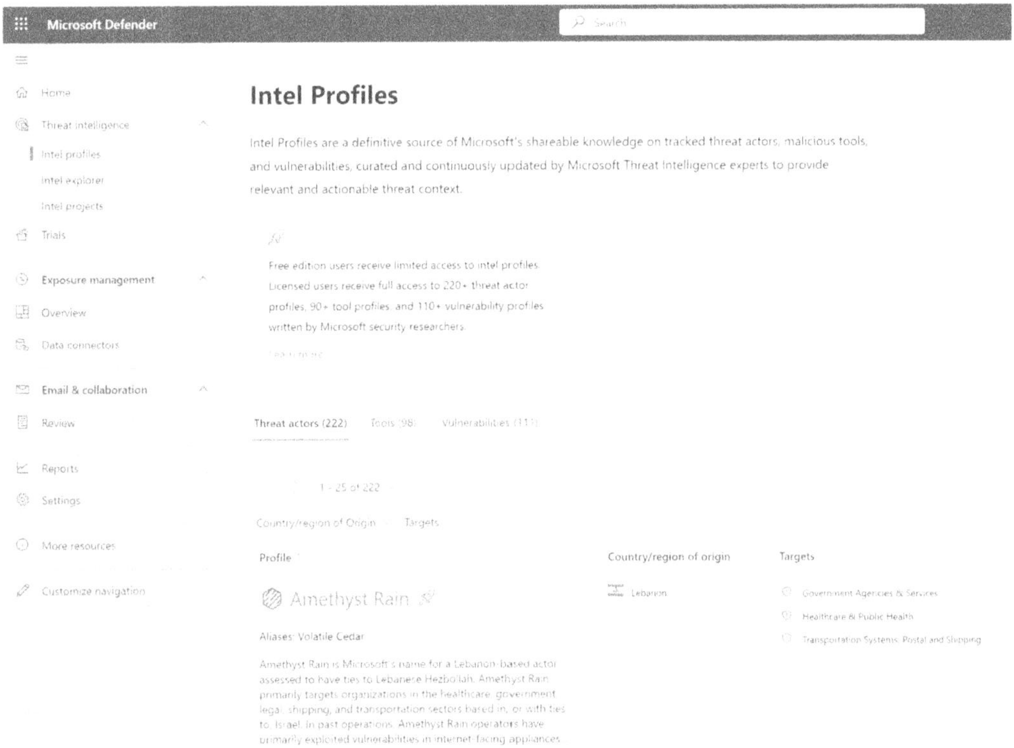

*Figure 9.20: Intel Profiles (for better visualization, refer to https://packt.link/gbp/9781836204879)*

To see the **Tactics, Techniques, and Procedures** (**TTPs**) used by a threat actor, you can open the threat actor's profile and read more about it. *Figure 9.21* shows an example of the TTPs from threat actor Aqua Blizzard.

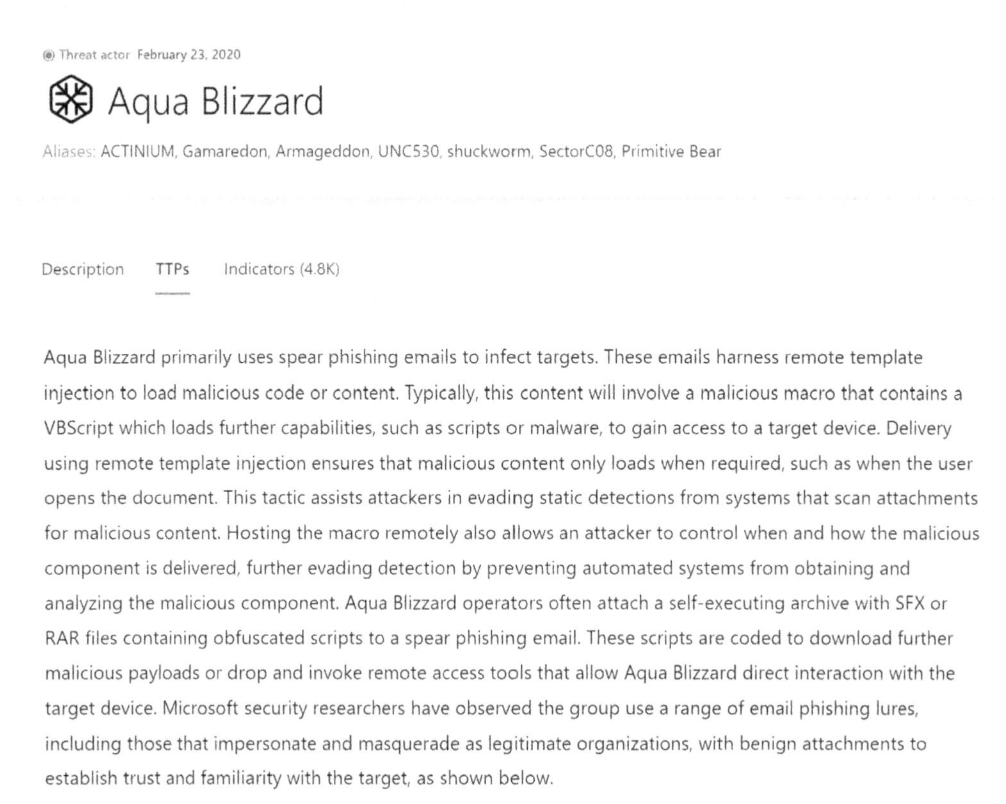

⊙ Threat actor  February 23, 2020

❋ Aqua Blizzard

Aliases: ACTINIUM, Gamaredon, Armageddon, UNC530, shuckworm, SectorC08, Primitive Bear

Description    TTPs    Indicators (4.8K)

Aqua Blizzard primarily uses spear phishing emails to infect targets. These emails harness remote template injection to load malicious code or content. Typically, this content will involve a malicious macro that contains a VBScript which loads further capabilities, such as scripts or malware, to gain access to a target device. Delivery using remote template injection ensures that malicious content only loads when required, such as when the user opens the document. This tactic assists attackers in evading static detections from systems that scan attachments for malicious content. Hosting the macro remotely also allows an attacker to control when and how the malicious component is delivered, further evading detection by preventing automated systems from obtaining and analyzing the malicious component. Aqua Blizzard operators often attach a self-executing archive with SFX or RAR files containing obfuscated scripts to a spear phishing email. These scripts are coded to download further malicious payloads or drop and invoke remote access tools that allow Aqua Blizzard direct interaction with the target device. Microsoft security researchers have observed the group use a range of email phishing lures, including those that impersonate and masquerade as legitimate organizations, with benign attachments to establish trust and familiarity with the target, as shown below.

*Figure 9.21: TTPs from a threat actor*

This type of information can be useful to help you create queries to search for indications of compromise or identify areas in your environment that are susceptible to the type of attack that is usually launched by a specific threat actor.

## Summary

In this chapter, we discussed how to leverage the insights collected by Defender for Cloud to perform proactive hunting. You learned how to use Cloud Security Explorer to create queries and how to use the available templates. In addition to that, you also learned how to use ARG to create queries using KQL and how to access ARG via Defender for Cloud. Lastly, you learned the advantages of using threat intel while doing proactive hunting.

In the next chapter, you will learn how to implement workload protection.

# Notes

1. For a historical timeline about Log4J, visit `https://bit.ly/CNAPPBook36`.

2. To learn more about KQL to use with ARG, visit `https://bit.ly/CNAPPBook37`.

3. For more information about the Microsoft.Security resource provider and the options available to query, visit `https://bit.ly/CNAPPBook38`.

# Additional resources

- Read more about the proactive approach to posture management in this article written by the author Yuri Diogenes: `https://bit.ly/CNAPPBook40`

- There is a demonstration of Cloud Security Explorer in this episode of *Defender for Cloud in the Field*: `https://bit.ly/CNAPPBook41`

# Join our community on Discord

Read this book alongside other users. Ask questions, provide solutions to other readers, and much more.

Scan the QR code or visit the link to join the community.

`https://packt.link/SecNet`

# 10

# Implementing Workload Protection

While security posture improvement is imperative for every business, part of this continuous improvement lifecycle also includes threat detection. If posture management is a proactive task to ensure that you are elevating a security posture, which decreases the likelihood of successful compromise, threat detection is a more reactive task. To ensure that your workloads are protected, you need to ensure that you have threat protection enabled for them.

Workload protection is a pillar originally from a **Cloud Workload Protection Platform** (**CWPP**) that is aggregated to CNAPP. This aggregation of capabilities helps to expand the impact of threat detections, while allowing the use of signals originating from CWPP to be consumed by other CNAPP components, with the intent of enhancing insights and enabling security administrators to make better decisions.

Defender for Cloud has different workload protection plans that can be enabled separately according to your needs.

This chapter covers:

- The need for tailored workload protection
- Threat detection in Defender for Cloud
- Workload protection plans

# The need for tailored workload protection

Threat actors constantly monitor current themes that take place worldwide and use them to create attack campaigns. In February 2024, Microsoft[1] reported that Iranian government-aligned actors launched a series of cyberattacks and **Influence Operations (IOs)** intended to help Hamas. Another example of attacks that took place as international events were unfolding was documented in the Microsoft Digital Defense Report 2023[2], which states that there was a 50 percent increase in destructive Russian attacks against Ukrainian networks during the first six weeks of the war between the two countries.

The threat landscape is continuously changing, and threat actors will continue to monitor current themes to craft new attack campaigns. While the examples above were very targeted actions against specific regions, the lessons learned from those actions need to be incorporated into defense mechanisms, as the techniques that were used will likely be used again in future campaigns against regular business targets.

This ongoing change in the threat landscape also affects how detections are created to protect workloads. Since the threat vectors will vary according to the type of workload, it becomes important that the creation of new detections is tailored for each type of workload. For example, the threat detections that are created for a storage account take into consideration the threat vectors for a storage account, which are not the same as the threat detections for a virtual machine.

Microsoft security researchers are also investing in creating threat matrices for different types of cloud workloads. For example, the threat matrix for Kubernetes (which you can access from `http://aka.ms/KubernetesThreatMatrix`) has the different attack phases based on the MITRE ATT&CK Framework (`attack.mitre.org`) and the different methods threat actors are using to attack Kubernetes. This type of matrix helps security developers create threat detections that are relevant for this type of workload (Kubernetes).

Tailored threat detections are a critical part of the CWP pillar within the CNAPP, not only ensuring that attack attempts are rapidly identified but also enriching SOC analysts to rapidly respond to these threats. The diagram shown in *Figure 10.1* shows a generic representation of this lifecycle:

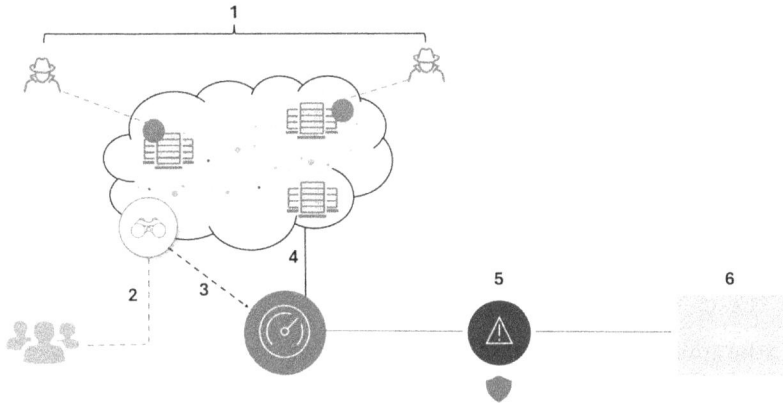

*Figure 10.1: Threat detection and response*

The numbers in *Figure 10.1* are documented below:

1.  Threat actors attack different cloud workloads, using specific methods according to the workload type.

2.  Security researchers constantly look at evolving threats to build new analytics, specific to the type of workload. This research happens in collaboration with developers who write the code behind the detection. Part of this work also includes:

    a.  Log analysis to understand the behavior of an activity

    b.  Cross-checks with **threat intelligence** (**TI**) to verify if such behavior is known in the TI database

3.  Once a new detection is created, validated, and tested, it can be deployed to a threat detection platform (a CWP component of CNAPP). The deployment happens in the backend (cloud service), which means it doesn't require customers to deploy anything on their environment.

4.  The CWP module within the CNAPP solution has a traffic match for the analytic that was created.

5.  An alert is triggered based on this activity, which can be visualized in the CNAPP dashboard.

6.  The CNAPP solution is configured to stream the alert to the **Security Information and Event Management** (**SIEM**) platform. This will help the SOC analyst triage the alert and respond.

In the following section, we will see how to apply threat detection in Defender for Cloud.

# Threat detection in Defender for Cloud

Threat detection in Defender for Cloud uses advanced security analytics that surpass traditional signature-based methods. By harnessing breakthroughs in big data and machine learning technologies, it can analyze events across the entire cloud infrastructure. This enables the detection of threats that would be impossible to identify manually. Alerts can be generated based on atomic events (audit events are an example of atomic actions), which look for known signatures and patterns within a single data source. An example of an atomic alert is a known bad process that was executed. "Known" in this case refers to being known by the signals that previously detected this process performing malicious operations.

Defender for Cloud threat detection also uses behavioral analytics, which is a technique that examines and compares data against a set of known patterns. Unlike simple signatures, these patterns are derived from complex machine learning algorithms applied to large datasets and through detailed analysis of malicious behaviors. Defender for Cloud leverages behavioral analytics to identify compromised resources by analyzing different types of logs, including virtual network device logs, fabric logs, and other sources.

Defender for Cloud employs anomaly detection to recognize potential threats in the environment. Unlike behavioral analytics, which relies on known patterns from large datasets, anomaly detection is more tailored and focuses on baselines specific to your deployments. Machine learning is used to establish normal activity for your cloud workloads, and rules are then generated to identify outlier conditions that could indicate a security event. For example, an RDP rate for a specific virtual machine is usually 5 times a day, and today, it happened 100 times. That's a deviation from the normal pattern of access.

## Alert dashboard

All alerts in Defender for Cloud surface in the **Security alerts** dashboard, located under the **General** section, as shown in *Figure 10.2*:

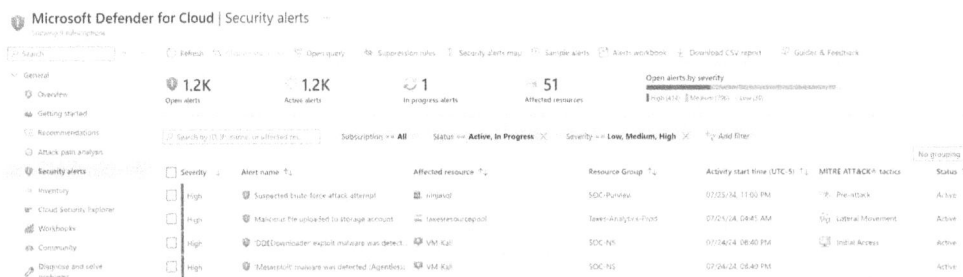

*Figure 10.2: Security alerts in Defender for Cloud (for better visualization, refer to https://packt.link/gbp/9781836204879)*

The **Security alerts** dashboard is very intuitive, and it enables you to create different filters, just like the **Recommendation** dashboard, as you've seen previously in this book. The upper part of the dashboard has a menu of options, which includes the following buttons:

- **Refresh** forces a page refresh.

- **Change status** enables you to change the alert's status to **Active**, **In Progress**, **Dismissed**, or **Resolved**.

- **Open query** opens **Azure Resource Graph** (**ARG**), using the context of the alert. In the ARG dashboard, you can make changes to customize your query using KQL.

- **Suppression rules** enables you to create rules to suppress alerts that you believe are false positives.

- **Security alerts map** presents a new dashboard with a map, where there will be marks representing the sources of the attack on your resources. This is available for alerts that contain IP addresses targeting your resources.

- **Sample alerts** allows you to trigger a set of sample alerts for different Defender for Cloud plans.

- **Alerts workbook** enables you to access a built-in alert workbook with a different visualization to see alerts.

- **Download CSV report** enables you to download the current view of alerts to a CSV file.

- **Guides & Feedback** opens up a new blade where you can learn more about security alerts and provide feedback to the Defender for Cloud team.

The middle part of the screen has some numbers to represent (in left to right order) the number of open alerts, the number of active alerts, the number of alerts in progress, and the number of affected resources. In addition, it also has a bar graph representing the open alerts by severity.

The bottom part of the screen is where the alerts will appear. This grid, which contains the alerts, has the following columns:

- **Severity** shows the severity (**High**, **Medium**, or **Low**) of the alert.

- **Alert name** shows the friendly name of the alert, which is part of the field **alertDisplay-Name**.

- **Affected resource** shows the target resource (**compromisedEntity** field) that was affected by this potential attack.

- **Resource Group** shows the resource group of the affected entity that triggered the alert.

- **Activity start time** presents the time (the **startTimeUtc** field) in which the activity (event) responsible for this alert started.

- **MITRE ATT&CK tactics** displays the MITRE ATT&CK tactic (`https://attack.mitre.org/tactics`) in which this type of attack fits.

- **Status** displays the current status (**Active, In Progress, Dismissed,** or **Resolved**) of the alert.

When you click on one alert from the list, a blade opens that shows more information about the alert, as shown in the example in *Figure 10.3*:

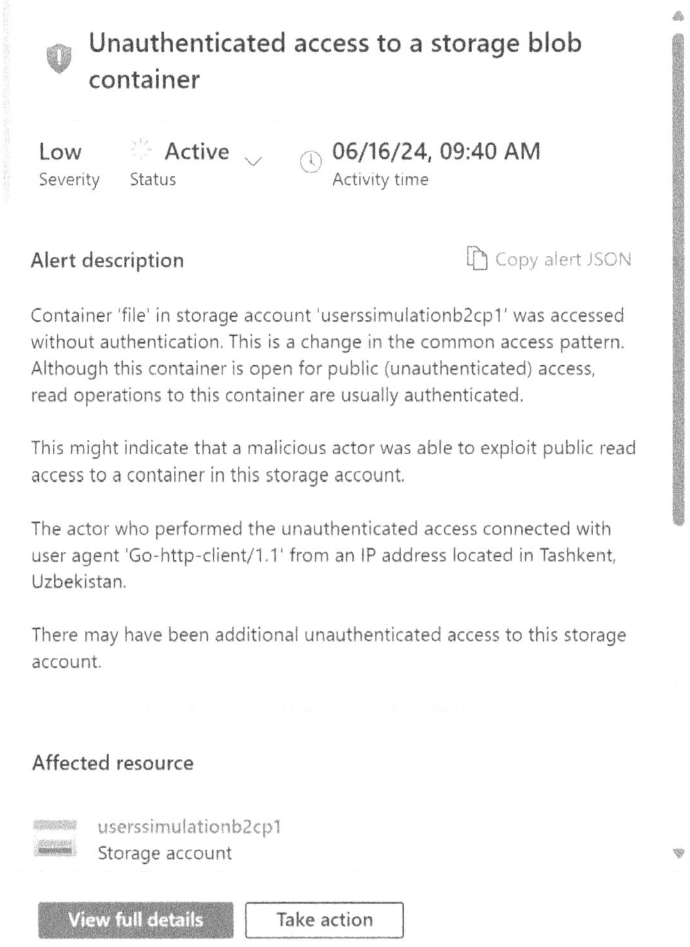

> ## Unauthenticated access to a storage blob container
>
> | Low | Active ⌄ | 06/16/24, 09:40 AM |
> |-----|----------|--------------------|
> | Severity | Status | Activity time |
>
> **Alert description**                             Copy alert JSON
>
> Container 'file' in storage account 'userssimulationb2cp1' was accessed without authentication. This is a change in the common access pattern. Although this container is open for public (unauthenticated) access, read operations to this container are usually authenticated.
>
> This might indicate that a malicious actor was able to exploit public read access to a container in this storage account.
>
> The actor who performed the unauthenticated access connected with user agent 'Go-http-client/1.1' from an IP address located in Tashkent, Uzbekistan.
>
> There may have been additional unauthenticated access to this storage account.
>
> **Affected resource**
>
> userssimulationb2cp1
> Storage account
>
> [ **View full details** ]   [ Take action ]

*Figure 10.3: Alert summary*

On this page, you can see a full description of the alert, and you can also access the JSON (raw data) for the alert by clicking on the **Copy alert JSON** button. You can then open Notepad and paste the content. You should be able to see all the details about the alert in a raw format, as shown in the example below (note that some fields were replaced with XXXX, for privacy reasons):

```
Copied alert from Microsoft Defender for Cloud on 07/27/24, 10:06 AM (UTC-
5)
https://ms.portal.azure.com/#blade/Microsoft_Azure_Security_
AzureDefenderForData/AlertBlade/location/centralus/subscriptionId/XXXXXXX/
alertId/ XXXX/referencedFrom/copyAlertButton
{
  "id": "/subscriptions/XXXXXXXXXX/resourceGroups/soc-ns/providers/
Microsoft.Security/locations/centralus/alerts/XXXX",
  "name": " XXXX",
  "type": "Microsoft.Security/Locations/alerts",
  "properties": {
    "status": "Active",
    "timeGeneratedUtc": "2024-07-15T16:55:57.971Z",
    "processingEndTimeUtc": "2024-07-15T16:55:57.719658Z",
    "version": "2022-01-01.0",
    "vendorName": "Microsoft",
    "productName": "Microsoft Defender for Cloud",
    "productComponentName": "AppService",
    "alertType": "AppServices_AnomalousPageAccess",
    "startTimeUtc": "2024-07-15T15:28:52.358376Z",
    "endTimeUtc": "2024-07-15T15:28:52.358376Z",
    "severity": "Low",
    "isIncident": false,
    "systemAlertId": " XXXX",
    "intent": "InitialAccess",
    "resourceIdentifiers": [
      {
        "$id": "centralus_1",
        "azureResourceId": "/subscriptions/XXXX/resourceGroups/soc-ns/
providers/Microsoft.Web/sites/owaspdirect",
        "type": "AzureResource",
        "azureResourceTenantId": "XXXX"
      },
```

```
        {
          "$id": "centralus_2",
          "aadTenantId": "XXXX",
          "type": "AAD"
        }
    ],
    "compromisedEntity": "owaspdirect",
    "alertDisplayName": "AppServices_AnomalousPageAccess",
    "description": "AppServices_AnomalousPageAccess",
    "extendedProperties": {
       "sampleUserAgents": "Mozilla/5.0+(Windows+NT+10.0;+Win64;+x64)+Apple
WebKit/537.36+(KHTML,+like+Gecko)+Chrome/126.0.0.0+Safari/537.36+Edg/
126.0.0.0",
       "sampleReferer": "-",
       "sampleURIs": "/rest/user/login",
       "sampleIps": "37.228.251.181",
       "resourceType": "App Service",
       "effectiveAzureResourceId": "/subscriptions/XXXX/resourceGroups/soc-
ns/
providers/Microsoft.Web/sites/owaspdirect",
       "compromisedEntity": "owaspdirect",
       "productComponentName": "AppService",
       "effectiveSubscriptionId": "XXXX"
    },
    "entities": [
       {
         "$id": "centralus_3",
         "resourceId": "/subscriptions/XXXX/resourceGroups/soc-ns/
providers/
Microsoft.Web/sites/owaspdirect",
         "resourceType": "App Service",
         "resourceName": "owaspdirect",
         "metadata": {
           "isGraphCenter": true
         },
```

```
        "asset": true,
        "type": "azure-resource"
      },
      {
        "$id": "centralus_4",
        "sourceAddress": {
          "$id": "centralus_5",
          "address": "XXXX",
          "location": {
            "countryCode": "IE",
            "countryName": "Ireland",
            "state": "Dublin",
            "city": "XXXX",
            "longitude":XXXX,
            "latitude":XXXX,
            "asn": 6830,
            "carrier": "XXXX",
            "organization": "Infrastructure"
          },
          "asset": false,
          "type": "ip"
        },
        "friendlyName": "XXXX",
        "asset": false,
        "type": "network-connection"
      },
      {
        "$ref": "centralus_5"
      }
    ],
    "alertUri": "https://portal.azure.com/#blade/Microsoft_Azure_Security_
AzureDefenderForData/AlertBlade/alertId/XXXX/subscriptionId/XXXX/
resourceGroup/soc-ns/referencedFrom/alertDeepLink/location/centralus"
  }
}
```

The raw data can be useful in some investigation scenarios, or if you are consuming the alert in a different platform, such as a SIEM. If you need to see more details about the alert, you can click the **View full details** button, and you will have a full page dedicated to the alert, as shown in *Figure 10.4*:

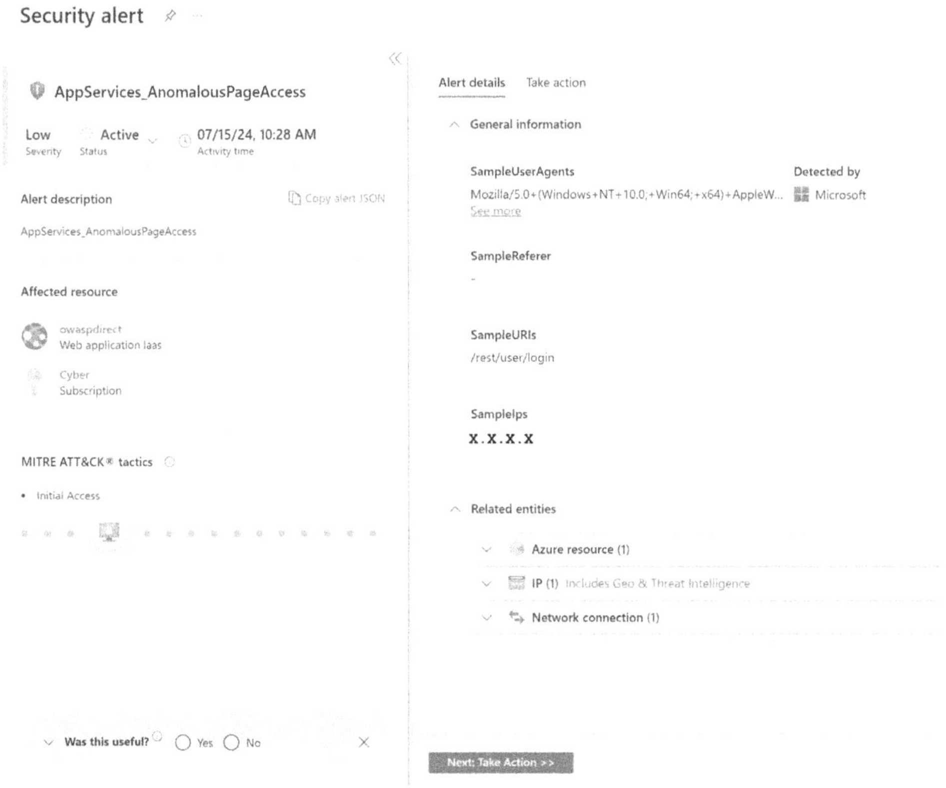

*Figure 10.4: More information about the alert (for better visualization, refer to https://packt. link/gbp/9781836204879)*

On this page, you have an expansion of the blade on the left side, with additional details, and on the right side, you have tailored information according to the type of alert. You will learn more about the different options available in the next chapters, where each Defender for Cloud plan will be covered. The last tab in the alert is the **Take action** tab, which has insights on actions that you can take to remediate the alert.

# Alert correlation

Triage and investigating security alerts can be time-consuming for even the most skilled SOC analysts, and for many, it is hard to even know where to begin. By using analytics to connect the information between distinct security alerts, Defender for Cloud provides you with a single view of an attack campaign and all related alerts, which is called security alert correlation[3]. When this correlation happens, Defender for Cloud creates a security incident. A security incident is an aggregation of all alerts for a resource that align with kill chain patterns. With this capability, SOC analysts can quickly understand what actions the threat actor took and what resources were impacted. For example, a brute-force attack, followed by a later suspicious VM activity, may indicate a potential breach, so these two alerts (brute force + suspicious VM activity) will be part one of a security incident.

A security incident will appear in the same **Security alerts** dashboard in Defender for Cloud, but it will have a different icon (three dots), as shown on the left side of *Figure 10.5*, and when you open the incident, the right side has the individual alerts that are part of this incident.

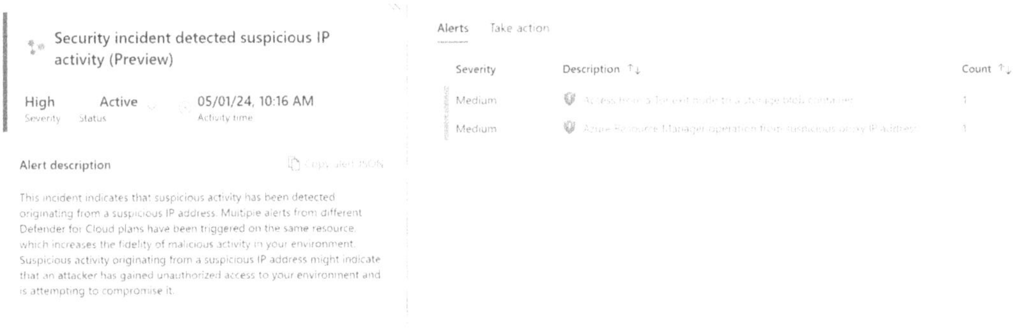

Figure 10.5: Security incident in Defender for Cloud (for better visualization, refer to https://packt.link/gbp/9781836204879)

# Sample alerts

Defender for Cloud enables you to trigger sample alerts for different plans to ensure that you can validate workflows, as well as better understand how an alert will look. You don't have to have the Defender for Cloud plan enabled in order to trigger an alert.

The sample alert will not have any data about a workload because it is an empty alert, so it will only contain the fields. This can be useful for SOC analysts who create playbooks and need to understand which fields they should expect from an alert.

To trigger a sample alert, you just need to click on the **Sample alert** button, and the blade shown in *Figure 10.6* will appear, allowing you to select which plans you want to trigger a sample alert:

## Create sample alerts (Preview)                                    ✕

Try Defender for Cloud alerts by creating sample alerts from our different Defender for Cloud plans. Learn more >>

**Subscriptions**

| Azure for Students | ⌄ |

**Defender for Cloud plans**

| 11 selected | ⌄ |

Create sample alerts

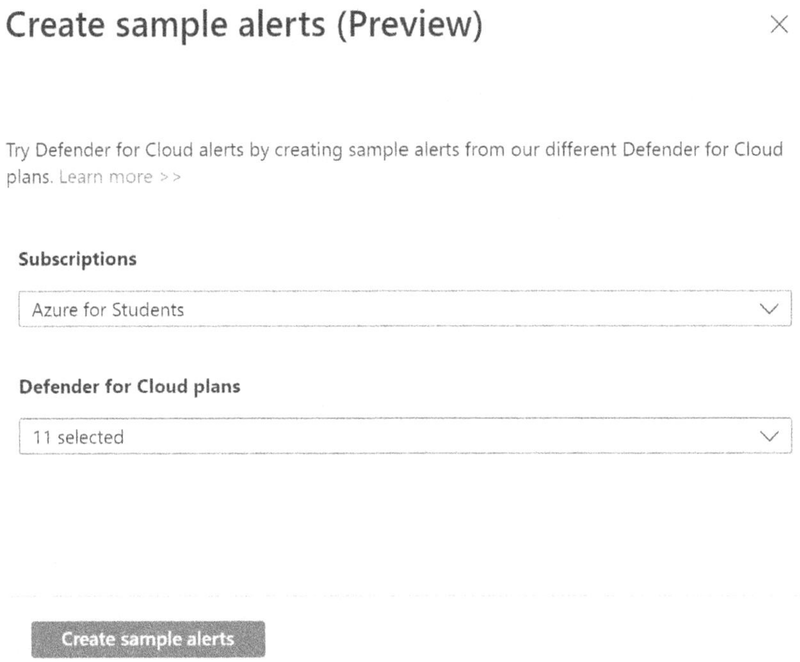

*Figure 10.6: Sample alert blade*

You can select the subscription that you want to use to trigger the alert and, in the **Defender for Cloud plans** drop-down list, select the plans that you want to use for this test. Below, you have an example of the main section of the raw sample alert:

```
"type": "Microsoft.Security/Locations/alerts",
"properties": {
  "status": "Active",
  "timeGeneratedUtc": "2024-07-26T22:07:02.248Z",
  "processingEndTimeUtc": "2024-07-26T22:07:01.965615Z",
  "version": "2022-01-01.0",
  "vendorName": "Microsoft",
  "productName": "Microsoft Defender for Cloud",
  "productComponentName": "KeyVault",
  "alertType": "SIMULATED_KV_OperationVolumeAnomaly",
  "startTimeUtc": "2024-07-26T22:06:55.965615Z",
```

```
    "endTimeUtc": "2024-07-26T22:06:55.965615Z",
    "severity": "Medium",
    "isIncident": false,
    "systemAlertId": "XXX",
    "intent": "Unknown",
    "resourceIdentifiers": [
      {
        "$id": "centralus_1",
        "azureResourceId": "/SUBSCRIPTIONS/XXX/RESOURCEGROUPS/Sample-RG/
providers/microsoft.keyvault/vaults/Sample-KV",
        "type": "AzureResource",
        "azureResourceTenantId": "XXX"
      },
      {
        "$id": "centralus_2",
        "aadTenantId": "XXX",
        "type": "AAD"
      }
    ],
    "compromisedEntity": "Sample-KV",
    "alertDisplayName": "[SAMPLE ALERT] User accessed high volume of Key
Vaults",
    "description": "THIS IS A SAMPLE ALERT: While may be benign it could
also indicate that a larger volume of Key Vault operations has been
performed compared to past historical data. Key Vaults typical exhibit the
same behavior over time. This may be a legitimate change in activity but
may also indicate that your Key Vault infrastructure has been compromised
warranting further investigation.",
    "remediationSteps": [
      "Please review your activity logs to determine if the access
attempts that triggered this alert were legitimate. If you are concerned
that these access attempts may not have been legitimate, please contact
your security administrator and disable access policies to the user or
application and rotate the secrets, keys, and passwords stored in this key
vault."
    ],
```

# Alert suppression

There will be scenarios where the SOC analyst may request to suppress an alert that they know is a false positive in their environment. Some other times, it may be that the Red Team is doing some exercises, and they know that specific alerts will be generated and can safely be ignored. In situations like that, you can leverage the alert suppression feature in Defender for Cloud. To access the alert suppression, click the **Suppression rules** button in the **Security alerts** dashboard. If you don't have any rule set yet, you will see a page like *Figure 10.7*:

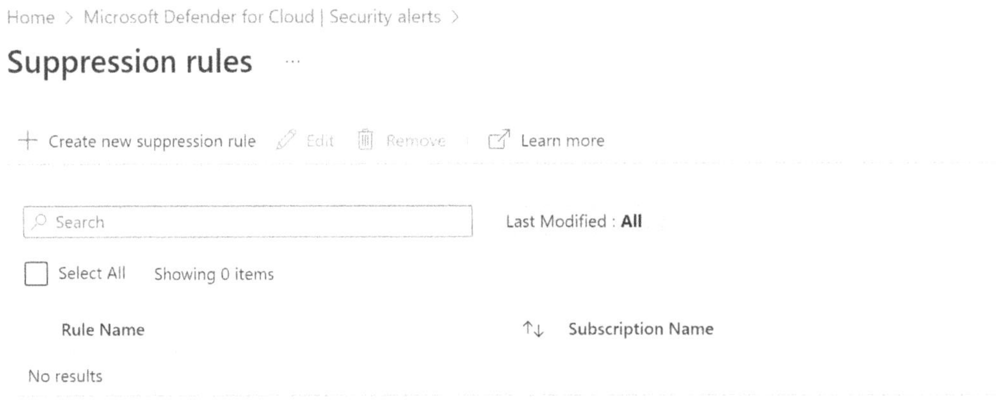

Home > Microsoft Defender for Cloud | Security alerts >

## Suppression rules   ...

+ Create new suppression rule   ✏ Edit   🗑 Remove   ⤢ Learn more

| 🔍 Search | | Last Modified : **All** |
|---|---|---|

☐ Select All    Showing 0 items

| Rule Name | ↑↓ Subscription Name |
|---|---|

No results

*Figure 10.7: Suppression rules page (for better visualization, refer to https://packt.link/ gbp/9781836204879)*

As new rules are created, they will be populated on this page. To create a new rule, click the **+ Create new suppression rule** button, and the **New suppression rule** blade will appear, as shown in *Figure 10.8*:

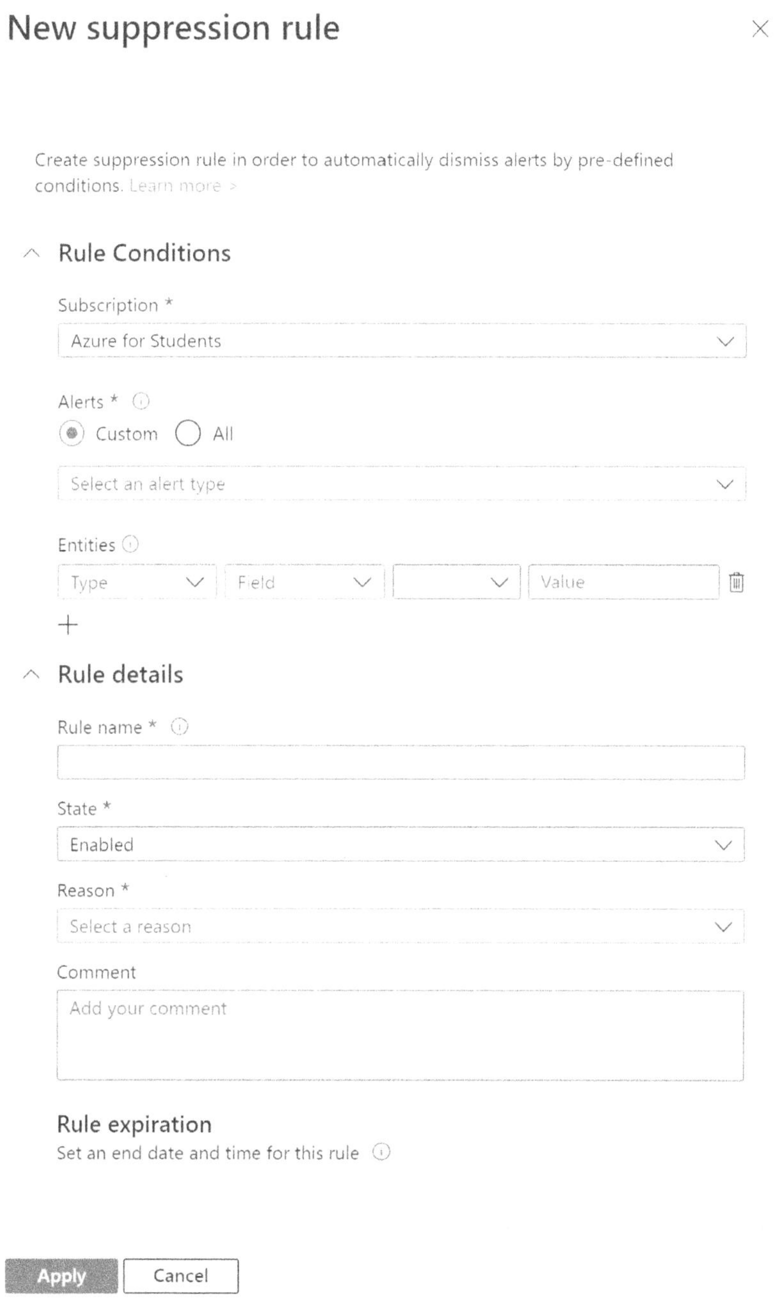

# New suppression rule ✕

Create suppression rule in order to automatically dismiss alerts by pre-defined conditions. Learn more >

## ⌃ Rule Conditions

Subscription *

| Azure for Students | ⌄ |

Alerts * ⓘ
◉ Custom ◯ All

| Select an alert type | ⌄ |

Entities ⓘ

| Type ⌄ | Field ⌄ | ⌄ | Value | 🗑 |

+

## ⌃ Rule details

Rule name * ⓘ

| |

State *

| Enabled | ⌄ |

Reason *

| Select a reason | ⌄ |

Comment

| Add your comment |

### Rule expiration
Set an end date and time for this rule ⓘ

**Apply**  Cancel

*Figure 10.8: New suppression rule blade*

Here, you can select the subscription that you want this rule to be created on; this also means that it will only apply the subscription against alerts that were generated in this subscription. In the **Alerts** section, you will usually select the **Custom** option to customize which alert you want to suppress. In the drop-down list for **Select an alert type**, you select which alert you want to suppress.

Under the **Entities** section, you select specific entities to refine your suppression rule; when choosing entities, the rule will be matched only for the alerts containing them. When choosing multiple entities, the relation between the entities will always have an AND logical operator, ensuring that all conditions are true in order to match. If the entity is an IP address, you can use **Classless Inter-Domain Routing** (CIDR) notation to represent an IP range.

Under the **Rule details** section, you should add a name for this rule, select the state (**Enabled** or **Disabled**) of the rule, select the reason (rationale) behind creating this suppression, and optionally, add a more detailed description about this rule. For example, you could explain why you created this rule and who authorized it.

The last part of the rule is the expiration, which is very important to always establish. The reason why this is important is that you don't want to create permanent blind spots in your detection. A suppression should have an expiration date, forcing you to reevaluate the need for the suppression moving forward. Rules that expire will change their state to **Expired**. The type of alert that you will receive will vary, according to the Defender for Cloud plan that is enabled, which is what you will learn about in the next section.

# Defender for Cloud plans

In Defender for Cloud, workload protection is done by leveraging different Defender for Cloud paid plans. These plans are the ones responsible for using the different types of threat detection that you learned about earlier in this chapter. At the time of writing, the workload protection plans are:

- Defender for Servers:

  - Virtual machine (cloud or on-premises) protection. Also covers bare metal server on-premises protection

- Defender for Databases:

  - Protection for Azure SQL Databases, SQL servers on machines (on-premises or in a different cloud provider), open-source relational databases, and Azure Cosmos DB

- Defender for Containers

- Defender for Storage

- Defender for App Services

- Defender for APIs

- Defender for Key Vault

- Defender for Resource Manager

You will learn more about these plans in more detail in the upcoming chapters of this book.

# Summary

In this chapter, we discussed the need to have a tailored approach to workload protection. You learned how threat detection works in Defender for Cloud and the different types of detection. You also learned about the **Security alerts** dashboard, alert correlation, sample alerts, and how alert suppression works. Lastly, you learned the different Defender for Cloud plans available.

In the next chapter, you will learn how to implement workload protection.

# Notes

1. You can read the entire article at `https://bit.ly/CNAPPBook42`.

2. You can download this digital report from `https://bit.ly/CNAPPBook43`.

3. Watch author Yuri Diogenes, a security researcher who worked on this project, discuss this type of correlation with Microsoft: `https://bit.ly/CNAPPBook44`.

# Additional resources

- All security alerts in Defender for Cloud: `https://bit.ly/CNAPPBook45`

- Learn more about alert simulation: `https://bit.ly/CNAPPBook15`

# Join our community on Discord

Read this book alongside other users. Ask questions, provide solutions to other readers, and much more.

Scan the QR code or visit the link to join the community.

https://packt.link/SecNet

# 11

# Protecting Compute Resources (Servers and Containers)

**Virtual machines (VMs)** are still a significant component of cloud workload footprints, but they are no longer the largest or most dominant. In 2024, the focus has shifted significantly toward other cloud services, such as containers. However, both VMs and containers are common targets for threat actors. The threat landscape for compute-based workloads includes attacks such as cryptocurrency mining[1] to abuse compute resources.

For this reason, it becomes imperative to have visibility and control across these types of workloads to not only reduce the likelihood of compromise but also rapidly detect potential malicious operations, including **Living off the Land Binaries (LoLBins)** types of attacks. **Living off the Land (LoL)** is a tactic where threat actors exploit existing tools and utilities that are already present within an operating system to execute their attacks. These tools, known as LoLBins, are legitimate binaries and scripts found in Windows or Linux environments. Though typically used for benign administrative tasks, attackers can manipulate them on compromised systems to carry out malicious activities.

Defender for Cloud has two plans to cover VMs and Containers, which are Defender for Servers and Defender for Containers. These plans provide security posture management and threat protection for these workloads.

This chapter covers:

- Defender for Containers
- Defender for Servers

# Defender for Containers

Defender for Containers provides security posture management and threat protection for the following types of Kubernetes:

- Azure:

    - **Azure Kubernetes Service (AKS)**

- **Amazon Web Services (AWS):**

    - **Amazon Elastic Kubernetes Service (EKS):**

        - In this scenario, you need to have the Defender for Cloud multicloud connector for the AWS account enabled (read *Chapter 6* for more information)

- **Google Cloud Platform (GCP):**

    - **Google Kubernetes Engine (GKE):**

        - In this scenario, you need to have the Defender for Cloud multicloud connector for the GCP project enabled (read *Chapter 6* for more information)

- Hybrid / On-premises
- An unmanaged Kubernetes distribution using Azure Arc-enabled Kubernetes[2]

Defender for Containers provides security posture management using an agentless approach. In other words, there is no need to deploy any agent to your Containers. The agentless capability is also part of Defender **Cloud Security Posture Management (CSPM)**, which means that even if you don't have the Defender for Containers plan enabled, but you do have Defender CSPM, you will have access to the following features that are related to Container posture management:

- Agentless discovery for Kubernetes
- Inventory capabilities
- Vulnerability assessment
- Control plane hardening

In addition to that, when you have Defender CSPM enabled, you will be able to obtain these insights from the Attack Path and Cloud Security explorer features, as shown in *Chapter 5*. While the security posture management capabilities are like the ones available when you enable Defender CSPM, when it comes to workload protection, the features are exclusively available only when you enable the Defender for Containers plan.

To be able to closely monitor your containers and bring awareness of potential attacks, Defender for Containers needs to collect audit logs, security events, cluster configuration information from the control plane, workload configuration from Azure Policy, security signals, and events from the node level. The approach to performing these tasks will vary according to which environment (Azure, AWS, GCP, or on-premises) the container resides in. The table below has the details about each environment:

| Components | Environment | | | |
| --- | --- | --- | --- | --- |
| | Azure | AWS | GCP | On-prem-ises |
| Defender sensor | X<br><br>(The AKS Defender sensor only supports AKS clusters that have RBAC enabled) | X | X | X |
| Azure Policy for Kubernetes | X | X | X | X |
| Azure Arc-enabled Kubernetes | Not applicable | Not applicable | Not applicable | X |
| Kubernetes audit logs using the AWS account's Cloud-Watch | Not applicable | X | Not applicable | Not applicable |
| Kubernetes audit logs using GCP Cloud Logging | Not applicable | Not applicable | X | Not applicable |

As you can see, the Defender sensor and the Azure Policy for Kubernetes are the only two components that are applicable to all platforms. The Defender sensor is a **DaemonSet**[4] that is deployed on each node. This **DaemonSet** is responsible for collecting signals from hosts using eBPF technology (see ebpf.io) and also for providing runtime protection. The Defender sensor has the following components:

- Two Kubernetes DaemonSets:
  - Pod name: `microsoft-defender-collector-ds-*`
    - This is responsible for collecting inventory and security events.

- Pod name: `microsoft-defender-publisher-ds-*`
  - This is responsible for collecting data for the Microsoft Defender for Containers backend service
- One Kubernetes Deployment[5]
  - Pod name: `microsoft-defender-collector-misc-*`
    - This is responsible for collecting inventory and security events that aren't limited to a specific node.

The Azure Policy for Kubernetes is a pod that extends the open-source Gatekeeper v3. Behind the scenes, this component will register as a webhook to Kubernetes admission control, which enables you to create at-scale enforcements and safeguards. Now that you understand Defender for Containers components, it is time to enable this plan.

# Enabling Defender for Containers

To enable Defender for Containers and all recommended extensions in your Azure subscription, you will need Owner or Contributor privileges in the subscription. To enable this plan, open the Defender for Cloud dashboard, click **Environment settings** under the **Management** section, click the subscription in which you want to enable Defender for Containers, click the **Defender plans** option, and click **On** besides the **Containers** plan, as shown in *Figure 11.1*:

*Figure 11.1: Enabling Defender for Containers (for better visualization, refer to https://packt. link/gbp/9781836204879)*

After switching the toggle to **On**, a hyperlink called **Settings** will appear; click on it to see the settings that will be enabled with this plan, as shown in *Figure 11.2*:

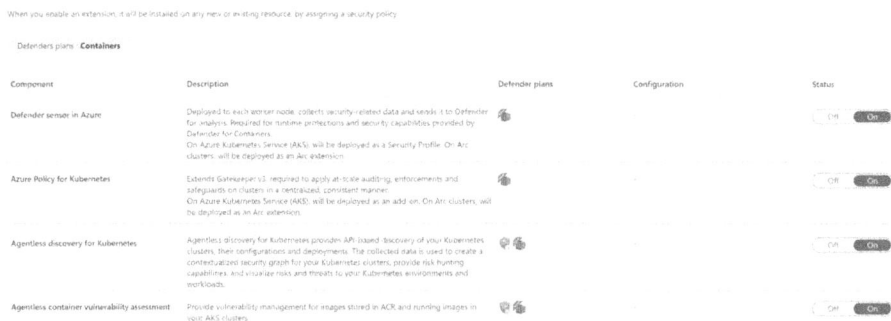

*Figure 11.2: The Defender for Containers components (for better visualization, refer to https:// packt.link/gbp/9781836204879)*

All these components will be enabled by default, and you should keep them enabled to ensure that all the operations that are handled by those components are working properly. After reviewing this page, click the **Continue** button and then click **Save** on the **Defender plans** page.

Since all components were selected, they will automatically be deployed to the Kubernetes cluster that are available in the subscription. As a result, some recommendations will not have unhealthy resources. For example, the recommendation **Azure Kubernetes Service clusters should have the Azure Policy add-on for Kubernetes** installed (shown in *Figure 11.3*) shows only the healthy resources.

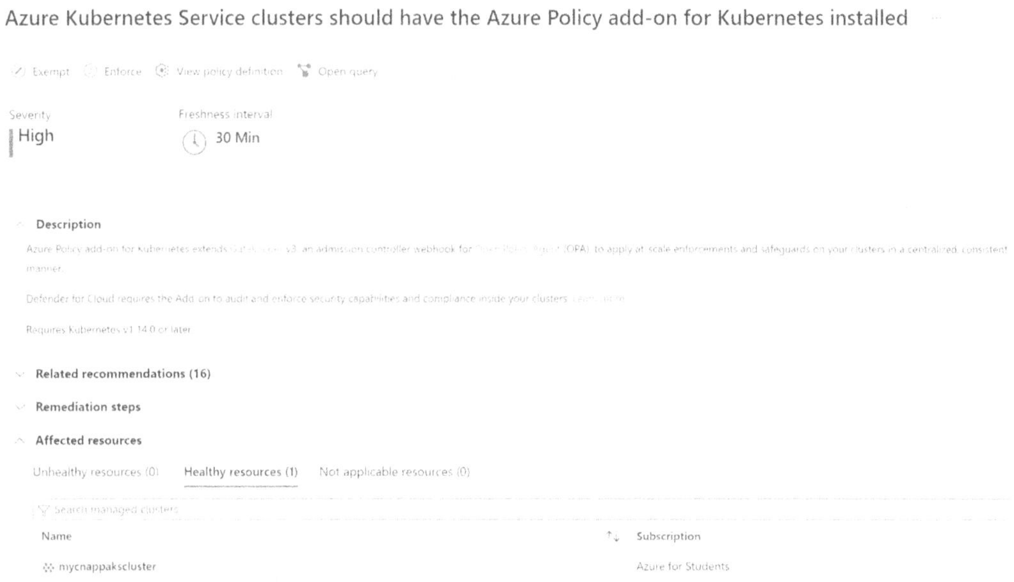

*Figure 11.3: Azure Kubernetes recommendation (for better visualization, refer to https://packt. link/gbp/9781836204879)*

You can find the list of Container security recommendations for Azure, AWS, and GCP at `https://bit.ly/CNAPPBook88`. This list also includes Kubernetes data plane hardening[6]-related recommendations and **Azure Container Registry** (**ACR**)[7]-related recommendations.

## Vulnerability assessment

Container registries can be used to assist you with container development and deployment pipelines. **ACR** tasks can be used to build container images in Azure. Defender for Containers will leverage the **Microsoft Defender Vulnerability Management** (**MDVM**) capability to scan your registry for commonly known vulnerabilities (**Common Vulnerabilities and Exposures** (**CVEs**)) and provide a detailed vulnerability report for each image.

The result of this scan will surface in a recommendation that shows the list of vulnerabilities that were found. *Figure 11.4* has an example of this recommendation:

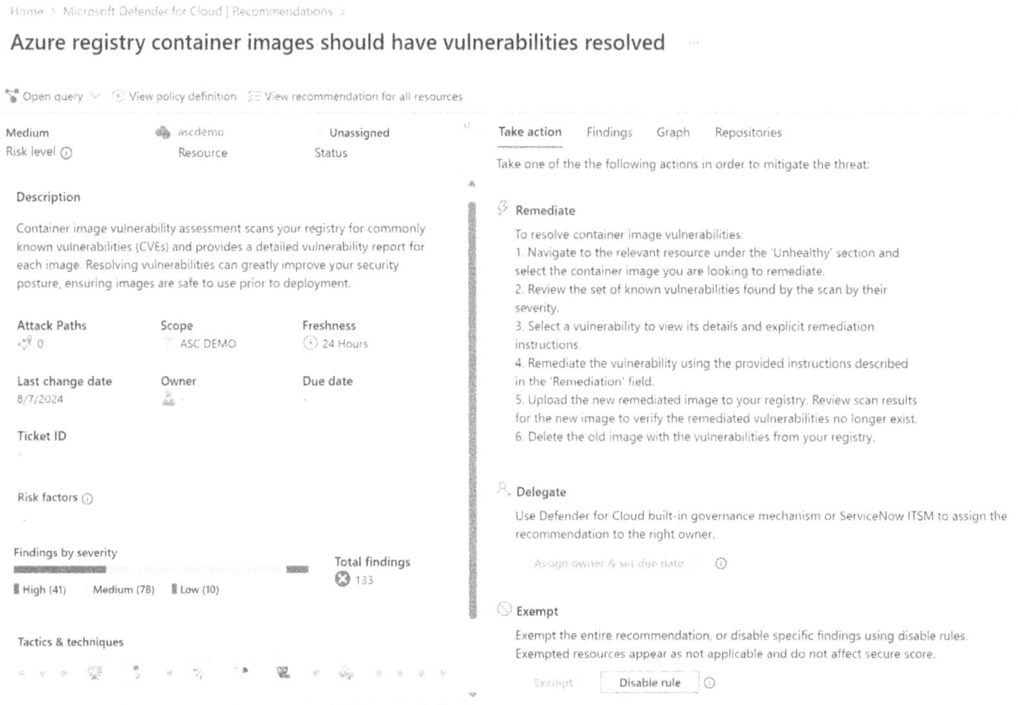

Home > Microsoft Defender for Cloud | Recommendations >

## Azure registry container images should have vulnerabilities resolved

☆ Open query ∨   ⊙ View policy definition   ☷ View recommendation for all resources

| | | |
|---|---|---|
| Medium | 🐳 ascdemo | Unassigned |
| Risk level ⓘ | Resource | Status |

**Description**

Container image vulnerability assessment scans your registry for commonly known vulnerabilities (CVEs) and provides a detailed vulnerability report for each image. Resolving vulnerabilities can greatly improve your security posture, ensuring images are safe to use prior to deployment.

| | | |
|---|---|---|
| Attack Paths | Scope | Freshness |
| ⚡ 0 | ⎙ ASC DEMO | ⊙ 24 Hours |
| Last change date | Owner | Due date |
| 8/7/2024 | 👤 · | · |
| Ticket ID | | |
| · | | |

Risk factors ⓘ
·

Findings by severity
▮▮▮▮▮▮▮▮▮▮▮▮▮▮▮▮▮▮▮▮▮▮  ▬▬
▮ High (41)   ▮ Medium (78)   ▮ Low (10)

Total findings
✖ 133

Tactics & techniques
⌃ ⌄ ⊙ 🖳 ⦿ ⚓ ⚗ ⬆ 🖾 ✦ 🕸 ✦ ✦ ✦ ⊳

**Take action**    Findings    Graph    Repositories

Take one of the the following actions in order to mitigate the threat:

⚡ **Remediate**

To resolve container image vulnerabilities:
1. Navigate to the relevant resource under the 'Unhealthy' section and select the container image you are looking to remediate.
2. Review the set of known vulnerabilities found by the scan by their severity.
3. Select a vulnerability to view its details and explicit remediation instructions.
4. Remediate the vulnerability using the provided instructions described in the 'Remediation' field.
5. Upload the new remediated image to your registry. Review scan results for the new image to verify the remediated vulnerabilities no longer exist.
6. Delete the old image with the vulnerabilities from your registry.

👤 **Delegate**

Use Defender for Cloud built-in governance mechanism or ServiceNow ITSM to assign the recommendation to the right owner.

Assign owner & set due date    ⓘ

◌ **Exempt**

Exempt the entire recommendation, or disable specific findings using disable rules. Exempted resources appear as not applicable and do not affect secure score.

Exempt    Disable rule    ⓘ

*Figure 11.4: Azure registry vulnerabilities recommendation (for better visualization, refer to https://packt.link/gbp/9781836204879)*

Just like many other recommendations that you've seen so far in this book, this recommendation follows the same standard. The left side shows the overall details about the recommendation, and the right side is composed of different tabs. To see the list of vulnerabilities, click the **Findings** tab and the list will appear, as shown in the example of *Figure 11.5*:

Take action  **Findings**  Graph  Repositories

🔍 Search to filter items...                    Status : **Unhealthy**

| Severity | ID | Security check |
|----------|-----|----------------|
| Critical | CVE-2022-4883 | CVE-2022-4883 |
| Critical | CVE-2022-27782 | CVE-2022-27782 |
| Critical | CVE-2021-46848 | CVE-2021-46848 |
| Critical | CVE-2022-40674 | CVE-2022-40674 |
| High | CVE-2022-37434 | CVE-2022-37434 |
| High | CVE-2022-2868 | CVE-2022-2868 |
| High | CVE-2021-3999 | CVE-2021-3999 |
| High | CVE-2022-4450 | CVE-2022-4450 |
| High | CVE-2023-0215 | CVE-2023-0215 |
| High | CVE-2023-38545 | CVE-2023-38545 |
| High | CVE-2023-4911 | CVE-2023-4911 |
| High | CVE-2022-22576 | CVE-2022-22576 |
| High | CVE-2022-2509 | CVE-2022-2509 |

*Figure 11.5: List of CVEs*

To obtain more details about the CVE, click on the CVE in the list, and the CVE blade appears, as shown in the examples of *Figure 11.6* (part 1) and *Figure 11.7* (part 2):

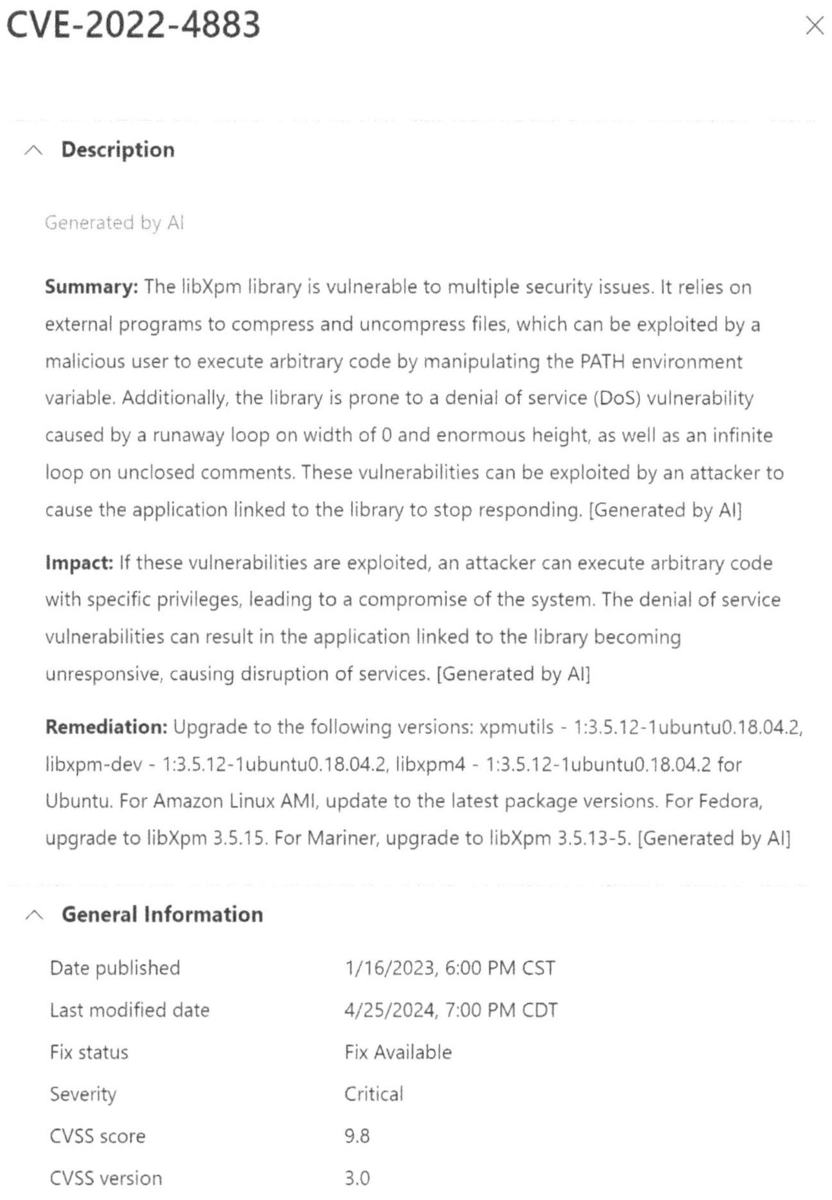

## CVE-2022-4883　　　　　　　　　　　　　　　　　　✕

### ∧　Description

Generated by AI

**Summary:** The libXpm library is vulnerable to multiple security issues. It relies on external programs to compress and uncompress files, which can be exploited by a malicious user to execute arbitrary code by manipulating the PATH environment variable. Additionally, the library is prone to a denial of service (DoS) vulnerability caused by a runaway loop on width of 0 and enormous height, as well as an infinite loop on unclosed comments. These vulnerabilities can be exploited by an attacker to cause the application linked to the library to stop responding. [Generated by AI]

**Impact:** If these vulnerabilities are exploited, an attacker can execute arbitrary code with specific privileges, leading to a compromise of the system. The denial of service vulnerabilities can result in the application linked to the library becoming unresponsive, causing disruption of services. [Generated by AI]

**Remediation:** Upgrade to the following versions: xpmutils - 1:3.5.12-1ubuntu0.18.04.2, libxpm-dev - 1:3.5.12-1ubuntu0.18.04.2, libxpm4 - 1:3.5.12-1ubuntu0.18.04.2 for Ubuntu. For Amazon Linux AMI, update to the latest package versions. For Fedora, upgrade to libXpm 3.5.15. For Mariner, upgrade to libXpm 3.5.13-5. [Generated by AI]

### ∧　General Information

| | |
|---|---|
| Date published | 1/16/2023, 6:00 PM CST |
| Last modified date | 4/25/2024, 7:00 PM CDT |
| Fix status | Fix Available |
| Severity | Critical |
| CVSS score | 9.8 |
| CVSS version | 3.0 |

*Figure 11.6: Detailed explanation of the CVE (part 1)*

Usually, this is a long list, which means that you need to continue scrolling down to see the rest of the details, including the affected resource, as shown in *Figure 11.7*:

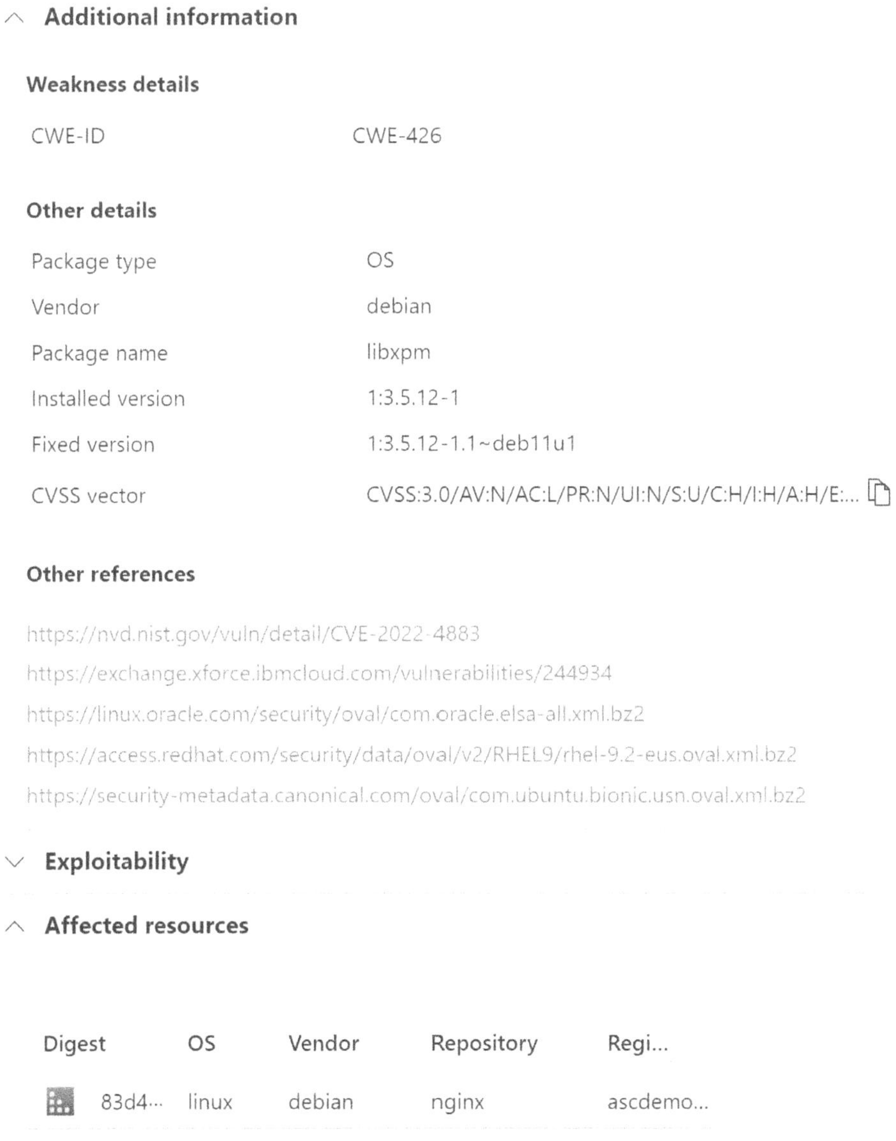

∧  **Additional information**

**Weakness details**

CWE-ID                                CWE-426

**Other details**

Package type                     OS
Vendor                              debian
Package name                  libxpm
Installed version              1:3.5.12-1
Fixed version                   1:3.5.12-1.1~deb11u1
CVSS vector                    CVSS:3.0/AV:N/AC:L/PR:N/UI:N/S:U/C:H/I:H/A:H/E:... 🗋

**Other references**

https://nvd.nist.gov/vuln/detail/CVE-2022-4883
https://exchange.xforce.ibmcloud.com/vulnerabilities/244934
https://linux.oracle.com/security/oval/com.oracle.elsa-all.xml.bz2
https://access.redhat.com/security/data/oval/v2/RHEL9/rhel-9.2-eus.oval.xml.bz2
https://security-metadata.canonical.com/oval/com.ubuntu.bionic.usn.oval.xml.bz2

∨  **Exploitability**

∧  **Affected resources**

| Digest | OS | Vendor | Repository | Regi... |
|--------|-----|--------|-----------|---------|
| 🔲 83d4⋯ | linux | debian | nginx | ascdemo... |

*Figure 11.7: Detailed explanation of the CVE (part 2)*

Now that you know the affected resource, you can click on it to open the image and see which packages are vulnerable, as shown in the example of *Figure 11.8*:

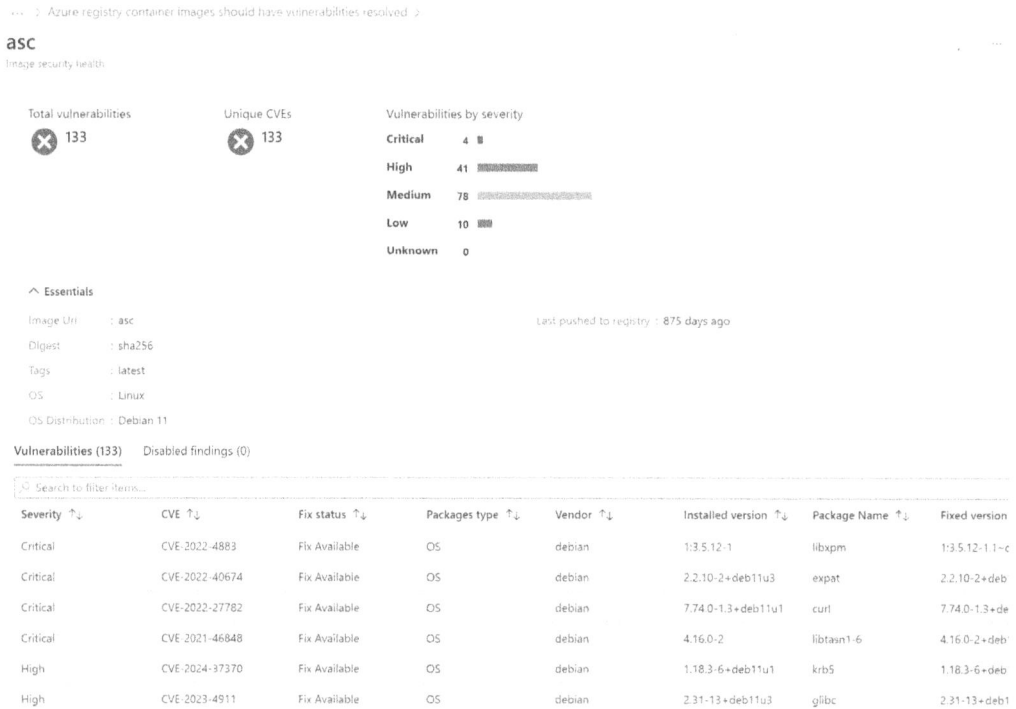

*Figure 11.8: Detailed explanation of the vulnerable packages within the image (for better visualization, refer to https://packt.link/gbp/9781836204879)*

With this information, you can start working on the remediation, which in this case is shown in *Figure 11.6* under the **Remediation** section. After performing those steps, you need to upload the new remediated image to your registry and wait for the next scan to validate the results and show the resources as healthy. It is important to mention that brand-new images are scanned shortly after being added to a registry and new cycles are performed once every 24 hours.

Although you can't trigger a manual scan, a scan is automatically performed for each image pushed or imported to a container registry. In addition, Defender for Containers provides a capability called **continuous rescan triggering**, which is a re-scan once a day for images that were pushed in the last 30 days, and in the last 90 days. Images that are deleted from the registry will be removed from Defender for Cloud within one hour, however, if there are communication (unable to reach due to network connectivity or any other reason) issues and Defender for Cloud is not notified that the image was deleted, it may take up to three days for the image to completely disappear from Defender for Cloud.

# Binary drift detection

Binary drift in a container environment refers to the unintended or unauthorized changes that occur to the binaries or software components within a running container. This could include modifications to the container's file system, binaries, or libraries that deviate from its original, defined state. Since containers are meant to be immutable, in other words, they should not change once deployed, binary drift can create security risks and operational challenges.

Ideally, Containers should not change once they are running. However, administrators or automated processes may inadvertently or intentionally make changes to binaries or configurations, and this action will lead to a binary drift scenario. While it can be a valid operation, and a benign change, it is still necessary to have visibility of those cases. Defender for Containers enables you to create binary drift detection policies to alert you when binary drift occurs. The list below has the prerequisites for this feature:

- Defender for Containers sensors must be enabled on the subscriptions and connectors.

    - The sensor is available in AWS, GCP, and AKS in versions 1.29 or higher.

- The Defender for Containers sensor needs to be running

- You need **Security Admin** or higher permissions on the tenant to create or modify drift policies

- You need **Security Reader** or higher permissions on the tenant to view drift policies

To create a new Container drift policy, open the Defender for Cloud dashboard, click **Environment Settings** under the **Management** section in the left navigation pane, and click the **Containers drift policy** tile, located on the far right side of the page, as shown in *Figure 11.9*:

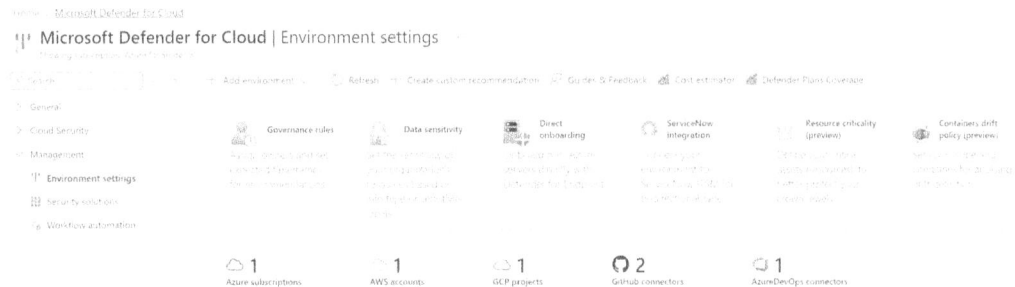

*Figure 11.9: Accessing the container drift policy (for better visualization, refer to https://packt. link/gbp/9781836204879)*

Once you click on this tile, you will see the **Binary drift policy** page, which has, by default, two rules, as shown in *Figure 11.10*:

*Figure 11.10: Accessing the container drift policy (for better visualization, refer to https:// packt.link/gbp/9781836204879)*

Before talking about these default rules, it is important to understand how these rules are processed. The rules are processed from top to bottom, based on the priority order (just like firewall rules are usually processed). This means that if the conditions match rule number one, this rule gets processed, and rule number two won't be processed. That's why it is very important that when you are creating rules, you put the more specific ones at the top (higher priority) and the more generic ones at the bottom (lower priority).

The first default rule is called **Alert on Kube-System namespace**, and the goal of this rule is to trigger an alert in case there is a binary drift in the **Kube-System** namespace. The second rule is called **Default binary drift**, and the action it takes when this rule is processed is to ignore the drift. The rules that you will create should be a higher priority since they will be more specific for your business needs, which means that if the condition doesn't match your custom rules, the default rules will be processed.

While you can edit both rules, you cannot delete the **Default binary drift**, and this is an expected behavior. To create a custom binary drift policy rule, click the + **Add rule** button, and the **Add new containers** drift rule appears, as shown in *Figure 11.11*:

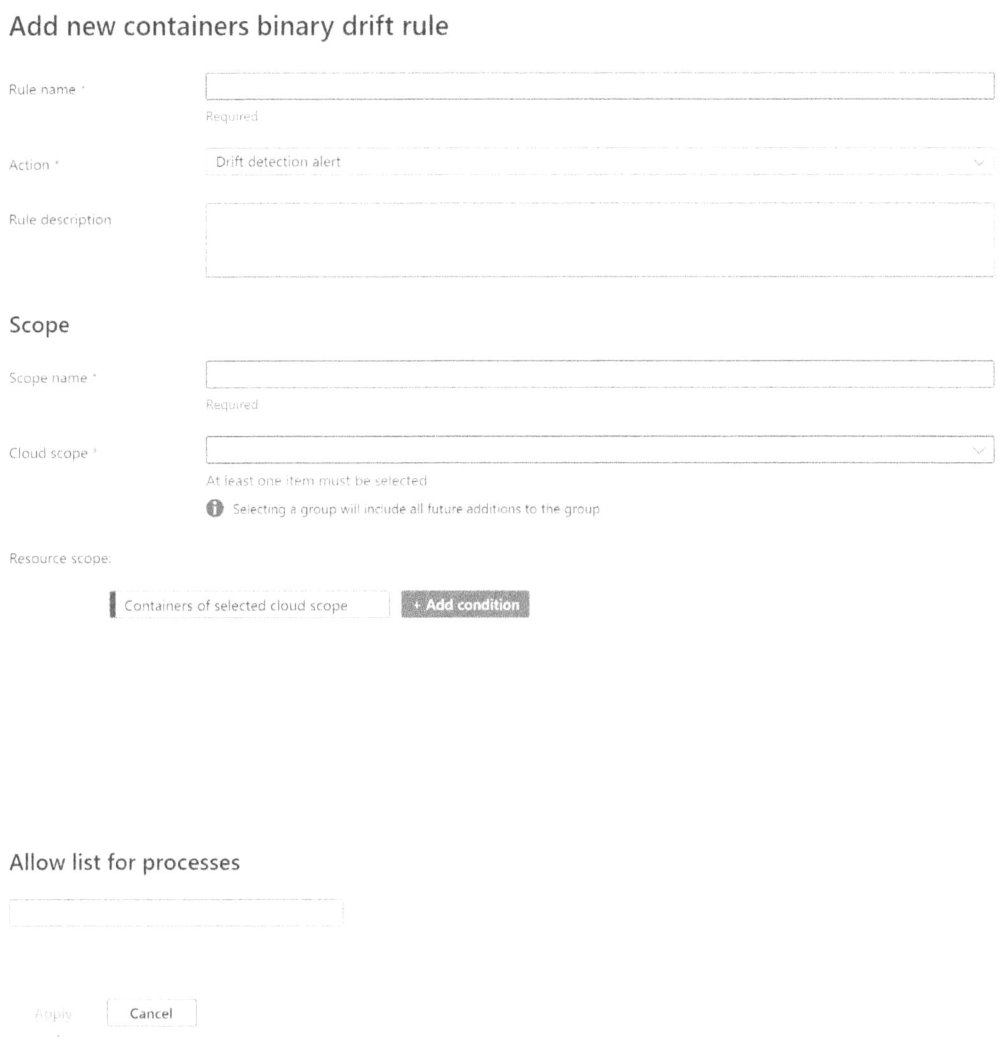

## Add new containers binary drift rule

Rule name *

Required

Action *          Drift detection alert

Rule description

## Scope

Scope name *

Required

Cloud scope *

At least one item must be selected

ⓘ  Selecting a group will include all future additions to the group

Resource scope:

Containers of selected cloud scope     + **Add condition**

## Allow list for processes

Apply       Cancel

*Figure 11.11: Creating a new binary drift rule (for better visualization, refer to https://packt.*
*link/gbp/9781836204879)*

On this page, you will need to provide some information to create the rule, which includes:

- **Rule name:** A comprehensive name for the rule
- **Action:** There are two actions available; you either select the option to trigger an alert or the action to ignore the drift
- **Rule description:** Although this is an optional field, it is recommended to write a description that documents the rationale behind the rule. This will help others in your team to better understand the motivation behind creating this rule
- **Scope name:** A comprehensive name to describe the scope of this rule. The actual scope is selected in the next two fields
- **Cloud scope:** Select which cloud service (Azure, AWS, or GCP) this rule will be applicable to
- **Resource scope:** This is the main part of the rule since, here, you will select the condition that will be processed, and once the match occurs, the drift detection will be triggered. The visual experience is very similar to the cloud security explorer (see example in *Figure 11.12*), and you can select the scope based on container name, image name, namespace, pod label, pod name, or cluster name.
- **Allow list for processes:** Here, you can write a list of processes that are allowed to run in the container; in other words, if the process is not on the list, an alert will be triggered.

In the example shown in *Figure 11.12*, you have a scope with two conditions, with AND as a logical operator, which means both conditions must be true for the rule to be successfully processed. In addition, there is only one process in the list of allowed processes to run.

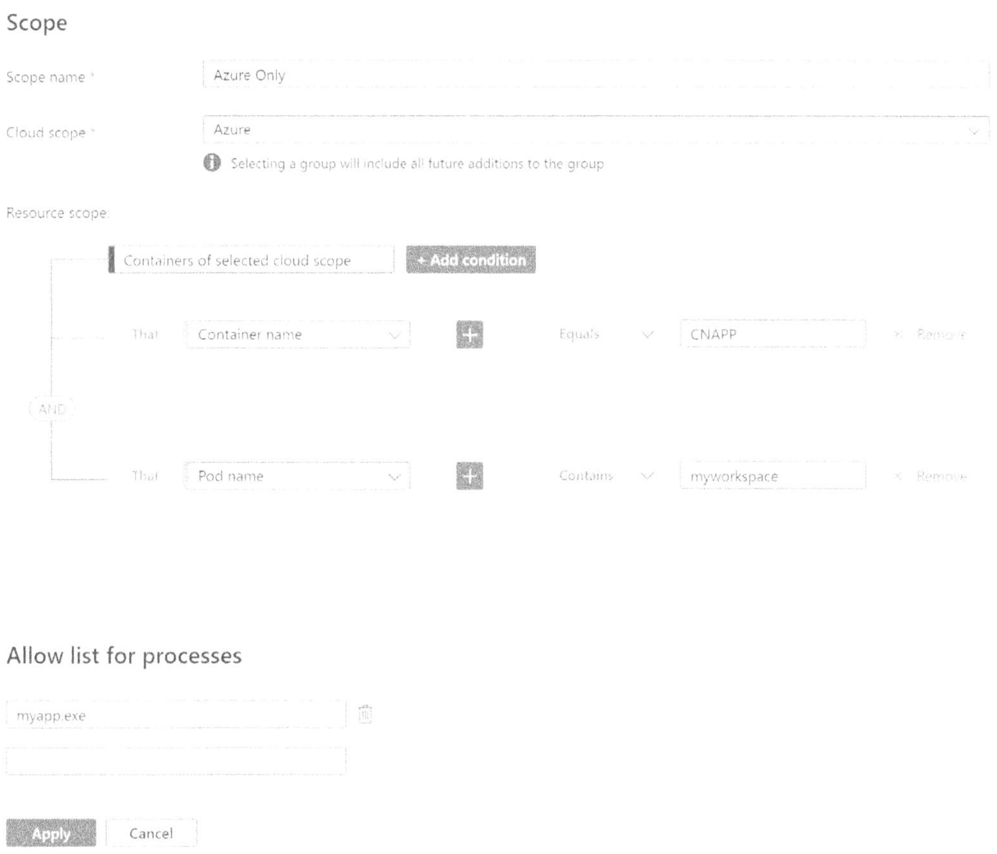

Scope

| | |
|---|---|
| Scope name * | Azure Only |
| Cloud scope * | Azure |

ℹ Selecting a group will include all future additions to the group

Resource scope:

Containers of selected cloud scope    **+ Add condition**

That    Container name ⌄    [+]    Equals ⌄    CNAPP    ✕ Remove

AND

That    Pod name ⌄    [+]    Contains ⌄    myworkspace    ✕ Remove

Allow list for processes

myapp.exe    🗑

**Apply**    Cancel

*Figure 11.12: Configuring the condition (for better visualization, refer to https://packt.link/gbp/9781836204879)*

Once you finish configuring these options, click the **Apply** button to commit the changes and click **Save** to create the rule. The new policy rule will be in effect immediately for the existing clusters, however, if a new cluster is added, it may take one hour for this new cluster to receive the policy. This happens because the discovery process takes place every hour, so the delay is not so much on the policy itself, but on the discovery of new clusters.

At any moment, if you want to delete or disable a policy, all you need to do is select the policy and choose either the **Delete rule** button or the **Disable** button, as shown in *Figure 11.13*:

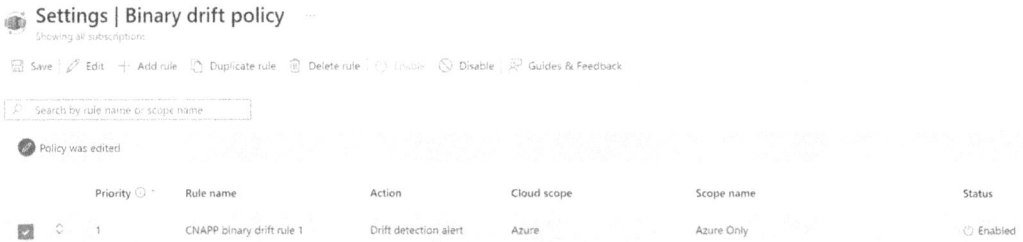

*Figure 11.13: Disabling or deleting a rule (for better visualization, refer to https://packt.link/ gbp/9781836204879)*

If the conditions match the rule, an alert will be triggered, and you will see this alert on the **Security alert** dashboard in Defender for Cloud. *Figure 11.14* has an example of this type of alert:

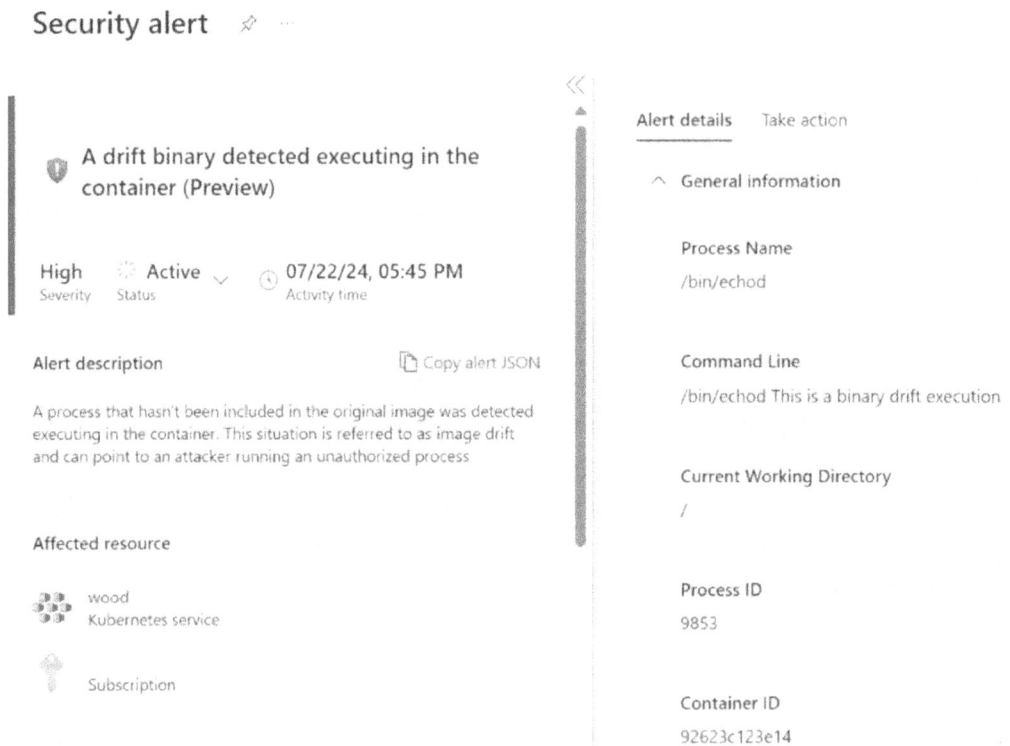

*Figure 11.14: Drift detection alert (for better visualization, refer to https://packt.link/ gbp/9781836204879)*

This alert follows the same schema as any other alert in Defender for Cloud, with an explanation about the alert on the left side, and the details about the resource on the right side. It is important to emphasize that a drift detection alert will always be a high-severity alert, due to the criticality of this type of operation.

While drift detection is something that you have control over because you need to create the rule, there are other out-of-the-box detections available in Defender for Containers that will automatically trigger alerts based on abnormal activities and unusual and malicious operations. To see a list of all alerts available for this plan, visit `https://bit.ly/CNAPPBook94`.

While container usage in cloud computing continues to grow, VMs are still heavily used as an **Infrastructure as a Service** (**IaaS**) workload. Next, let's see how Defender for Servers can help protect this type of workload.

# Defender for Servers

In *Chapter 4*, you learned how to create an adoption plan for CNAPP, and part of this adoption plan was to learn more about the Defender for Servers capabilities available in Plan 1 and Plan 2. Before proceeding with reading this section, you should revisit *Chapter 4* to understand the difference between Defender for Servers Plan 1 versus Plan 2 in terms of feature sets, and when to use one over the other.

Once you decide which plan is the most appropriate for your environment, you need to decide if you are going to enable the plan for the entire subscription or at the resource level. The advantage of enabling at the subscription level is that all VMs that are in the subscription will be protected, and future VMs that are provisioned in the subscription will also be automatically protected. If you have valid reasons to enable it at the resource level only (for example, budget constraints), you will need to ensure that you are documenting in which resources you are enabling Defender for Servers. In addition, the enablement at the resource level is only available via REST APIs[8] or PowerShell script[9]; in other words, it is not available in the Defender for Cloud dashboard.

Defender for Servers leverages agents to collect different types of data, and this data will be used to improve security posture management improvement, as well as threat detection and attack surface reduction.

The table below has the different agents and their functionalities:

| Agent | Scope | Functionalities |
|---|---|---|
| Azure Arc* | Leveraged in on-premises and multicloud scenarios | • Connectivity to Azure<br>• Automatic installation of other agents<br>• System-managed identity<br>• Azure Policy |
| **Microsoft Defender for Endpoint (MDE)** | Applicable to all servers protected by Defender for Servers plans (1 or 2) | • Next-generation antivirus<br>• Endpoint detection and response<br>• Threat and vulnerability management<br>• File integrity monitoring |
| **Azure Monitor Agent (AMA)** | | • Security data collection<br>• Additional threat detection |
| Guest Configuration[10] | | • Operating system hardening recommendations |

**\*** Hybrid servers need to have the Azure ARC agent installed first to support auto-provisioning and onboarding.

To enable Defender for Servers at the subscription level, you need to have the **Security Admin** privilege. Once you open the Defender for Cloud dashboard, click **Environment settings** under the **Management** section, click the subscription in which you want to enable Defender for Servers, click the **Defender plans** option, and click **On** beside the plan, as shown in *Figure 11.15*:

| Plan | Pricing* | Resource quantity | Monitoring coverage | Status |
|---|---|---|---|---|
| Servers | Plan 2 ($15/Server/Month)  Change plan > | 1 servers | | Off ( ) On |

*Figure 11.15: Enabling Defender for Servers (for better visualization, refer to https://packt.link/gbp/9781836204879)*

After you enable it, you will see that the **Change plan** link becomes available, and you can click on it to choose Plan 1 or Plan 2. For this section of this chapter, the assumption is that you will use Plan 2. Another option that is available after enabling the plan is the **Settings** option, under the **Monitoring coverage** column. Click on it to ensure that all extensions are enabled, as shown in *Figure 11.16*:

*Figure 11.16: Defender for Servers onboarding settings (for better visualization, refer to https://packt.link/gbp/9781836204879)*

It is important to mention that at the time this chapter was written, the Log Analytics agent was already on the deprecation path[11], hence the warning message in the first row on the left of *Figure 11.16*, and by the time you are reading this chapter, this page may not have this message anymore. Moving forward in this section of the chapter, the Log Analytics agent is out of scope and won't be covered. For this scenario, you will leave the **File Integrity Monitoring** toggle set to **Off**, as you will learn more about this feature later in the chapter. Click the **Continue** button, and then click the **Save** button.

The enablement of the plan triggers an automatic deployment of MDE to associated VMs on the subscription. If a VM already has MDE deployed, it won't be redeployed. Keep in mind that if you disable the Defender for Servers plan on the subscription, MDE will not be uninstalled. In other words, you will need to manually uninstall MDE[12].

While MDE is being deployed, Defender for Servers will still be able to provide an initial level of protection for your VMs using the agentless machine scanning capabilities (available with Defender CSPM or Defender for Servers Plan 2). Agentless VM scanning (an extension that was enabled during the onboarding process—see *Figure 11.16*) leverages cloud APIs to gather data without requiring an agent on the VM. In contrast, agent-based scanning utilizes the VM's operating system APIs in real time to continuously monitor and collect security data. Microsoft Defender for Cloud captures VM disk snapshots and performs a thorough analysis of the operating system configuration and file system based on the snapshot.

This process is conducted out-of-band, ensuring that the snapshot remains in the same region as the VM, and the VM itself is not impacted by the scanning process. Once the necessary metadata is retrieved from the copied disk, Defender for Cloud promptly deletes the disk snapshot. The collected metadata is then sent to Microsoft's analysis engines to identify configuration vulnerabilities and potential threats.

If for some reason you want to exclude VMs from being scanned using agentless scanning capability, you can add the VM name to the exclusion list. To do that, go to the extension page (see *Figure 11.16*), and click the **Edit configuration** option in the same row as **Agentless scanning for machines**. The **Agentless scanning configuration** page appears as shown in *Figure 11.17*:

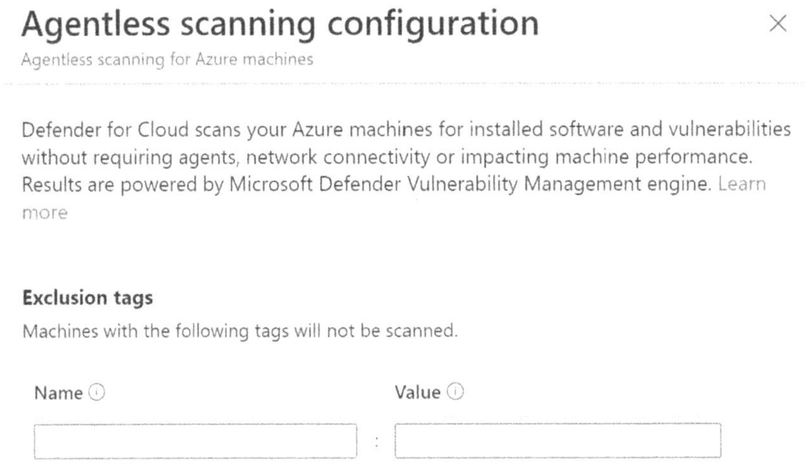

## Agentless scanning configuration        ✕

Agentless scanning for Azure machines

Defender for Cloud scans your Azure machines for installed software and vulnerabilities without requiring agents, network connectivity or impacting machine performance. Results are powered by Microsoft Defender Vulnerability Management engine. Learn more

**Exclusion tags**

Machines with the following tags will not be scanned.

Name ⓘ                                  Value ⓘ

[                    ]    :    [                    ]

*Figure 11.17: Creating an exclusion list for VMs (for better visualization, refer to https://packt. link/gbp/9781836204879)*

On this page, you should enter the tag name and the value that applies to the machines that you want to exclude. If you click the **Name** field, a drop-down list will automatically appear as Defender for Cloud retrieves the available names so you can select from the list; the same applies to the **Value** field. Once you finish configuring, click **Apply, Continue,** and then **Save** to commit the changes.

## Agentless malware scanning

Leveraging the same agentless extension technology, a new functionality was added to Defender for Servers Plan 2 in early 2024 that allows Defender for Servers to perform malware scanning without the need to have agents installed. The advantage of this feature is that it is completely frictionless onboarding and there is no performance impact on workloads.

Agentless malware scanning provides automatic quick and full scans using heuristic and signature-based threat detection using the MDE antivirus engine. The scanning process is not customizable; in other words, it will be automatically performed behind the scenes, and if a piece of malware is detected, a security alert is triggered, as shown in the example of *Figure 11.18*:

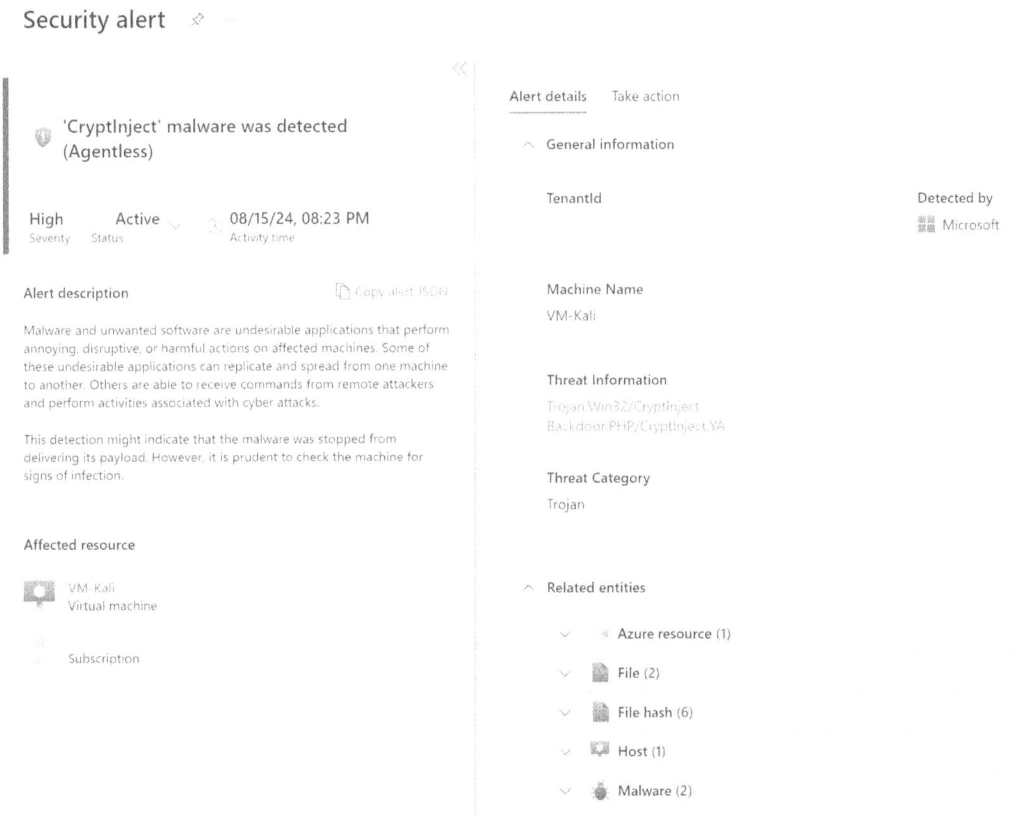

*Figure 11.18: Agentless malware scanning alert (for better visualization, refer to https://packt.link/gbp/9781836204879)*

Keep in mind that the alert is there to bring you awareness that a piece of malware was detected; however, it will not automatically clear the malware. If you click the **Take action** tab, you will see the options to remediate, as shown in *Figure 11.19*:

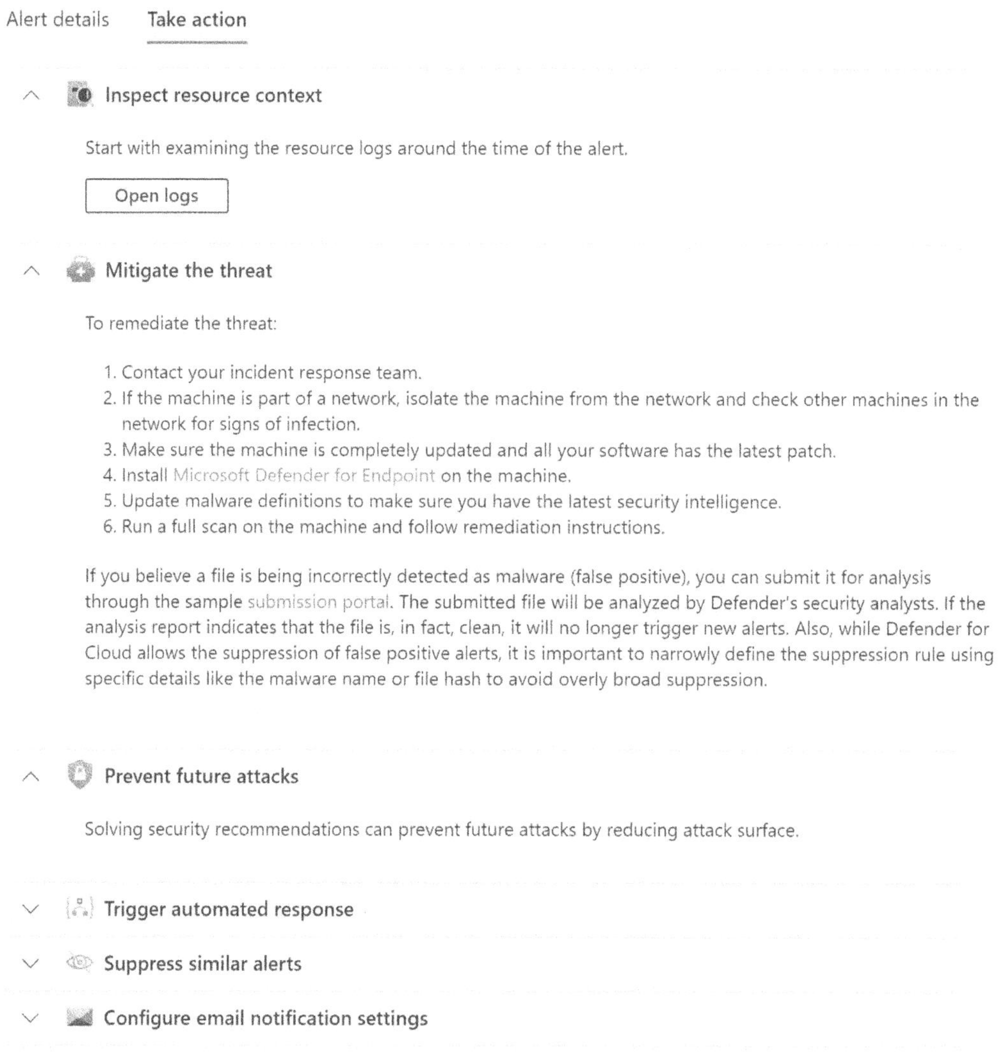

Alert details        Take action

∧     🔲 Inspect resource context

       Start with examining the resource logs around the time of the alert.

       [ Open logs ]

∧     ⚙ Mitigate the threat

       To remediate the threat:

          1. Contact your incident response team.
          2. If the machine is part of a network, isolate the machine from the network and check other machines in the network for signs of infection.
          3. Make sure the machine is completely updated and all your software has the latest patch.
          4. Install Microsoft Defender for Endpoint on the machine.
          5. Update malware definitions to make sure you have the latest security intelligence.
          6. Run a full scan on the machine and follow remediation instructions.

       If you believe a file is being incorrectly detected as malware (false positive), you can submit it for analysis through the sample submission portal. The submitted file will be analyzed by Defender's security analysts. If the analysis report indicates that the file is, in fact, clean, it will no longer trigger new alerts. Also, while Defender for Cloud allows the suppression of false positive alerts, it is important to narrowly define the suppression rule using specific details like the malware name or file hash to avoid overly broad suppression.

∧     🛡 Prevent future attacks

       Solving security recommendations can prevent future attacks by reducing attack surface.

∨     ⦿ Trigger automated response

∨     👁 Suppress similar alerts

∨     ✉ Configure email notification settings

*Figure 11.19: Take action tab (for better visualization, refer to https://packt.link/ gbp/9781836204879)*

Optionally, you can automate a response for this type of alert using the **Trigger automated response** option. You will need to build an automation using Azure Logic Apps and select this automation to link to this alert.

While this feature is powerful for monitoring potential malware in a system, there are other types of malicious operations beyond malware, for example, threat actors, that infiltrate the system trying to make changes to the registry. For this type of scenario, File Integrity Monitoring can help.

# File Integrity Monitoring (FIM)

FIM is a capability that is exclusively available as part of Defender for Servers Plan 2, and it leverages MDE to collect data from machines according to the configured rules. With FIM, you can monitor changes made to critical files and Windows registries based on a predefined list that you can customize. FIM will trigger alerts upon the creation or deletion of files and registry keys. It also covers alerts for changes in the file's size, name, location, hash of its content, and alterations of the registry.

To access the FIM dashboard, click **Workload protections**, under the **Cloud Security** section, as shown in *Figure 11.20*:

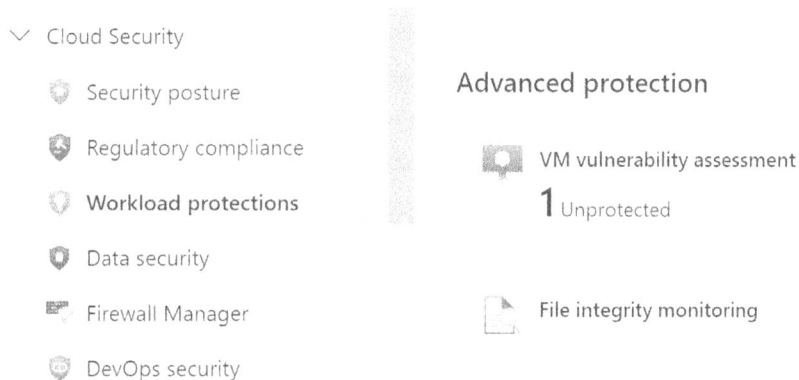

*Figure 11.20: The FIM tile*

If this is the first time you are accessing the FIM dashboard, you may see the message shown in *Figure 11.21*:

Home > Microsoft Defender for Cloud | Workload protections >

# File Integrity Monitoring   ···

Showing subscription 'Visual Studio Ultimate with MSDN'

⟳ Refresh   🕓 Change to previous experience

ℹ Toward Log Analytics agent retirement, new File Integrity Monitoring (FIM) is now available through   →
Microsoft Defender for Endpoint integration in Defender for Servers Plan 2. Enable it on now to benefit from
the new improved capabilities and ensure your environment is protected.

## No subscriptions are onboarded

Click here to start enabling **File Integrity Monitoring** over your scopes

**Onboard subscriptions**

Learn more about service ↗

*Figure 11.21: The File Integrity Monitoring page when the subscription is not onboarded*

To enable this feature in the subscription, you need to go back to the Defender for Servers extension page (see *Figure 11.16*) and switch the **File Integrity Monitoring** toggle to **On**. When you do this, the FIM configuration page appears as shown in *Figure 11.22*:

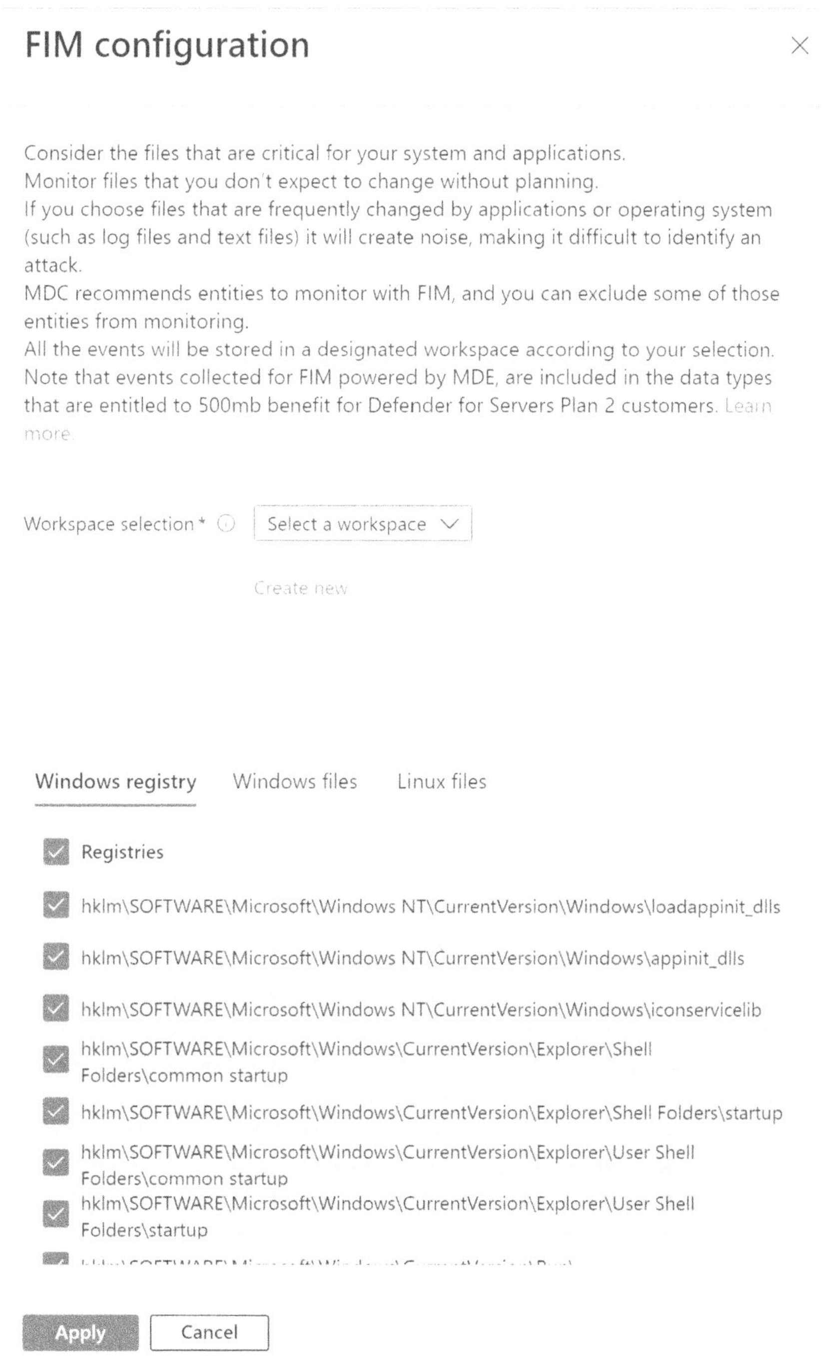

# FIM configuration ✕

Consider the files that are critical for your system and applications.
Monitor files that you don't expect to change without planning.
If you choose files that are frequently changed by applications or operating system
(such as log files and text files) it will create noise, making it difficult to identify an
attack.
MDC recommends entities to monitor with FIM, and you can exclude some of those
entities from monitoring.
All the events will be stored in a designated workspace according to your selection.
Note that events collected for FIM powered by MDE, are included in the data types
that are entitled to 500mb benefit for Defender for Servers Plan 2 customers. Learn
more.

Workspace selection * ⓘ   Select a workspace ∨

Create new

**Windows registry**   Windows files   Linux files

☑ Registries

☑ hklm\SOFTWARE\Microsoft\Windows NT\CurrentVersion\Windows\loadappinit_dlls

☑ hklm\SOFTWARE\Microsoft\Windows NT\CurrentVersion\Windows\appinit_dlls

☑ hklm\SOFTWARE\Microsoft\Windows NT\CurrentVersion\Windows\iconservicelib

☑ hklm\SOFTWARE\Microsoft\Windows\CurrentVersion\Explorer\Shell
Folders\common startup

☑ hklm\SOFTWARE\Microsoft\Windows\CurrentVersion\Explorer\Shell Folders\startup

☑ hklm\SOFTWARE\Microsoft\Windows\CurrentVersion\Explorer\User Shell
Folders\common startup

☑ hklm\SOFTWARE\Microsoft\Windows\CurrentVersion\Explorer\User Shell
Folders\startup

☑ hklm\SOFTWARE\Microsoft\Windows\CurrentVersion\Run\

**Apply**   Cancel

*Figure 11.22: Configuring the FIM settings (for better visualization, refer to https://packt.link/
gbp/9781836204879)*

The first step to configure FIM is to ensure that you select the Log Analytics workspace that will retain the data collected by FIM. To do that, click the **Workspace selection** drop-down list and select the proper workspace. Keep in mind that events collected by FIM (which is powered by MDE) are included in the data types eligible for the 500 MB benefit for Defender for Servers Plan 2[13].

The FIM configuration is divided into three tabs: Windows registry, Windows files, and Linux files. By default, all the main registry and files are configured. Keep in mind that at the time this chapter was written, the capability to add other registry keys and files was not available. You can review the current selection and uncheck the ones that you believe are not applicable. After doing this, click the **Apply** button, click **Continue**, and then click **Save**.

Once FIM is fully configured and monitoring your subscription, you can access the dashboard via **Workload protections**, under the **Cloud Security** section, and now you will be able to see the machines that are being monitored, as well as some usage and metrics regarding the changes, as shown in *Figure 11.23*:

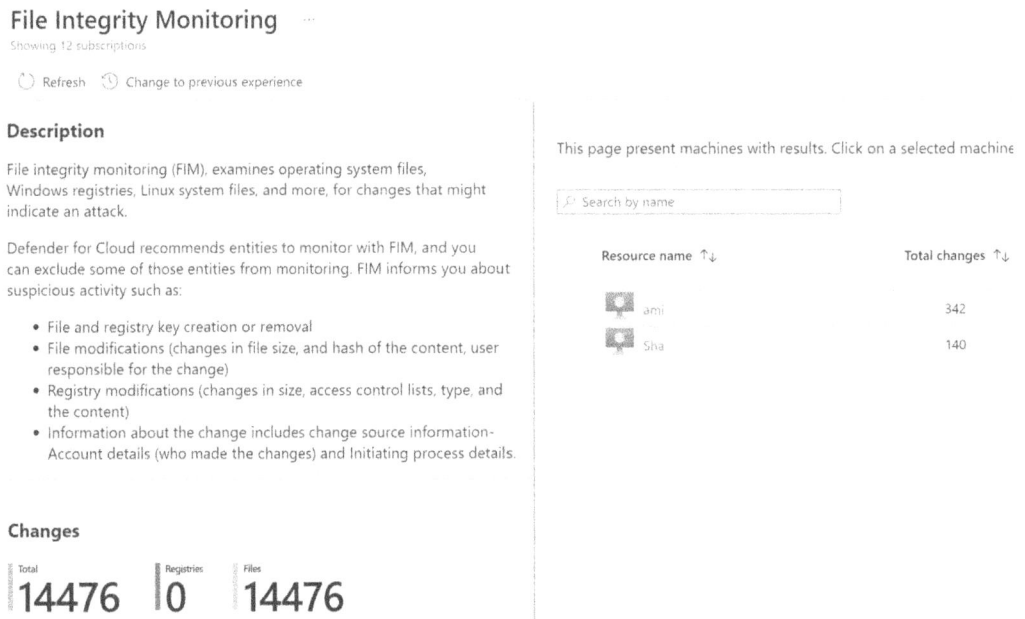

*Figure 11.23: The FIM dashboard (for better visualization, refer to https://packt.link/ gbp/9781836204879)*

To see more details about which files changed, you can click on the machine on the right side, and the Log Analytics workspace page appears with a predefined query that will search for the changes in the workspace and will show them to you, as shown in the example from *Figure 11.24*:

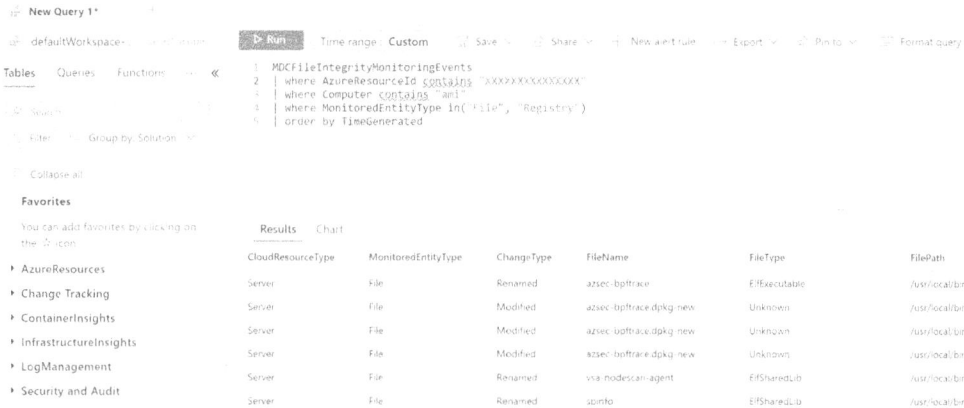

*Figure 11.24: Log Analytics workspace with the query results (for better visualization, refer to https://packt.link/gbp/9781836204879)*

From this page, you can navigate to each row in the table results to see more details about the changes. Aside from monitoring files and registry, you need to ensure that you are reducing the attack surface of your VMs, mainly for the ones that are publicly exposed to the internet. To address this scenario, you will leverage Just-in-Time VM access, another exclusive feature of Defender for Servers Plan 2.

# JIT VM access

With the growth of people working from home since COVID-19 changed the industry, the use of remote desktop technology also grew, and with that the number of attacks against **Remote Desktop Protection (RDP)**. The RDP brute-force attack is a very common type of attack against VMs in the cloud. This means that if you have VMs that are publicly exposed to the internet with the default RDP port (TCP port 3389) open, the likelihood that these VMs will constantly be attacked is very high. While Defender for Cloud will trigger an alert (see example in *Figure 11.25*) for a potential RDP brute-force attempt, you should always aim to reduce the attack surface for public internet resources.

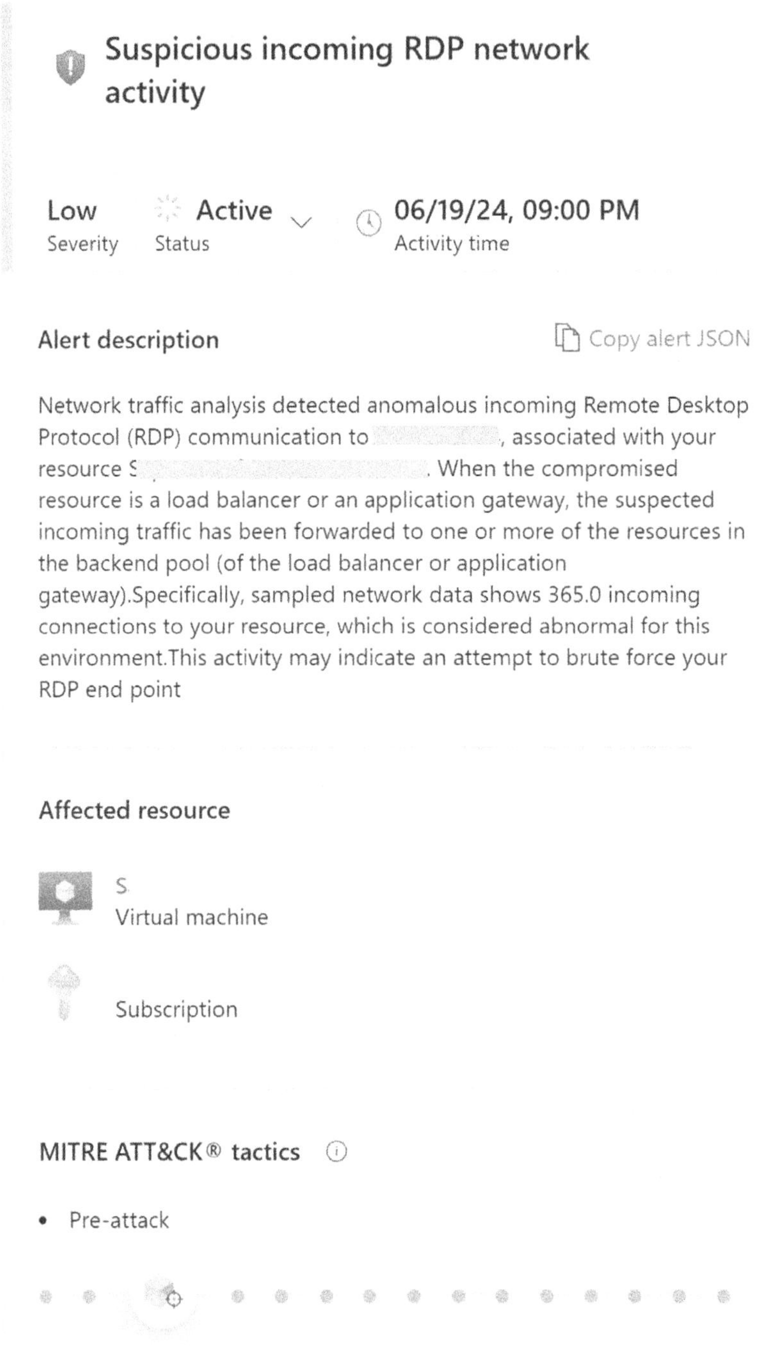

## Suspicious incoming RDP network activity

**Low**    ☀ **Active** ⌄    🕐 **06/19/24, 09:00 PM**
Severity    Status                    Activity time

### Alert description                    📋 Copy alert JSON

Network traffic analysis detected anomalous incoming Remote Desktop Protocol (RDP) communication to ▮▮▮▮▮▮▮▮, associated with your resource S▮▮▮▮▮▮▮▮▮▮▮▮▮▮. When the compromised resource is a load balancer or an application gateway, the suspected incoming traffic has been forwarded to one or more of the resources in the backend pool (of the load balancer or application gateway).Specifically, sampled network data shows 365.0 incoming connections to your resource, which is considered abnormal for this environment.This activity may indicate an attempt to brute force your RDP end point

### Affected resource

🖥 S
Virtual machine

🔑 Subscription

### MITRE ATT&CK® tactics   ⓘ

• Pre-attack

*Figure 11.25: RDP brute-force attack attempt*

To reduce the attack surface, you should limit the exposure of VMs, and with JIT VM access, you can do that by configuring specifically the time interval that this machine will be accepting RDP connections. For example, if you know that this machine will be accessed by remote users four hours a day, then configure the machine to be accessible via RDP coming from the internet for only four hours a day.

Defender for Servers will go through a process[14] to verify if the machines are exposed to the internet, and if they are eligible to have JIT. If the conditions are true, a new recommendation will be generated, as shown in *Figure 11.26*:

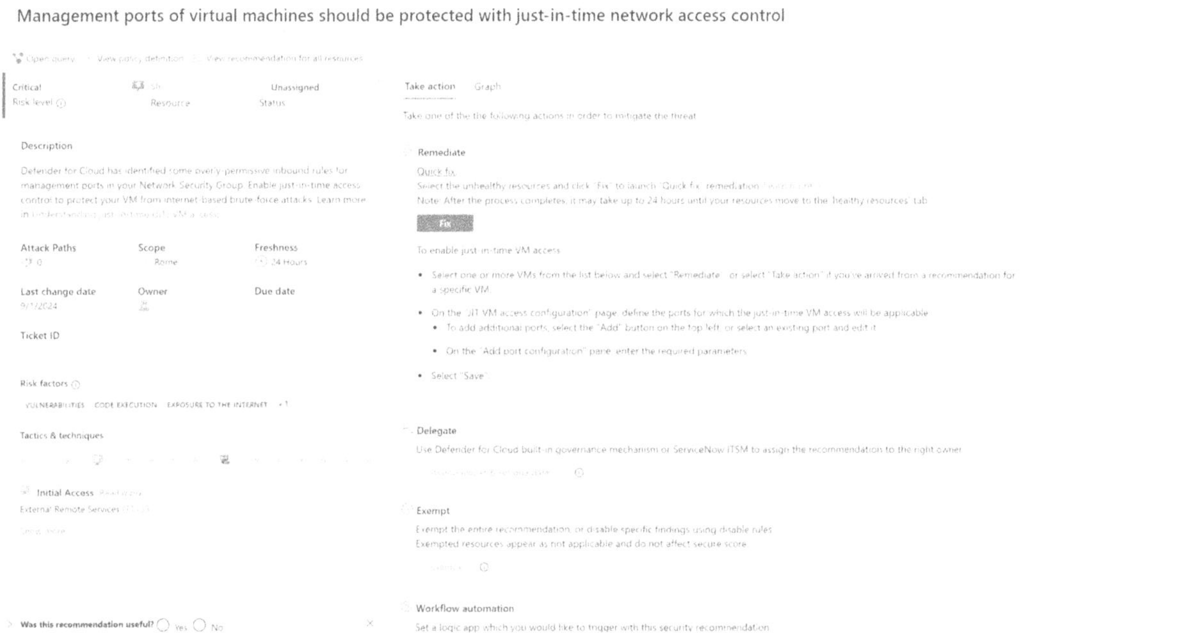

*Figure 11.26: Recommendation to enable JIT (for better visualization, refer to https://packt. link/gbp/9781836204879)*

This recommendation brings visibility about the VM (resource) that needs JIT enabled, and you can use the **Fix** button on the right to quickly expedite the enablement of JIT on this machine. Keep in mind that to remediate this recommendation, you will need the following privileges to enable JIT:

- On the scope of a subscription or resource group that is associated with the VM, you need **Microsoft.Security/locations/jitNetworkAccessPolicies/write**

- On the scope of a subscription or resource group of the VM, you need **Microsoft.Compute/ virtualMachines/write**

Click the **Fix** button, and you will see the JIT VM access configuration page, as shown in *Figure 11.27*:

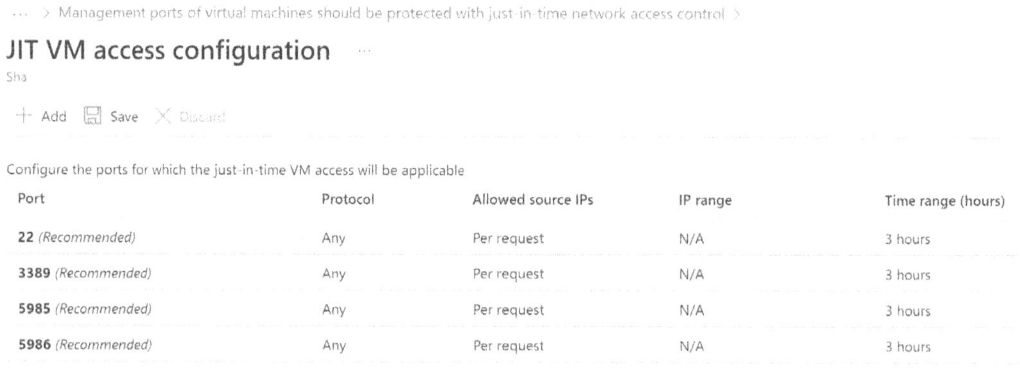

··· > Management ports of virtual machines should be protected with just-in-time network access control >

## JIT VM access configuration ···
Sha

+ Add    📄 Save   ✕ Discard

Configure the ports for which the just-in-time VM access will be applicable

| Port | Protocol | Allowed source IPs | IP range | Time range (hours) |
|------|----------|--------------------|----------|--------------------|
| **22** *(Recommended)* | Any | Per request | N/A | 3 hours |
| **3389** *(Recommended)* | Any | Per request | N/A | 3 hours |
| **5985** *(Recommended)* | Any | Per request | N/A | 3 hours |
| **5986** *(Recommended)* | Any | Per request | N/A | 3 hours |

*Figure 11.27: The JIT configuration page (for better visualization, refer to https://packt.link/ gbp/9781836204879)*

By default, there are four ports available, but you can customize this by leaving only the ports you need and customize each line by simply clicking on them. For example, if you click the line for 3389, a new blade opens, as shown in *Figure 11.28*:

## Add port configuration  ✕

Port *

3389

Protocol

( Any    TCP    UDP )

Allowed source IPs

( Per request   CIDR block )

IP addresses ⓘ

Max request time

●—○············································   3
                                            (hours)

Discard      OK

*Figure 11.28: Port configuration page (for better visualization, refer to https://packt.link/ gbp/9781836204879)*

In this blade, you can change the port number, the transport protocol, the source IP address, and the maximum time (in hours) that the VM will be available to be accessed remotely. Once you perform those customizations, you can click the **OK** button, and then click the **Save** button.

Once you enable a machine to be protected with JIT, you will need to request access to the machine. To request access, you need to assign these actions to the user on the scope of the subscription or resource group associated with the VM:

- `Microsoft.Security/locations/jitNetworkAccessPolicies/initiate/action.`
- `Microsoft.Security/locations/jitNetworkAccessPolicies/*/read.`
- `Microsoft.Compute/virtualMachines/read.`
- `Microsoft.Network/networkInterfaces/*/read.`

Assuming that the privileges are correct, you can request access from the VM dashboard in Azure, click the **Connect** option, and click the **Request access** option, located on the right side of the blade, as shown in the example of *Figure 11.29*:

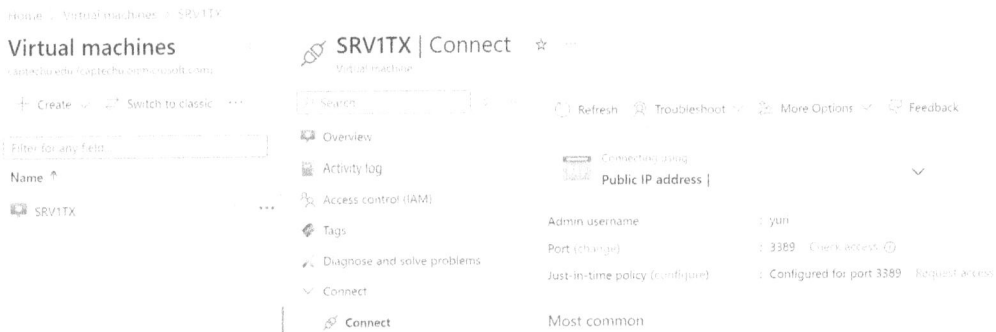

*Figure 11.29: Requesting VM access (for better visualization, refer to https://packt.link/ gbp/9781836204879)*

If you want to see the JIT configuration, click the **Configure** option, located beside the name JIT policy, and the **JIT VM access** page appears as shown in *Figure 11.30*:

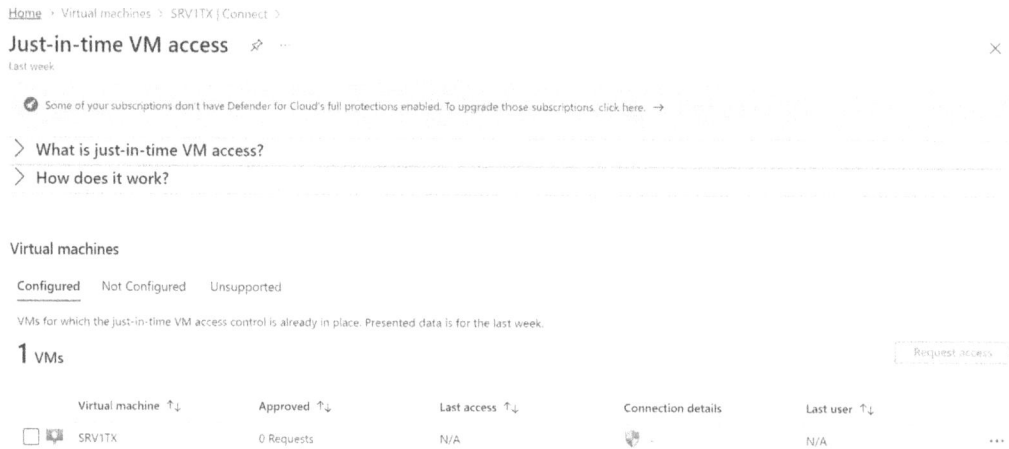

Home > Virtual machines > SRV1TX | Connect >

## Just-in-time VM access
Last week

✓ Some of your subscriptions don't have Defender for Cloud's full protections enabled. To upgrade those subscriptions. click here. →

> What is just-in-time VM access?
> How does it work?

**Virtual machines**

Configured    Not Configured    Unsupported

VMs for which the just-in-time VM access control is already in place. Presented data is for the last week.

**1** VMs                                                                                    Request access

| | Virtual machine ↑↓ | Approved ↑↓ | Last access ↑↓ | Connection details | Last user ↑↓ | |
|---|---|---|---|---|---|---|
| ☐ | SRV1TX | 0 Requests | N/A | | N/A | ... |

*Figure 11.30: JIT VM access page (for better visualization, refer to https://packt.link/gbp/9781836204879)*

On this page, you will see the machines that are configured, the ones that are not configured in the second tab, and the ones that are not supported in the third tab. Unsupported VMs include VMs deployed with a classic deployment model and VMs protected by Azure Firewall, which is controlled by Azure Firewall Manager. On this page, you will also find a grid that shows the number of requests approved under the **Approved** column, the last access, the connection details, and the last user that accessed via JIT.

By now, your VM is protected by MDE, which was deployed during the Defender for Servers on-boarding process, agentless malware scanning, FIM, and JIT. In addition to these security controls, it is important to keep the systems (both Windows and Linux) updated, and to ensure that you have the right level of visibility of potential vulnerabilities, you can leverage the vulnerability assessment capability from Defender for Servers.

## Vulnerability assessment

Due to the native integration with MDE, you automatically get the core version of **Microsoft Defender Vulnerability Management (MDVM)** for P1 and the premium capabilities for P2. All that without the need to install an additional agent. The vulnerabilities that are found by MDVM will surface within the **Machines should have vulnerability findings resolved** recommendation, as shown in *Figure 11.31*:

## Machines should have vulnerability findings resolved ✕

Open query  |  View policy definition  |  View recommendation for all resources

| High | SRV11X | Unassigned |
|------|--------|------------|
| Risk level | Resource | Status |

**Take action**   Findings   Graph

Take one of the the following actions in order to mitigate the threat:

### Description

Resolving vulnerability findings on virtual machines is a recommended step in maintaining a secure environment.
These findings, identified by vulnerability assessment solutions, highlight potential weaknesses that could be exploited by malicious actors.
If these vulnerabilities are not addressed, they could lead to unauthorized access, data breaches, or even system failure.
Therefore, it is important to resolve these findings promptly to ensure the security and integrity of the virtual machines.

**Remediate**

Review and remediate vulnerabilities discovered by the vulnerability assessment solutions

**Delegate**

Use Defender for Cloud built-in governance mechanism or ServiceNow ITSM to assign the recommendation to the right owner.

Assign owner & set due date

| Attack Paths | Scope | Freshness |
|--------------|-------|-----------|
| 1 | Azure for Stude... | 12 Hours |

| Last change date | Owner | Due date |
|------------------|-------|----------|
| 9/9/2024 | | |

Ticket ID

Risk factors

EXPOSURE TO THE INTERNET   VULNERABILITIES

**Exempt**

Exempt the entire recommendation, or disable specific findings using disable rules. Exempted resources appear as not applicable and do not affect secure score.

Exempt   Disable rule

**Workflow automation**

Set a logic app which you would like to trigger with this security recommendation

Trigger logic app

**Findings by severity**

High (2)   Medium (0)   Low (1)

**Total findings**
3

*Figure 11.31: Vulnerability findings recommendation (for better visualization, refer to https:// packt.link/gbp/9781836204879)*

On this page, you will see the description of the recommendation on the left and the details about the risk factors that are contributing to the criticality level (High) of this recommendation. To see the list of vulnerabilities, you will need to click on the **Findings** tab, as shown in the example of *Figure 11.32*:

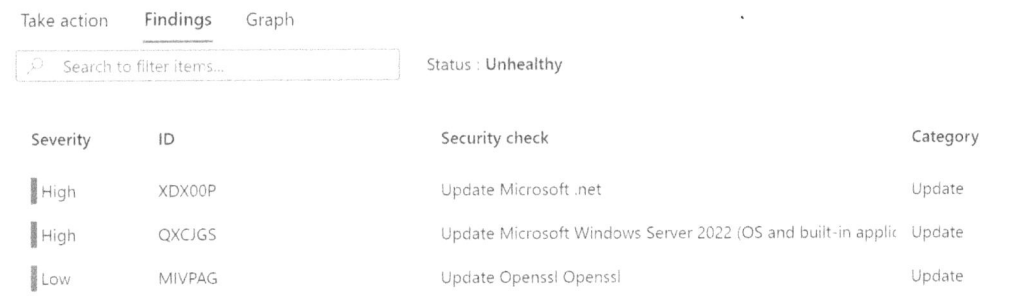

Take action   **Findings**   Graph

Search to filter items...        Status : **Unhealthy**

| Severity | ID | Security check | Category |
|----------|-----|----------------|----------|
| High | XDX00P | Update Microsoft .net | Update |
| High | QXCJGS | Update Microsoft Windows Server 2022 (OS and built-in applic | Update |
| Low | MIVPAG | Update Openssl Openssl | Update |

*Figure 11.32: List of vulnerabilities (for better visualization, refer to https://packt.link/ gbp/9781836204879)*

To see more details about the vulnerability itself, click on it and a new blade appears showing the details, including the remediation steps, as shown in *Figure 11.33*:

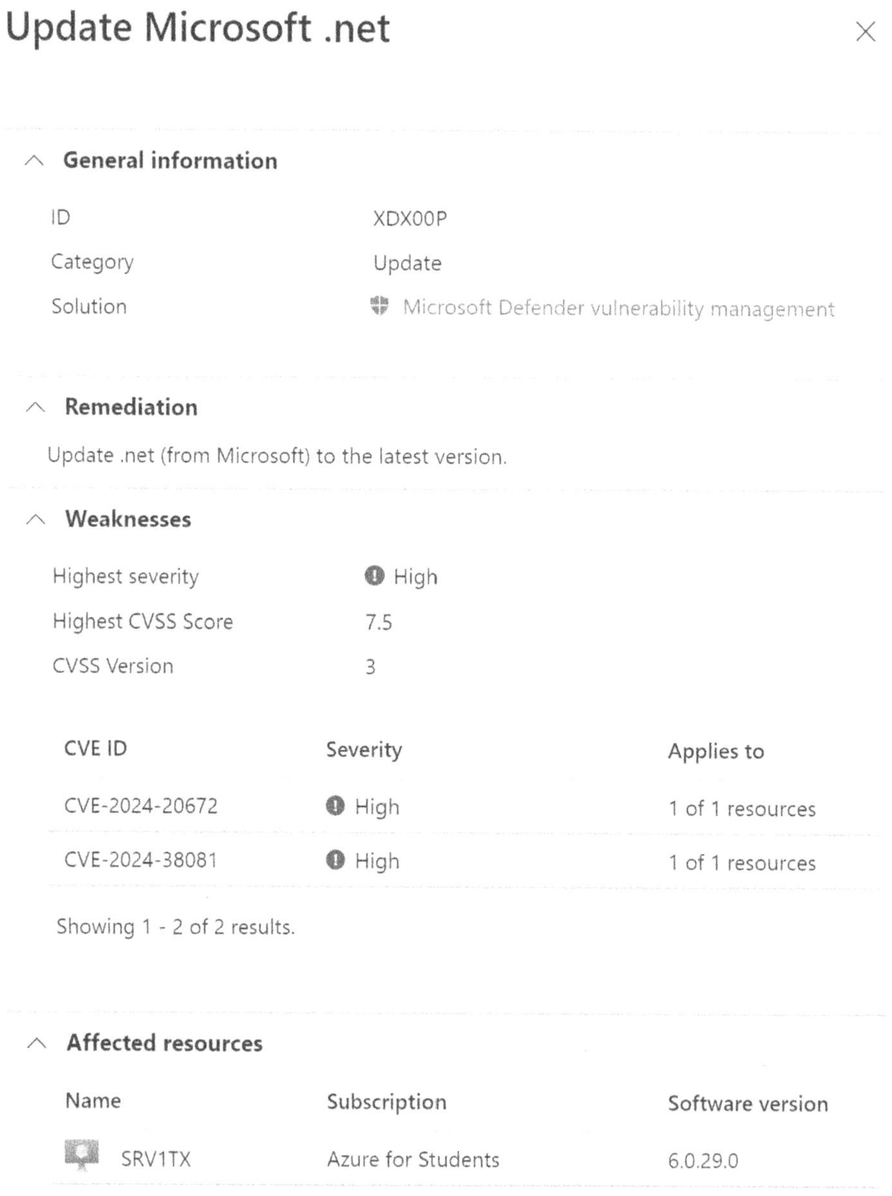

## Update Microsoft .net                                                    ✕

∧  **General information**

ID                                    XDX00P

Category                              Update

Solution                              🛡 Microsoft Defender vulnerability management

∧  **Remediation**

Update .net (from Microsoft) to the latest version.

∧  **Weaknesses**

Highest severity              ❗ High

Highest CVSS Score            7.5

CVSS Version                  3

| CVE ID | Severity | Applies to |
|--------|----------|------------|
| CVE-2024-20672 | ❗ High | 1 of 1 resources |
| CVE-2024-38081 | ❗ High | 1 of 1 resources |

Showing 1 - 2 of 2 results.

∧  **Affected resources**

| Name | Subscription | Software version |
|------|--------------|------------------|
| 🖥 SRV1TX | Azure for Students | 6.0.29.0 |

*Figure 11.33: Details about the vulnerability*

Keep in mind that while this feature will give you awareness of the vulnerabilities, you will need to use another tool, such as Microsoft Intune, to deploy the patches. The advantage of having the vulnerability assessment built into Defender for Servers is the fact that the insights that are generated by MDVM are shared across other elements within Defender for Cloud, for example, the attack path, which you learned more about in *Chapter 5* of this book.

## Summary

In this chapter, you learned how to protect Containers with Defender for Containers and machines with Defender for Servers. You learned about the supported scenarios in Defender for Containers and its capabilities and constraints. You also learned how to enable Defender for Containers, how to visualize vulnerabilities in your containers, and how to configure binary drift detection. You also learned about Defender for Servers capabilities, such as agentless malware scanning, FIM, JIT VM access, and vulnerability assessment powered by Microsoft Defender Vulnerability Management.

In the next chapter, you will learn how to protect the data, including storage and databases.

## Notes

1. You can read more about this type of attack at `https://bit.ly/CNAPPBook80`.

2. For more information about Azure Arc-enabled Kubernetes, visit `https://bit.ly/CNAPPBook83`.

3. For more information about Kubernetes RBAC, visit `https://bit.ly/CNAPPBook87`.

4. For more information about Kubernetes DaemonSet, visit `https://bit.ly/CNAPPBook85`.

5. Visit `https://bit.ly/CNAPPBook84` for more information about this component.

6. For more information about Kubernetes data plane hardening, visit `https://bit.ly/CNAPPBook89`.

7. For more information about ACR, visit `https://bit.ly/CNAPPBook90`.

8. For more information about Defender for Cloud REST APIs used to deploy Defender for Servers, visit `https://bit.ly/CNAPPBook82`.

9. Microsoft created a PowerShell script that you can download and use to deploy Defender for Servers at the resource level at scale. You can download this script from `https://bit.ly/CNAPPBook81`.

10. For more information about Azure Policy guest configuration, visit `https://bit.ly/CNAPPBook97`.

11. If you are in an environment that still uses Log Analytics, make sure to read this article at `https://bit.ly/CNAPPBook95` for more information about the migration path.

12. Use the steps from this article at `https://bit.ly/CNAPPBook96` to offboard MDE from devices.

13. For more information about the 500 MB allowance, read this article at `https://bit.ly/CNAPPBook98`.

14. To learn more about this workflow, see this diagram at `https://bit.ly/CNAPPBook99`.

## Additional resources

- You can watch the author Yuri Diogenes interviewing the Feature PM of Defender for Servers, where they talk about the agentless malware detection feature, in this episode: `https://bit.ly/CNAPPBook100`

- You can watch the author Yuri Diogenes interviewing the Feature PM of Defender for Servers, where they talk about the vulnerability assessment feature, in this episode: `https://bit.ly/CNAPPBook101`

## Join our community on Discord

Read this book alongside other users. Ask questions, provide solutions to other readers, and much more.

Scan the QR code or visit the link to join the community.

`https://packt.link/SecNet`

# 12

# Protecting Storage and Databases

Regardless of how threats actors will potentially get into the network and start moving laterally to perform their malicious operations, one of their goals may be to obtain sensitive information about the company. Protecting data is a challenging task because data can be, and usually is, spread across multiple workloads, the most common ones being cloud storage and databases.

According to **Thales 2024 Cloud Security Study**,[1] 44% of the organizations surveyed have experienced a cloud data breach. The most alarming data is the fact that 31% of the data breaches were caused by human error and misconfigurations. This scenario emphasizes the importance of adopting a CNAPP solution that can look across different datasets and proactively disrupt potential attacks even more.

When it comes to data, it is not only about protecting the data, but also discovering, classifying, and having threat detection in place to identify malicious operations that may be occurring. Defender for Cloud has native data protection for storage accounts and many options available for different types of databases.

This chapter covers:

- Defender for Storage
- Defender for Databases

# Defender for Storage

In *Chapter 5*, you learned about data security posture, which is a feature of Defender CSPM. While this feature gives you a great level of visibility that can help you uncover misconfigurations, those characteristics are part of the overall posture management; the gap that exists by only using this feature is the lack of threat detection. In other words, you need to gain knowledge about active threat actors trying to break into your storage account, and that's where Defender for Storage comes into play.

Cloud storage may be the entry point for your organization. For example, if you have an exposed storage account (due to valid reasons for your business), how can you identify if a piece of malware was uploaded? In addition to this scenario, you also need to be aware of potential data exfiltration, and the use of infrastructure tools to encrypt and delete data inside the storage account.

Defender for Storage conducts data plane and control plane analysis to detect threats, and these threats can come from malicious insiders and external attackers. Defender for Storage leverages Microsoft Threat Intelligence and behavioral modeling signals to identify early signs of a breach. Once it is identified, it will trigger an alert,[2] and this alert will appear on the **Security alerts** dashboard in Defender for Cloud. *Figure 12.1* has an example of a low-severity alert generated by Defender for Storage:

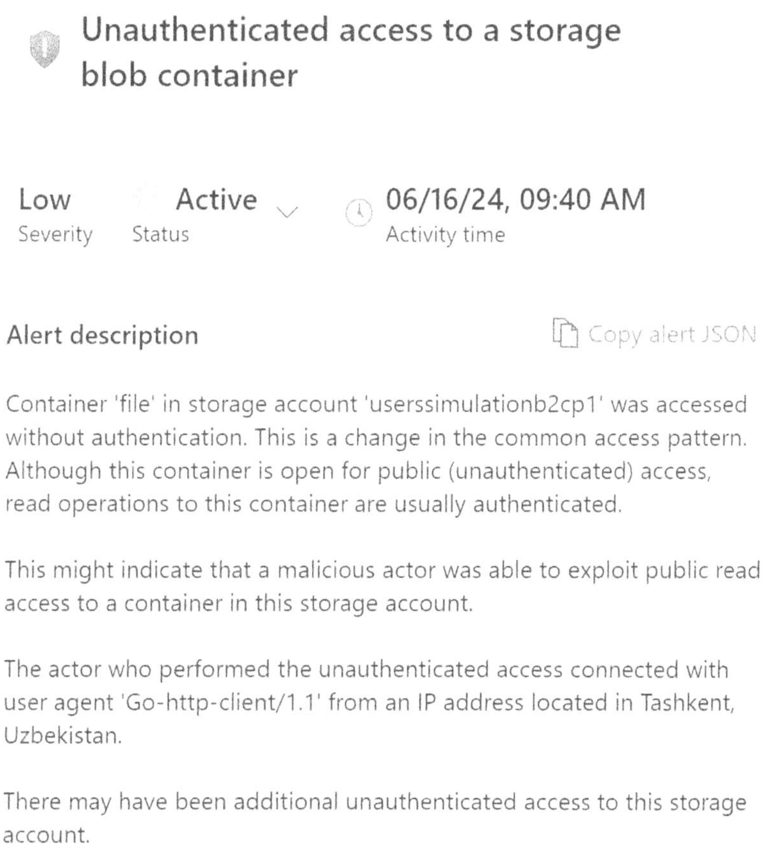

# Unauthenticated access to a storage blob container

**Low**
Severity

**Active** ⌄
Status

🕐 **06/16/24, 09:40 AM**
Activity time

## Alert description

📋 Copy alert JSON

Container 'file' in storage account 'userssimulationb2cp1' was accessed without authentication. This is a change in the common access pattern. Although this container is open for public (unauthenticated) access, read operations to this container are usually authenticated.

This might indicate that a malicious actor was able to exploit public read access to a container in this storage account.

The actor who performed the unauthenticated access connected with user agent 'Go-http-client/1.1' from an IP address located in Tashkent, Uzbekistan.

There may have been additional unauthenticated access to this storage account.

*Figure 12.1: Alert generated by Defender for Storage*

# Enabling Defender for Storage

Defender for Storage supports Azure Blob Storage, Azure Files, and Azure Data Lake Storage services. However, by the time this chapter was written, the malware scanning feature in Defender for Storage, which is enabled separately, does not support Azure Files. To enable Defender for Storage, you need to be a Security Admin in the subscription where the Azure storage resides, and if you also want to enable the malware scanning feature, you need to be a subscription owner or storage account owner.

You can enable Defender for Storage in the entire subscription, or at the storage account level only, which usually happens when you need to control costs and focus only on the most important storage accounts, or accept the risk. The most important storage accounts, for instance, are those that are publicly exposed and may have anonymous access allowed, and those that contain sensitive data. The disadvantage of this option is that when new storage accounts are provisioned in the subscription, they will not be automatically protected by Defender for Storage, since it is disabled at the subscription level. For this reason, enabling the plan at the subscription level is the recommended option. This ensures that you have continuous protection for current and future storage accounts.

To enable Defender for Storage in the subscription level, open the Defender for Cloud dashboard, click **Environment settings** under the **Management** section, click the subscription that you want to enable Defender for Storage, click the **Defender plans** option, and click **On** beside the storage plan, as shown in *Figure 12.2*:

| Storage | $10/Storage account/month $0.15/GB scanned for On-Upload Malware Sc   1 storage accounts   Details > | | Full Settings > | Off   On |

*Figure 12.2: Enabling Defender for Storage (for better visualization, refer to https://packt. link/gbp/9781836204879)*

Before enabling the plan, you can also click the **Details** hyperlink (shown in *Figure 12.2*) to see more information about pricing. The **Plan details** blade appears as shown in *Figure 12.3*:

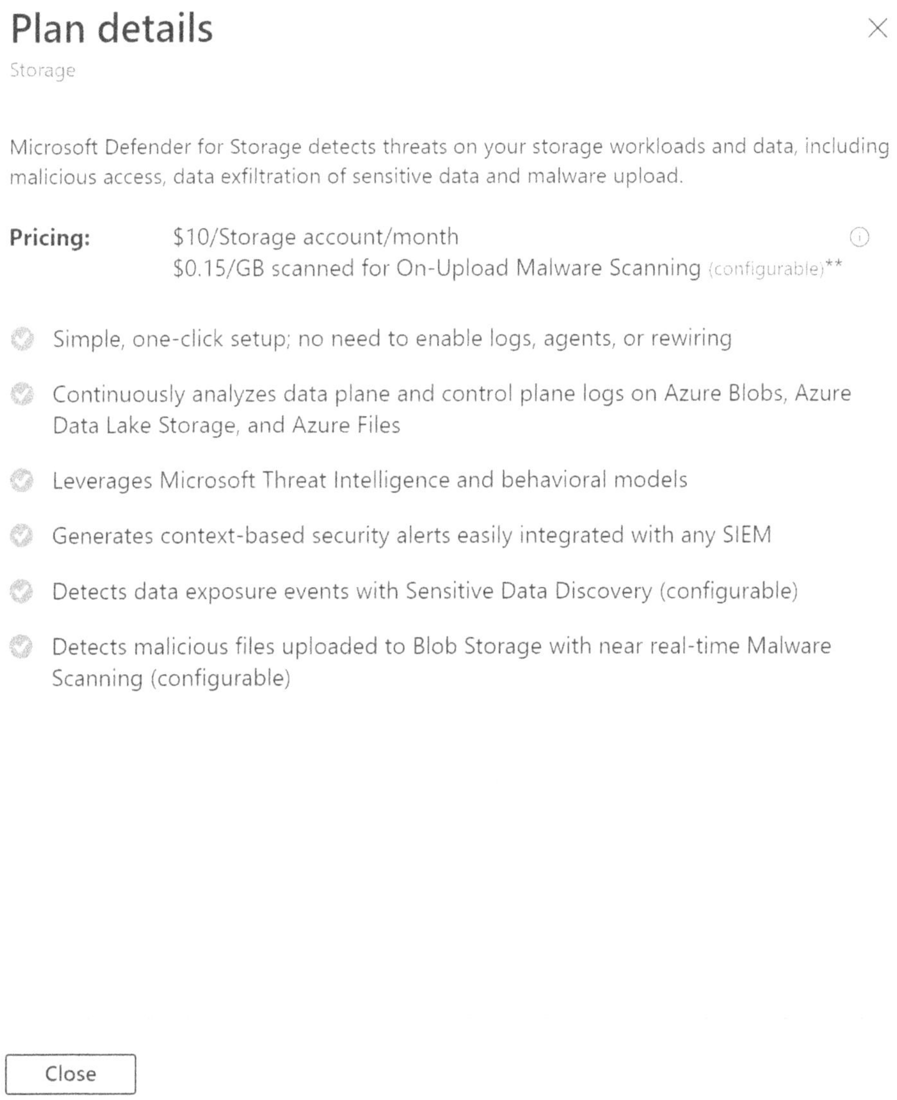

# Plan details

Storage

×

Microsoft Defender for Storage detects threats on your storage workloads and data, including malicious access, data exfiltration of sensitive data and malware upload.

**Pricing:**        $10/Storage account/month                             ⓘ
                      $0.15/GB scanned for On-Upload Malware Scanning (configurable)**

- ✅ Simple, one-click setup; no need to enable logs, agents, or rewiring

- ✅ Continuously analyzes data plane and control plane logs on Azure Blobs, Azure Data Lake Storage, and Azure Files

- ✅ Leverages Microsoft Threat Intelligence and behavioral models

- ✅ Generates context-based security alerts easily integrated with any SIEM

- ✅ Detects data exposure events with Sensitive Data Discovery (configurable)

- ✅ Detects malicious files uploaded to Blob Storage with near real-time Malware Scanning (configurable)

Close

*Figure 12.3: Reviewing the plan's pricing*

After finishing the review, you can click the **Close** button, and before saving the configuration, and if you want the malware scanning feature to be enabled, click the **Settings** hyperlink (see *Figure 12.2*), and the **Seatings & monitoring** page appears as shown in *Figure 12.4*:

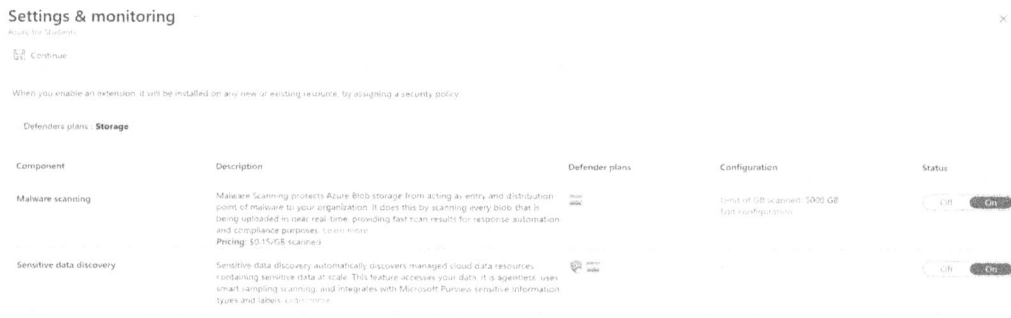

*Figure 12.4: Enabling malware scanning (for better visualization, refer to https://packt.link/gbp/9781836204879)*

As you can see, the default scanning limit is 5,000 GB per month, which works like a cap to control cost. You can lower or increase this number by clicking the **Edit configuration** hyperlink beside the malware scanning option. The **On-upload malware scanning configuration** blade is shown in *Figure 12.5*:

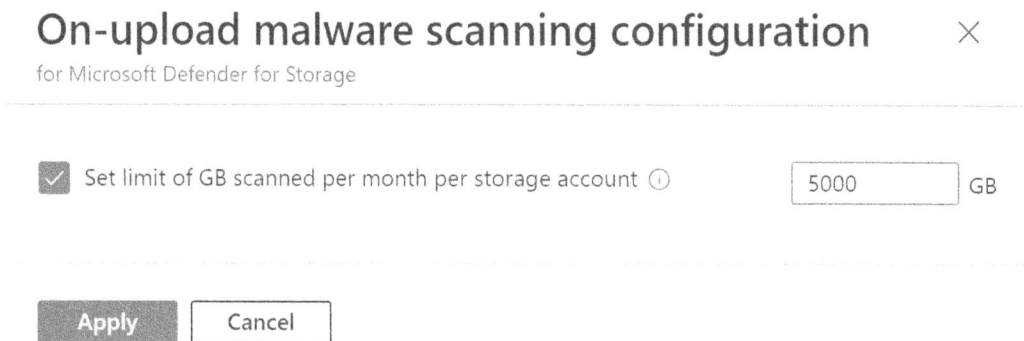

*Figure 12.5: Controlling the monthly limit*

It is important to mention that once it crosses this limit in a single billing period, files will not be scanned for malware, but you will receive an alert warning about this. After making the changes, click the **Apply** button, and you will return to the **Settings & monitoring** page (*Figure 12.4*). Now, you need to ensure that the malware scanning toggle is set to **On**.

You can see in *Figure 12.4* that you have another option called **Sensitive data discovery**, and **yes**, this is exactly the same feature that you learned about in *Chapter 5* when you were enabling Defender CSPM. The reason this feature appears in both plans is to cover scenarios where a customer has Defender for Storage enabled but doesn't have Defender CSPM. For example, an environment that only has storage accounts and doesn't have any other type of workload, and the organization just wants to monitor storage accounts. After you finish configuring, click the **Continue** button and then the **Save** button to commit the changes.

## Malware scanning

The malware scanning feature in Defender for Storage leverages Microsoft Defender Antivirus engine to scan, and since it is a **software-as-a-service** (**SaaS**) type of solution, it doesn't require any maintenance, or manual engine update. However, to scan your data, the malware scanning service needs access to it. This access is granted when you enable the feature (*Figure 12.4*); at that time, a new resource named **StorageDataScanner** is created in your Azure subscription and given a system-assigned managed identity. When enabling malware scanning at the subscription level, a new Security Operator resource named **StorageAccounts/securityOperators/DefenderForStorageSecurityOperator** is created in your Azure subscription and assigned a system-managed identity.

When users upload files to a storage account, and the storage account is protected by Defender for Storage with the malware scanning feature turned on, the file (every file type, including ZIP files) will be scanned "in-memory" (scanned files are immediately deleted after the scan) and the malware scanning will be performed in the same region of the storage account that is getting scanned.

By default, the malware scanning is triggered by any operation that results in a **BlobCreated** event. It is important to emphasize that incremental operations, such as **AppendFile** (Azure Data Lake Storage Gen2) and **PutBlock** (Azure BlockBlob), do not trigger malware scanning. This is because malware scanning is only triggered when the operation is committed, for example, the **FlushWithClose** commits and finalizes the **AppendFile** operation, which, in this case, will trigger the scan. It is also important to emphasize that when this chapter was written, the following limitations were present in this feature:

- The malware scanning feature doesn't allow on-demand scanning; in other words, it is not possible to trigger a manual malware scan of files located in a storage account.
- A file size that can be scanned is limited to 2 GB.

Once the malware scanning identifies a malicious file, a security alert is triggered, and you will be able to visualize it in the Security Alert dashboard. *Figure 12.6* has an example of a summarized version of this type of alert:

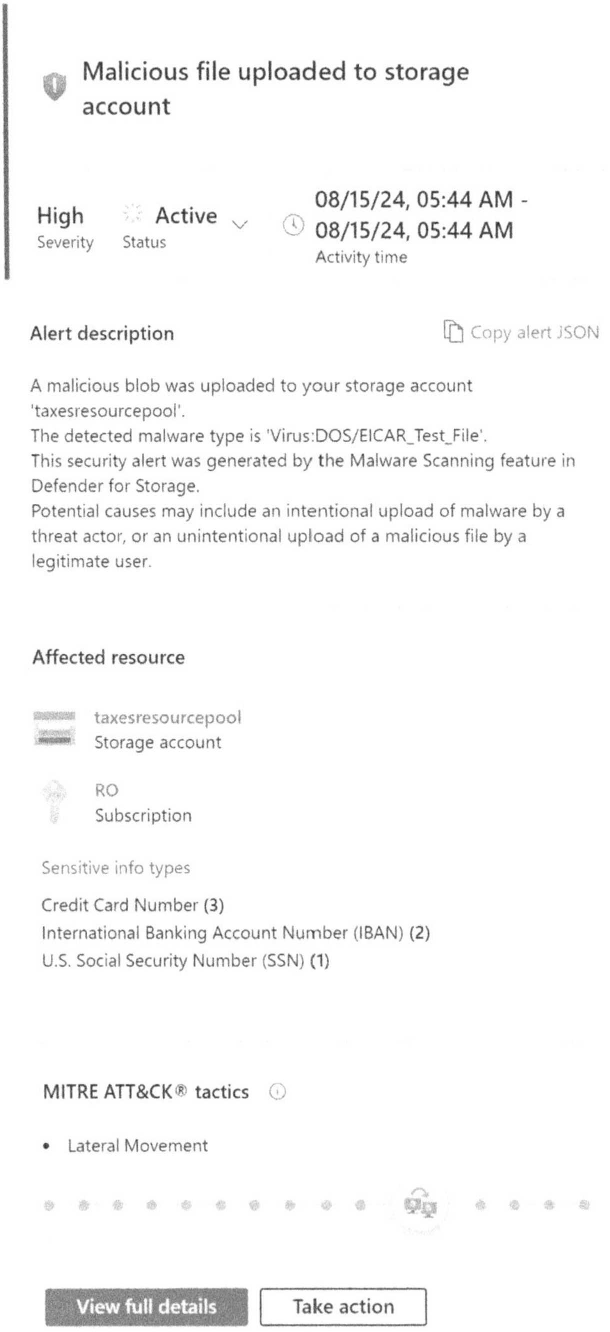

**Malicious file uploaded to storage account**

| High | Active ⌄ | 08/15/24, 05:44 AM - |
|------|----------|----------------------|
| Severity | Status | 08/15/24, 05:44 AM |
| | | Activity time |

**Alert description**　　　　　　　　　　　🗋 Copy alert JSON

A malicious blob was uploaded to your storage account
'taxesresourcepool'.
The detected malware type is 'Virus:DOS/EICAR_Test_File'.
This security alert was generated by the Malware Scanning feature in
Defender for Storage.
Potential causes may include an intentional upload of malware by a
threat actor, or an unintentional upload of a malicious file by a
legitimate user.

**Affected resource**

taxesresourcepool
Storage account

RO
Subscription

Sensitive info types

Credit Card Number (3)
International Banking Account Number (IBAN) (2)
U.S. Social Security Number (SSN) (1)

**MITRE ATT&CK® tactics** ⓘ

- Lateral Movement

| View full details | Take action |

*Figure 12.6: Malware scanning alert summary*

The summarized version of this alert, which is available once you click on the alert itself, in the Security Alert dashboard, has some very important information. While this was a test malware (from `https://www.eicar.org`) that was uploaded to a storage account, you can see in the lower part of the screen, under the **Sensitive information type** section, the type of data located in this storage account. This is super important information to help you prioritize the proper response for this attack; for example, if the sensitive info type data says that there is credit card information, this will likely increase the level of priority that you should give to this alert. This information is only available because of the sensitivity data discovery feature (*Figure 12.4*). Another important piece of information is the MITRE ATT&CK tactics, which show that the threat actor is using tactics that are mapped to lateral movement.

If you click the **View full details** button, you will see the additional information that is shown in *Figure 12.7*:

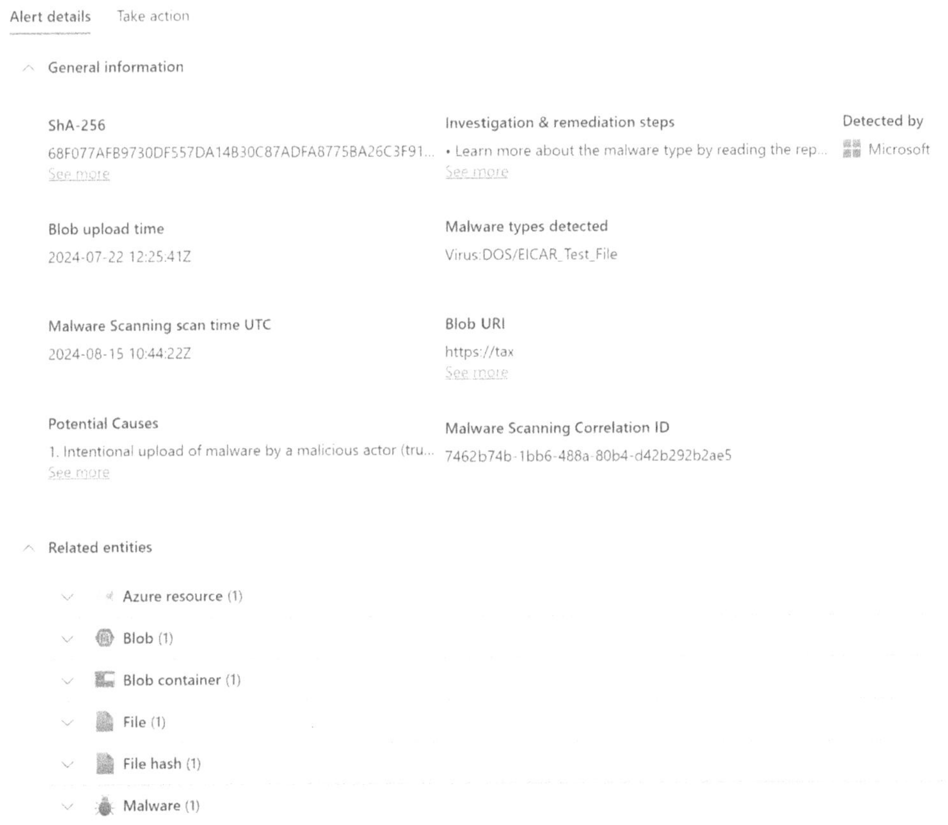

Alert details    Take action

⌄ General information

ShA-256
68F077AFB9730DF557DA14B30C87ADFA8775BA26C3F91...
See more

Investigation & remediation steps
• Learn more about the malware type by reading the rep...
See more

Detected by
▪▪ Microsoft

Blob upload time
2024-07-22 12:25.41Z

Malware types detected
Virus:DOS/EICAR_Test_File

Malware Scanning scan time UTC
2024-08-15 10:44:22Z

Blob URI
https://tax
See more

Potential Causes
1. Intentional upload of malware by a malicious actor (tru...
See more

Malware Scanning Correlation ID
7462b74b-1bb6-488a-80b4-d42b292b2ae5

⌄ Related entities

⌄  ◁ Azure resource (1)

⌄  ⊕ Blob (1)

⌄  ▥ Blob container (1)

⌄  ▤ File (1)

⌄  ▤ File hash (1)

⌄  ▦ Malware (1)

*Figure 12.7: Additional information about the alert (for better visualization, refer to https://packt.link/gbp/9781836204879)*

The lower part of this page has the details about the entities that were involved in this attack, including the malware information, which you can see when you expand the **Malware** dropdown on the page, which, in this scenario, will show information about the test malware, as shown in *Figure 12.8*:

| Name | Category | Files | Processes |
|---|---|---|---|
| Virus:DOS/EICAR_Test_File | Virus | EICAR receipt - compressed 3 times.zip | |

*Figure 12.8: Malware information (for better visualization, refer to https://packt.link/ gbp/9781836204879)*

While the Security Alert dashboard in Defender for Cloud will likely be the primary location where you will review the alerts related to malware scanning, the scan result will also surface at blob index tags.[3]

Just like any other alert in Defender for Cloud, you also have the **Take action** tab, which enables you to configure different ways to respond to this type of alert. For example, you can configure a Logic App[4] to be triggered to move a file with malware to another location or to delete an infected file. *Figure 12.9* has the example of the **Take action** tab for malware scanning alerts:

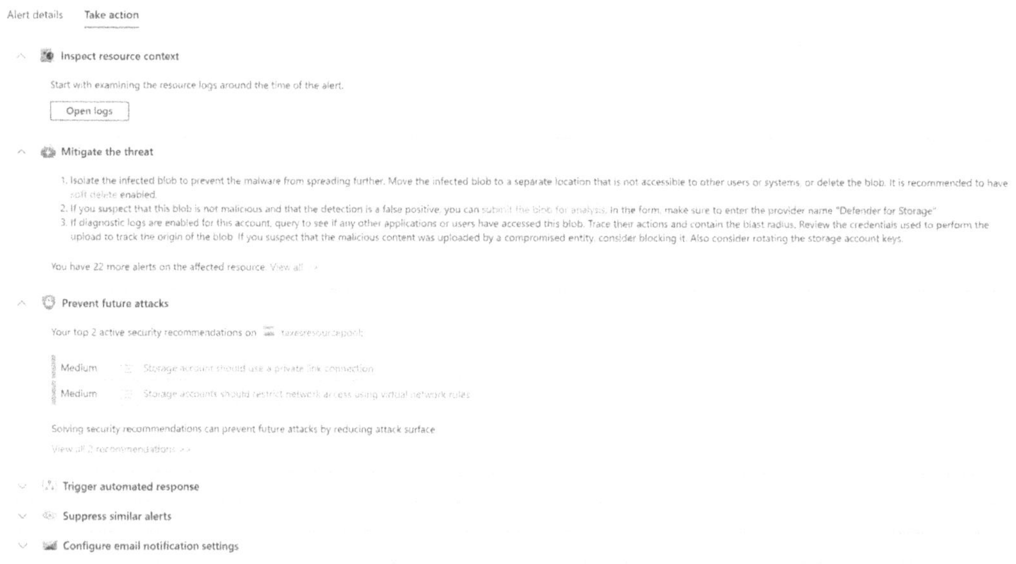

*Figure 12.9: Take action tab in a malware scanning alert (for better visualization, refer to https://packt.link/gbp/9781836204879)*

Some Defender for Cloud alerts will have the capability to examine resource logs around the time of the alert and the malware scanning alert is one of those. The first option on this page enables you to visualize the logs that were triggered around the time the alert was generated. This can help during your investigation and to better understand the circumstances that led to the generation of this alert. When you click **Open logs** on this page, you will be redirected to the **Activity log** blade, and the relevant operations will appear, as shown in *Figure 12.10*:

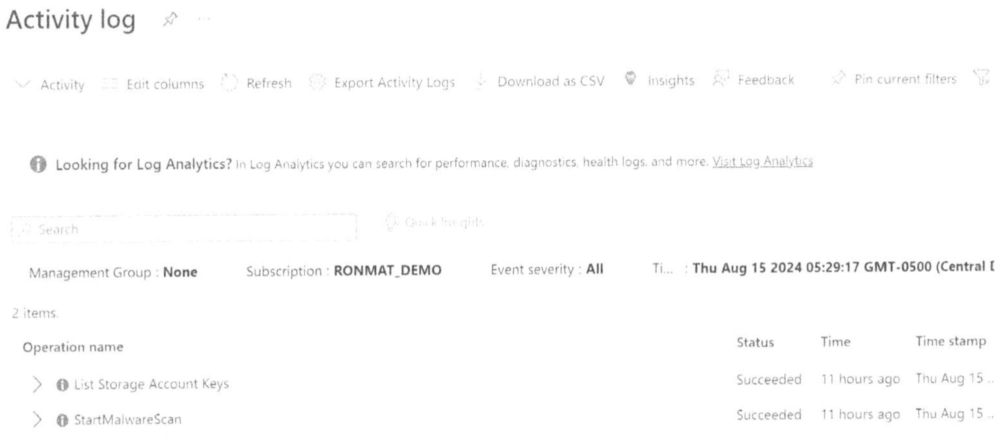

Activity log

☑ Activity ☰ Edit columns ⟳ Refresh ⚙ Export Activity Logs ⬇ Download as CSV ⚬ Insights ⚬ Feedback ⚬ Pin current filters ⚬

ⓘ **Looking for Log Analytics?** In Log Analytics you can search for performance, diagnostics, health logs, and more. Visit Log Analytics

| Search | Quick Insights |
| --- | --- |

Management Group : **None**     Subscription : **RONMAT_DEMO**     Event severity : **All**     Ti... : **Thu Aug 15 2024 05:29:17 GMT-0500 (Central [**

2 items.

| Operation name | Status | Time | Time stamp |
| --- | --- | --- | --- |
| > ⓘ List Storage Account Keys | Succeeded | 11 hours ago | Thu Aug 15 .. |
| > ⓘ StartMalwareScan | Succeeded | 11 hours ago | Thu Aug 15 .. |

*Figure 12.10: Azure Activity log (for better visualization, refer to https://packt.link/ gbp/9781836204879)*

From this blade, you can expand each operation to obtain more details about the operation, including the user who performed the action. However, storage is not the only location where data resides; sensitive data can also be located in databases, and Defender for Databases will add the necessary layer of protection for this type of workload.

# Defender for Databases

In 2023, SQL database attacks were a significant part of the broader cybersecurity landscape, with several high-profile incidents highlighting the growing threat. SQL injection attacks remain one of the most common methods used by attackers to exploit database vulnerabilities.[5] The FBI's **Internet Crime Complaint Center (IC3)**[6] reported that database-related attacks, including those targeting SQL databases, were a significant contributor to the $10.3 billion in losses reported in 2022, with trends continuing into 2023.

If your SQL Server is exposed to the internet using the default port, it becomes a prime target for brute-force attacks. Attackers often employ a technique that combines a list of commonly used usernames with a dictionary of frequently used passwords.

This approach allows them to guess weak credentials effectively, potentially granting them access to the database, often with high-level privileges.

Defender for Database has threat detections that can be divided into three major categories, as shown in the following table:

| Category | Type of alert |
|---|---|
| Brute-force attack | Potential brute-force |
| | Potential brute-force on a valid user |
| | Potential successful brute-force |
| Access anomalies | Access from an unusual location |
| | Access from a suspicious IP |
| | Data center anomaly |
| | Principal anomaly |
| | Domain anomaly |
| | Suspicious app |
| Query anomalies | Vulnerability to/potential SQL injection |
| | Anomalous amount/destination of data extraction |
| | Script execution with an anomalous external source |
| | Script execution under obfuscation |

You can find the full list of alerts available for Defender for Database according to the type of database using the list below:

- Alerts for SQL Database and Azure Synapse Analytics https://bit.ly/CNAPPBook52
- Alerts for open-source relational databases https://bit.ly/CNAPPBook53
- Alerts for Azure Cosmos DB https://bit.ly/CNAPPBook54

# Enabling Defender for Databases

Defender for Database is the overarching name of the plan, but you will enable the plan only for the type of databases that are available in your organization. The supported databases are as follows:

- Azure SQL Database: Single databases and Elastic pools, Azure SQL Managed Instances, and Azure Synapse Analytics
- SQL on Azure **virtual machines** (**VMs**), SQL servers on-premises, and Azure Arc-enabled SQL servers
- Open-source relational databases, such as Azure Database for PostgreSQL flexible servers, Azure Database for MySQL flexible servers, and non-basic tier Azure Database for PostgreSQL, MySQL, MariaDB single servers, and AWS RDS
- Azure Cosmos DB

You can enable Defender for Databases in the entire subscription, or at the database level only, which usually happens when you need to control costs and focus only on the most important databases (for example, databases where sensitive data is located). The disadvantage of this option is that when new databases are provisioned in the subscription, they will not be automatically protected by Defender for Databases, since it is disabled at the subscription level. For this reason, enabling the plan at the subscription level is the recommended option. This ensures that you have continuous protection for current and future databases.

To enable Defender for Database in the subscription level, open the Defender for Cloud dashboard, click **Environment settings** under the **Management** section, click the subscription that you want to enable Defender for Databases, click the **Defender plans** option, and click **On** beside the Databases plan, as shown in *Figure 12.11*:

*Figure 12.11: Enabling Defender for Databases (for better visualization, refer to https://packt. link/gbp/9781836204879)*

Next, click the **Select types >** hyperlink, and you will see the individual options to enable each database type, as shown in *Figure 12.12*:

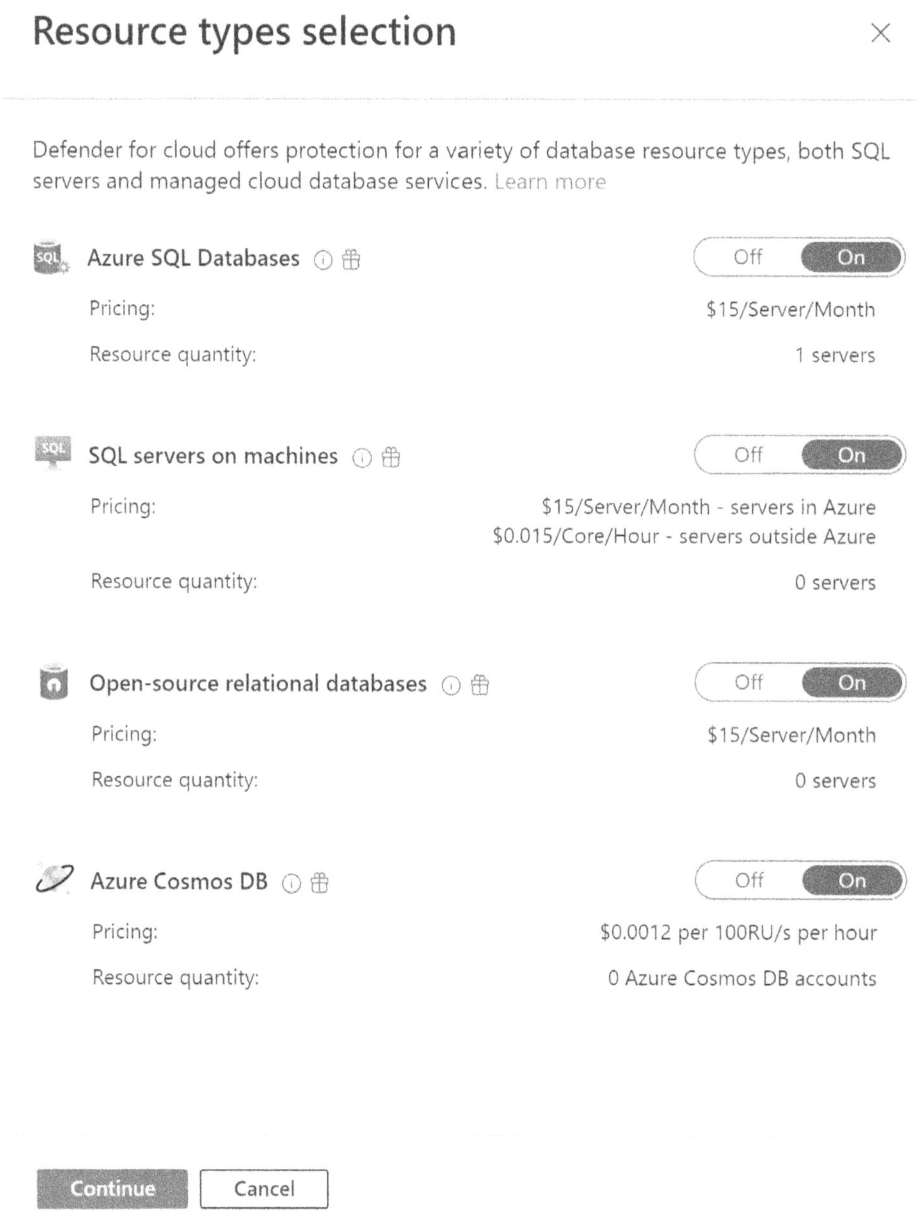

# Resource types selection                                      ✕

Defender for cloud offers protection for a variety of database resource types, both SQL servers and managed cloud database services. Learn more

**Azure SQL Databases** ⓘ ⊞                                    Off    **On**

Pricing:                                                        $15/Server/Month

Resource quantity:                                              1 servers

**SQL servers on machines** ⓘ ⊞                                Off    **On**

Pricing:                                    $15/Server/Month - servers in Azure
                                       $0.015/Core/Hour - servers outside Azure

Resource quantity:                                             0 servers

**Open-source relational databases** ⓘ ⊞                       Off    **On**

Pricing:                                                        $15/Server/Month

Resource quantity:                                             0 servers

**Azure Cosmos DB** ⓘ ⊞                                        Off    **On**

Pricing:                                              $0.0012 per 100RU/s per hour

Resource quantity:                                   0 Azure Cosmos DB accounts

[ Continue ]   [ Cancel ]

*Figure 12.12: Enabling individual database according to its type*

You can either leave all options selected or select only the ones that are applicable to your current needs. The advantage of leaving all enabled is that you are ensuring that in the future if any one of those types of databases are provisioned in the subscription, they will be automatically protected. After making the selection, click **Continue**, and then click **Save** to commit the changes.

While all databases that are based on **Platform as a Service (PaaS)** do not require any additional steps to enable it, the SQL Server on machines plan does require additional steps. When you enable the SQL Server component of the Defender for Databases plan, the auto-provisioning process is automatically triggered. This process installs and configures all necessary components for the plan, including the **Azure Monitor Agent (AMA)**, SQL IaaS extension, and Defender for SQL extensions. It also handles workspace configuration, data collection rules, identity setup (if required), and the SQL IaaS extension. Keep in mind that the SQL Server on machines is not only for on-premises and can also be an SQL Server in a VM running in another cloud provider, such as AWS or GCP. For more information on how to configure Defender for SQL Server on Machines, visit https://bit.ly/CNAPPBook55.

## Vulnerability assessment

SQL vulnerability assessment is a feature that gives you clear insight into your database's security status. It not only identifies security issues but also offers practical steps to address them and strengthen your database's defenses. Vulnerability assessment uses a knowledge base of rules to identify security vulnerabilities and highlight deviations from best practices, such as misconfigurations, excessive permissions, and unprotected sensitive data.

This feature becomes particularly useful for monitoring a dynamic database environment, where tracking changes can be challenging, helping you to continually improve your SQL security posture. Vulnerability assessment supports Azure SQL Database, Azure SQL Managed Instance, and Azure Synapse Analytics.

Vulnerability assessment is not enabled by default, and if you did not enable it manually, you will see a security recommendation that advises you to enable this feature, as shown in *Figure 12.13*:

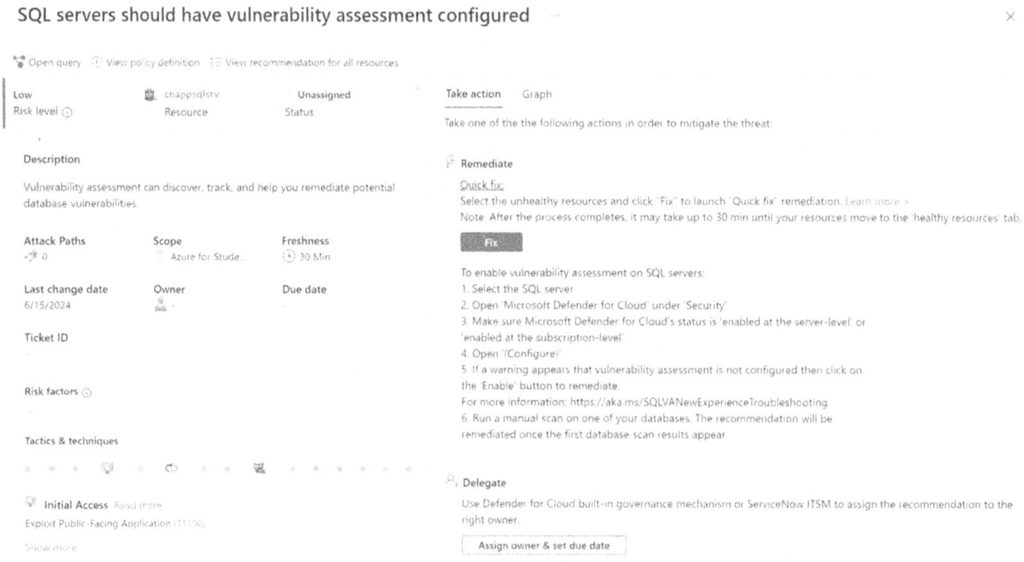

*Figure 12.13: Enabling SQL vulnerability assessment (for better visualization, refer to https://packt.link/gbp/9781836204879)*

You can easily enable this feature by using the **Fix** button shown in *Figure 12.13*; however, it is important to be aware of the different types of configurations before enabling it. Vulnerability assessment can be configured using the express configuration, which is the default enablement method, and allows you to configure vulnerability assessment without dependency on external storage to store baseline and scan result data. The other option is to use the classic configuration (also known as legacy configuration), which does require you to manage an Azure storage account to store baseline and scan result data. Aside from this difference (one requires storage, and the other does not), there are a few differences between the two configuration modes that you can review in the table described in this article: `https://bit.ly/CNAPPBook56`.

Now that you understand the two different types of vulnerability assessment configuration, you can click the **Fix** button in the recommendation, and the **Fixing resources** blade appears, as shown in *Figure 12.14*:

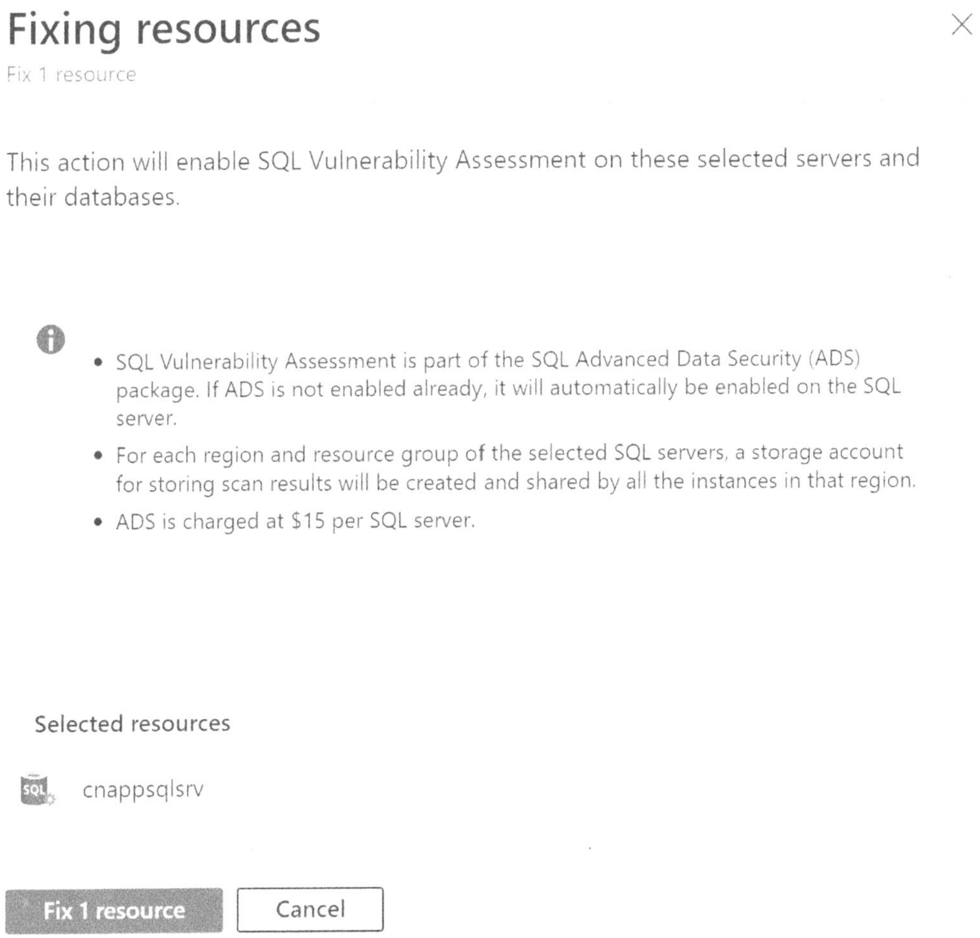

# Fixing resources      ✕

Fix 1 resource

This action will enable SQL Vulnerability Assessment on these selected servers and their databases.

🛈
- SQL Vulnerability Assessment is part of the SQL Advanced Data Security (ADS) package. If ADS is not enabled already, it will automatically be enabled on the SQL server.
- For each region and resource group of the selected SQL servers, a storage account for storing scan results will be created and shared by all the instances in that region.
- ADS is charged at $15 per SQL server.

Selected resources

🗄 cnappsqlsrv

| Fix 1 resource |    Cancel    |
|---|---|

*Figure 12.14: Using the Fix button to enable vulnerability assessment*

Notice that the lower part of the screen will specify which databases will be affected by this change, and to confirm, click the **Fix [number of databases selected] resource** button. Vulnerability scans will run on a weekly basis, and the results will surface in a new recommendation called **SQL databases should have vulnerability findings resolved**.

*Figure 12.15* has an example of this recommendation:

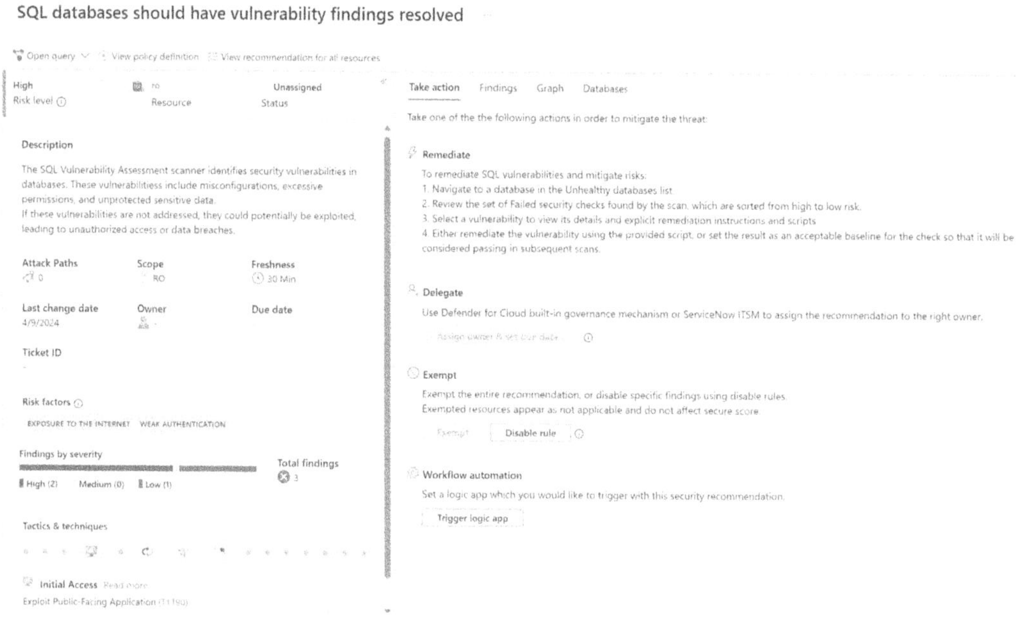

SQL databases should have vulnerability findings resolved

*Figure 12.15: SQL database vulnerability findings recommendation (for better visualization, refer to https://packt.link/gbp/9781836204879)*

Just like any other recommendation in Defender for Cloud, you have the details on the left part of the screen, and on the right, you have the potential actions that you can take. Click the **Findings** tab to see the complete list of vulnerabilities that were found in this database. *Figure 12.16* has an example of this:

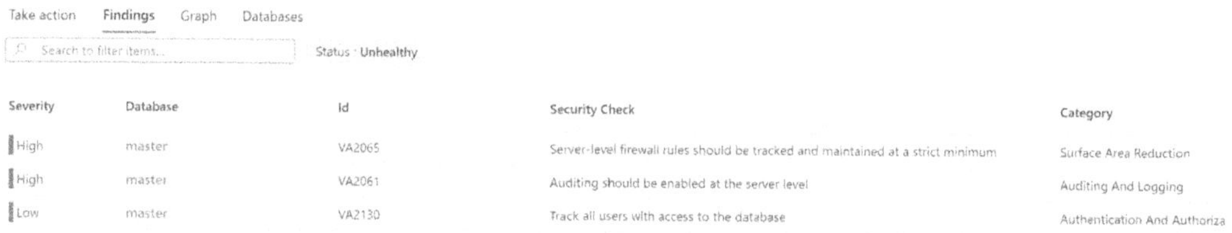

*Figure 12.16: List of findings (for better visualization, refer to https://packt.link/gbp/9781836204879)*

The list is organized by severity, with the database name, the ID of the security check,[7] the name of the security check, and the category. To see more details about each vulnerability, you can simply click on the vulnerability, and another blade appears, as shown in *Figure 12.17*:

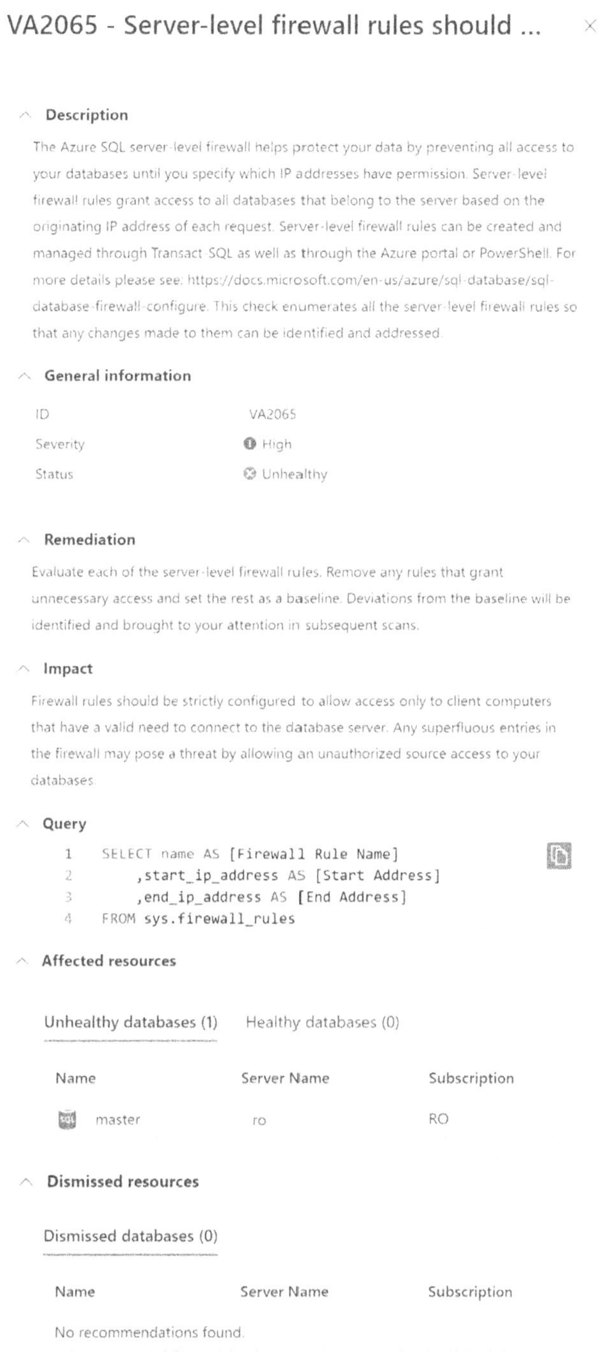

## VA2065 - Server-level firewall rules should ...   ✕

### ⌃ Description

The Azure SQL server-level firewall helps protect your data by preventing all access to your databases until you specify which IP addresses have permission. Server-level firewall rules grant access to all databases that belong to the server based on the originating IP address of each request. Server-level firewall rules can be created and managed through Transact-SQL as well as through the Azure portal or PowerShell. For more details please see: https://docs.microsoft.com/en-us/azure/sql-database/sql-database-firewall-configure. This check enumerates all the server-level firewall rules so that any changes made to them can be identified and addressed.

### ⌃ General information

| | |
|---|---|
| ID | VA2065 |
| Severity | ❶ High |
| Status | ⊗ Unhealthy |

### ⌃ Remediation

Evaluate each of the server-level firewall rules. Remove any rules that grant unnecessary access and set the rest as a baseline. Deviations from the baseline will be identified and brought to your attention in subsequent scans.

### ⌃ Impact

Firewall rules should be strictly configured to allow access only to client computers that have a valid need to connect to the database server. Any superfluous entries in the firewall may pose a threat by allowing an unauthorized source access to your databases.

### ⌃ Query

```
1    SELECT name AS [Firewall Rule Name]
2         ,start_ip_address AS [Start Address]
3         ,end_ip_address AS [End Address]
4    FROM sys.firewall_rules
```

### ⌃ Affected resources

**Unhealthy databases (1)**     Healthy databases (0)

| Name | Server Name | Subscription |
|---|---|---|
| 🗄 master | ro | RO |

### ⌃ Dismissed resources

Dismissed databases (0)

| Name | Server Name | Subscription |
|---|---|---|

No recommendations found.

*Figure 12.17: More details about a vulnerability (for better visualization, refer to https://packt.link/gbp/9781836204879)*

This page has a very comprehensive list of details about the vulnerability, and how to remediate it. If you are the cloud administrator, most likely this will be only good information to have, because the persona that will address this vulnerability is the database administrator. You can create a governance rule to assign this type of recommendation to the appropriate workload owner and educate this person on how to use this page to remediate the vulnerability.

## Summary

In this chapter, you learned about the importance of protecting the data located in storage and databases. You learned about the most common threat vectors for storage and databases and how Defender for Storage and Defender for Databases can help protect this type of workload. You learned how to enable Defender for Storage in your subscription, and how malware scanning can be used to protect your storage accounts from getting compromised with malware. You learned how to enable Defender for Databases, the different types of databases that are supported by this plan, and how to use the vulnerability assessment feature to improve your database security posture.

In the next chapter, you will learn how to implement Defender for APIs.

## Notes

1. You can read the entire report at https://bit.ly/CNAPPBook46.
2. To view the complete list of alerts that Defender for Azure Storage has, visit https://bit.ly/CNAPPBook47.
3. To learn how to use blob index tags to find data, visit https://bit.ly/CNAPPBook48.
4. To learn more about how to create an automated response for malware scanning alerts, visit https://bit.ly/CNAPPBook49.
5. This statistic took into consideration the data provided at https://bit.ly/CNAPPBook50.
6. More information at https://bit.ly/CNAPPBook51.
7. You can see the full list of vulnerability checks for SQL at https://bit.ly/CNAPPBook57.

## Additional resources

- Enable vulnerability assessment manually without using the recommendations: https://bit.ly/CNAPPBook58
- Learn more about open-source relational database support in AWS by watching this episode of *Defender for Cloud in the Field,* where the author, Yuri Diogenes, interviews the PM who manages this feature: https://bit.ly/CNAPPBook59

# Leave a Review!

Thank you for purchasing this book from Packt Publishing—we hope you enjoy it! Your feedback is invaluable and helps us improve and grow. Once you've completed reading it, please take a moment to leave an Amazon review; it will only take a minute, but it makes a big diff erence for readers like you. Scan the QR or visit the link to receive a free ebook of your choice.

https://packt.link/NzOWQ

# 13

# Protecting APIs

APIs are everywhere. Whether you're booking a flight online, checking the weather on your search engine, or purchasing groceries through a mobile app, an **Application Program Interface (API)** is likely driving those digital interactions. Despite all that, current API security measures are often insufficient. The complexity of mobile access, microservice design patterns, and hybrid on-premises/cloud environments make API security challenging, as there is rarely a single point where security can be enforced. According to the **State of API Security Report 2024** from Salt[1], API security incidents more than doubled between 2023 and 2024.

While tools like application gateways, **Web Application Firewalls (WAFs)**, and API gateways provide traffic control and rule-based detection, they often lack the behavioral insights needed to fully secure APIs. This gap leaves API security vulnerable to threats related to business logic abuse, such as inadequate authorization and authentication, and the exposure of excessive data. In 2022, Twitter (currently X) revealed that millions of users' personal information was leaked[2] due to an exploitation of API vulnerability. This could have been prevented with proper API security hygiene and implementation of proactive measures to identify these vulnerabilities, mainly for the APIs that are public-facing.

As a complete CNAPP solution, Defender for Cloud also has capabilities to protect APIs with the Defender for APIs plan. Defender for APIs provides security posture management and threat detection for APIs.

This chapter covers:

- Preparing the environment
- Enabling Defender for APIs
- Operationalizing Defender for APIs

# Preparing the environment

At the time this chapter was written, Defender for APIs protects APIs that are published using Azure API Management[3] (the Premium, Standard, Basic, and Developer tiers of Azure API Management). From the APIs that are published in Azure API Management, Defender for APIs will discover and analyze only REST APIs. Before enabling Defender for APIs, it is important to ensure that you have at least one API Management instance in your Azure subscription.

Defender for APIs can be enabled at the subscription level, and to enable this plan, you need **API Management Service Contributor** role access and Security Admin role access in the subscription in which the plan will be enabled. Another way to enable Defender for APIs is when you are creating a new Azure API Management instance. During the setup process, you will have an option (second checkbox in *Figure 13.1*) to enable Defender for APIs in the subscription.

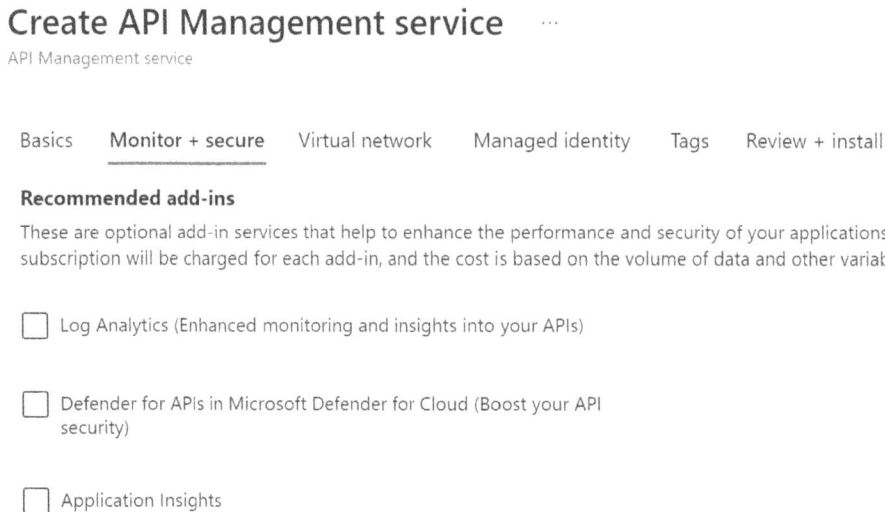

## Create API Management service

API Management service

Basics    **Monitor + secure**    Virtual network    Managed identity    Tags    Review + install

**Recommended add-ins**

These are optional add-in services that help to enhance the performance and security of your applications. Your Azure subscription will be charged for each add-in, and the cost is based on the volume of data and other variables.

☐ Log Analytics (Enhanced monitoring and insights into your APIs)

☐ Defender for APIs in Microsoft Defender for Cloud (Boost your API security)

☐ Application Insights

*Figure 13.1: Enabling Defender for APIs during Azure API Management creation (for better visualization, refer to https://packt.link/gbp/9781836204879)*

Defender for APIs has different pricing plans that will vary according to the API Management traffic profile. Therefore, before enabling the plan, it is important to review your historical Azure API Management API traffic usage. To do that, open the API Management service in the Azure portal, click the API Management instance for which you want to gather this data, and then, under the **Monitoring** section, click **Metrics**, and make the adjustments according to *Figure 13.2*:

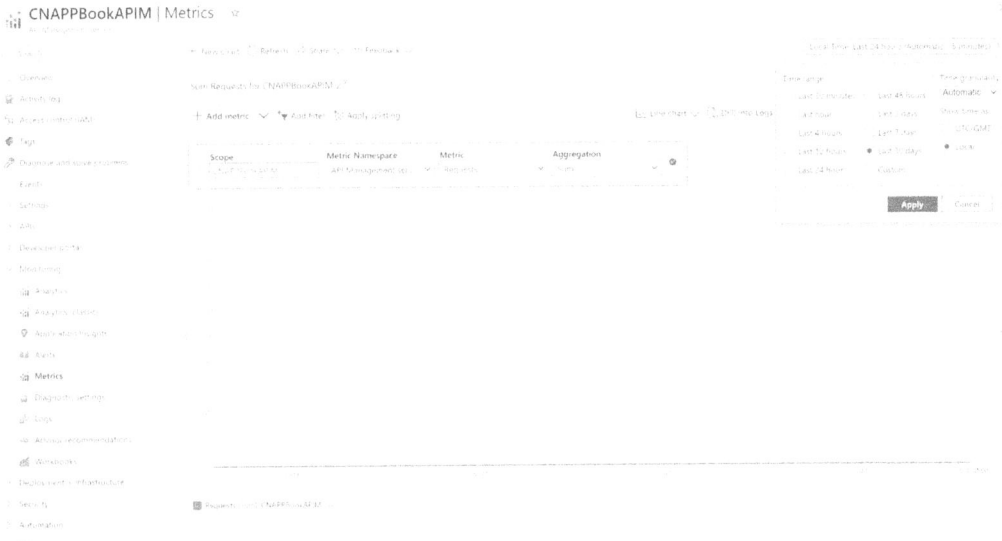

*Figure 13.2: Collecting historical API Management data (for better visualization, refer to https://packt.link/gbp/9781836204879)*

Notice that the metric in use is **Requests**, the aggregation is **Sum**, and the time range is 30 days. Once you apply these settings, the number of requests will appear at the bottom of the chart. You can use this number to guide you in selecting the best Defender for APIs plan. Keep in mind that Defender for APIs monitors any calls that make it to the Azure API Management instance, and they will be counted toward billing. Defender for APIs monitors and bills all API requests/response status codes, including unauthorized/blocked call attempts, primarily since failed calls often represent a security event, for example, a brute force attack where an attacker uses different API keys/token until a successful call is generated.

## Network architecture

While many organizations may already have invested in edge-based security controls such as firewalls and WAF, Defender for APIs adds another layer of protection by protecting against threats related to business logic abuse, broken authentication, and other OWASP Top 10 API security risks (see https://owasp.org/API-Security for more information). With that in mind, the network architecture with Defender for APIs will look like the diagram shown in *Figure 13.3*:

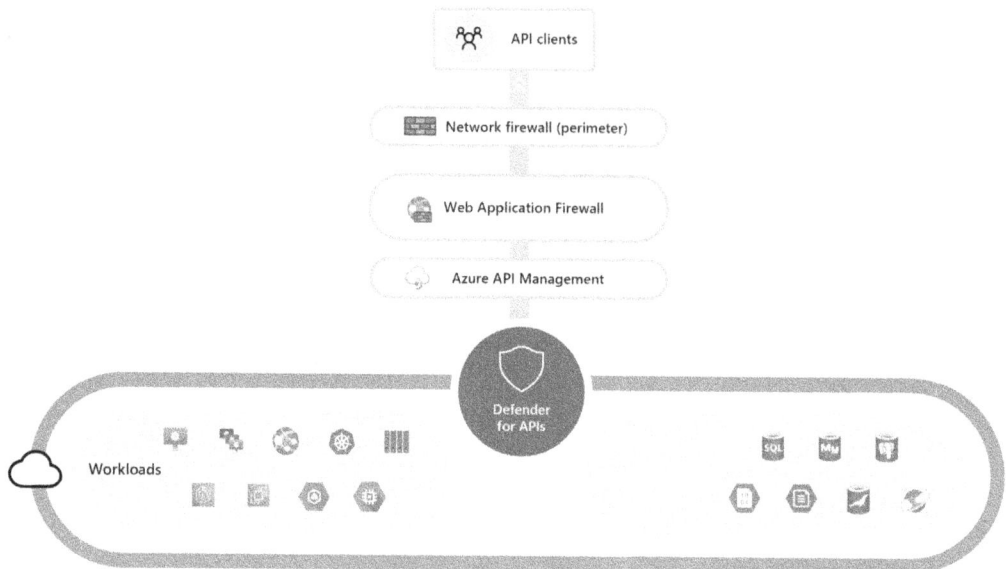

*Figure 13.3: Defender for APIs placement (for better visualization, refer to https://packt.link/ gbp/9781836204879)*

The addition of Defender for APIs in this architecture helps not only from the security posture perspective but also from the threat detection standpoint. For example, let's say a threat actor begins probing an application and gathering information about the target. During this process, the threat actor discovers that the API endpoints are accessible via the internet, without proper authentication, and sensitive data is in transit. This threat actor can exploit this vulnerability, using tactics such as **Broken Object Level Authorization**[4] to gain unauthorized access to the data. Defender for APIs would have prevented this attack by triggering a security recommendation showing that you have exposed APIs without proper authentication and with sensitive data. If you didn't remediate the recommendation, Defender for APIs has threat detection that will trigger an alert when suspicious activity occurs. To see the entire list of alerts that can be generated by Defender for APIs, see this article: `https://bit.ly/CNAPPBook62`. Now that you understand the rationale behind protecting APIs, let's see how to enable Defender for APIs in your Azure subscription.

# Enabling Defender for APIs

To enable Defender for APIs in the subscription, open the Defender for Cloud dashboard, click **Environment settings** in the **Management** section, click the subscription for which you want to enable Defender for APIs, click the **Defender plans** option, and click **On** beside the APIs plan, as shown in *Figure 13.4*:

APIs                    No plan selected                         1 Azure API Management services                              Off    On

Figure 13.4: Enabling Defender for APIs (for better visualization, refer to https://packt.link/gbp/9781836204879)

As soon as you switch the toggle to **On**, the **Plan selection** blade appears, as shown in *Figure 13.5*:

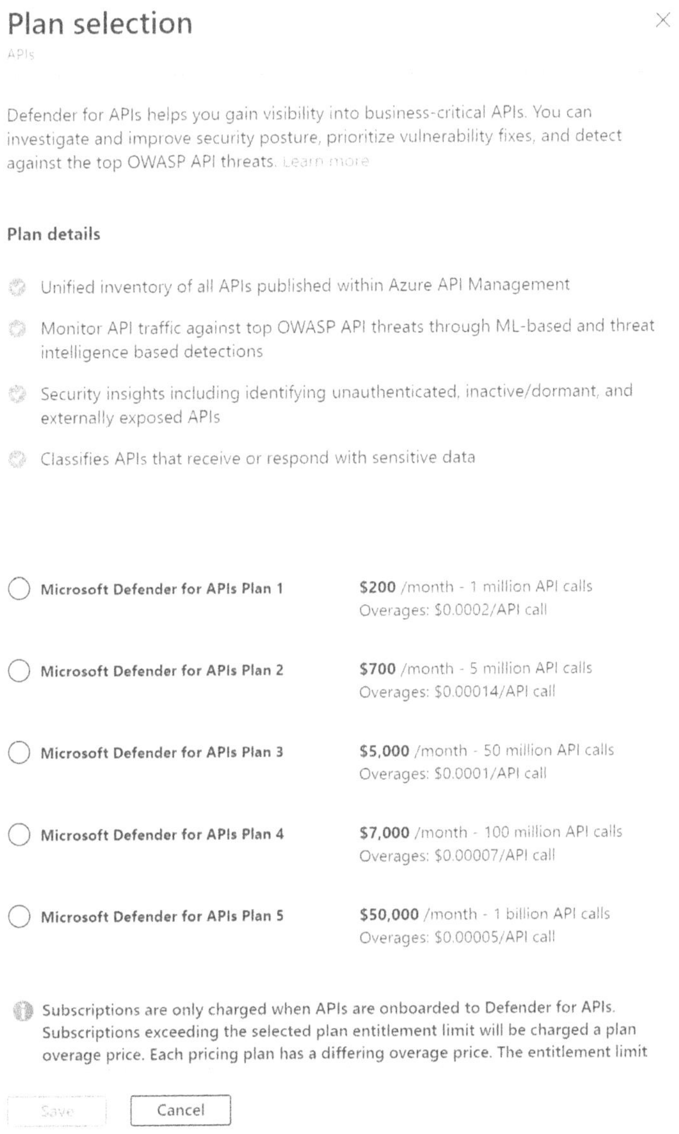

# Plan selection                                                                                        ✕

APIs

Defender for APIs helps you gain visibility into business-critical APIs. You can investigate and improve security posture, prioritize vulnerability fixes, and detect against the top OWASP API threats. Learn more

### Plan details

- Unified inventory of all APIs published within Azure API Management
- Monitor API traffic against top OWASP API threats through ML-based and threat intelligence based detections
- Security insights including identifying unauthenticated, inactive/dormant, and externally exposed APIs
- Classifies APIs that receive or respond with sensitive data

| ○ Microsoft Defender for APIs Plan 1 | **$200** /month - 1 million API calls<br>Overages: $0.0002/API call |
| ○ Microsoft Defender for APIs Plan 2 | **$700** /month - 5 million API calls<br>Overages: $0.00014/API call |
| ○ Microsoft Defender for APIs Plan 3 | **$5,000** /month - 50 million API calls<br>Overages: $0.0001/API call |
| ○ Microsoft Defender for APIs Plan 4 | **$7,000** /month - 100 million API calls<br>Overages: $0.00007/API call |
| ○ Microsoft Defender for APIs Plan 5 | **$50,000** /month - 1 billion API calls<br>Overages: $0.00005/API call |

Subscriptions are only charged when APIs are onboarded to Defender for APIs. Subscriptions exceeding the selected plan entitlement limit will be charged a plan overage price. Each pricing plan has a differing overage price. The entitlement limit

Save          Cancel

Figure 13.5: Selecting the appropriate Defender for APIs plan (for better visualization, refer to https://packt.link/gbp/9781836204879)

By now, you should already know which plan needs to be selected, since you already calculated the estimation traffic volume for API Management. Select the appropriate plan, click **Save**, and then click **Save** again to commit the changes. Notice that after saving, you will receive a warning (**Action required**), as shown in *Figure 13.6*:

*Figure 13.6: Action required warning (for better visualization, refer to https://packt.link/ gbp/9781836204879)*

This warning appears because to complete the Defender for APIs onboarding process, you need to access the security recommendation **Azure API Management APIs should be onboarded to Defender for APIs** and then select which individual APIs you want to protect. Keep in mind that if you don't do this step (remediate the security recommendation), the APIs won't be protected using Defender for APIs, even though the plan is enabled. An example of this recommendation is shown in *Figure 13.7*:

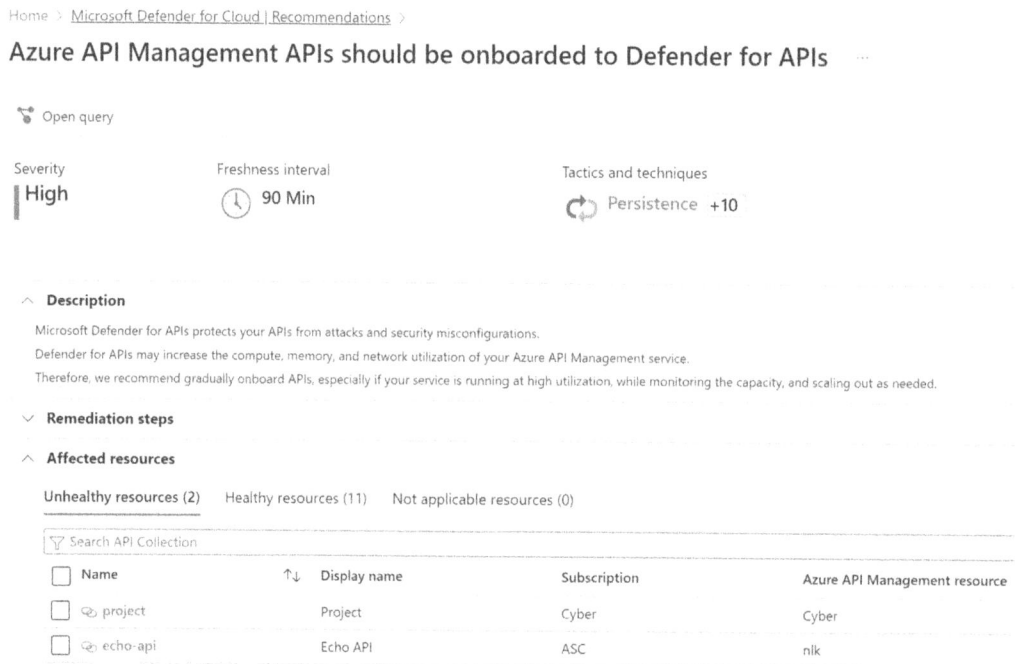

*Figure 13.7: Recommendation to onboard Defender for APIs (for better visualization, refer to https://packt.link/gbp/9781836204879)*

To remediate this recommendation, you need to select the API Management resource that you want to monitor and click the **Fix** button. Once you do that, the **Fixing resources** blade appears, as shown in *Figure 13.8*:

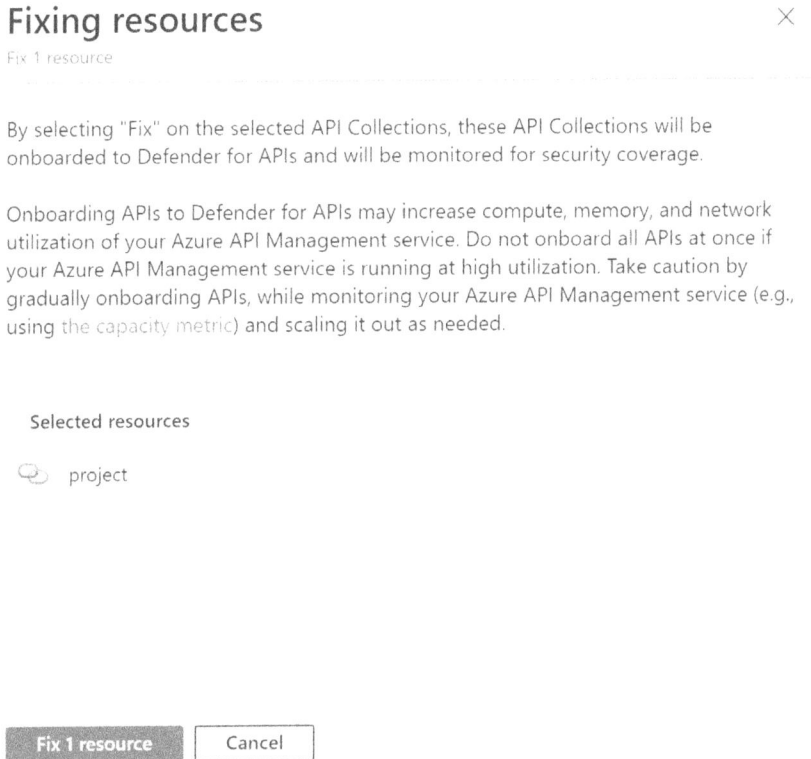

## Fixing resources

Fix 1 resource

By selecting "Fix" on the selected API Collections, these API Collections will be onboarded to Defender for APIs and will be monitored for security coverage.

Onboarding APIs to Defender for APIs may increase compute, memory, and network utilization of your Azure API Management service. Do not onboard all APIs at once if your Azure API Management service is running at high utilization. Take caution by gradually onboarding APIs, while monitoring your Azure API Management service (e.g., using the capacity metric) and scaling it out as needed.

Selected resources

project

Fix 1 resource     Cancel

*Figure 13.8: Remediating the recommendation with the Fix button*

To complete the onboarding process, click the **Fix [number of resources] resource** button. Once the onboarding process is finished, you will see a notification confirming the remediation, as shown in *Figure 13.9*:

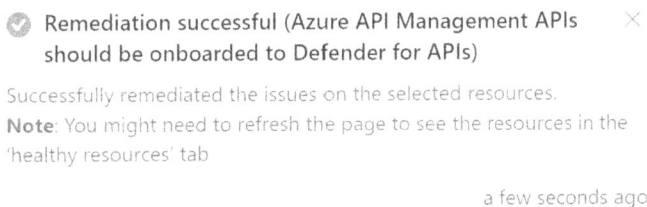

Remediation successful (Azure API Management APIs should be onboarded to Defender for APIs)

Successfully remediated the issues on the selected resources.
**Note**: You might need to refresh the page to see the resources in the 'healthy resources' tab

a few seconds ago

*Figure 13.9: Remediation notification*

The Azure policy to onboard Defender for APIs at scale was deprecated, but you can use a script created by the product team to onboard at scale. You can download this script from https://bit.ly/CNAPPBook64.

# Operationalizing Defender for APIs

Once Defender for APIs is enabled and operational, it will perform an API inventory and then start accessing the security posture of the APIs that were onboarded to generate security recommendations. Defender for APIs will check if API endpoints published within Azure API Management are enforcing authentication to help minimize security risk, among other API security best practices. You can view all Defender for APIs security recommendations in this article: https://bit.ly/CNAPPBook63. The approach that you will use to elevate the security posture of your APIs is the same as any other workload; in other words, you will use the risk-based approach and address the ones that are more critical to your environment.

Defender for APIs will also add insights in Cloud Security Explorer, and you can use these insights to do proactive hunting, as you learned in *Chapter 9*. When you open Cloud Security Explorer and create a new query, you will be able to search for APIs and select the different parameters, as shown in *Figure 13.10*:

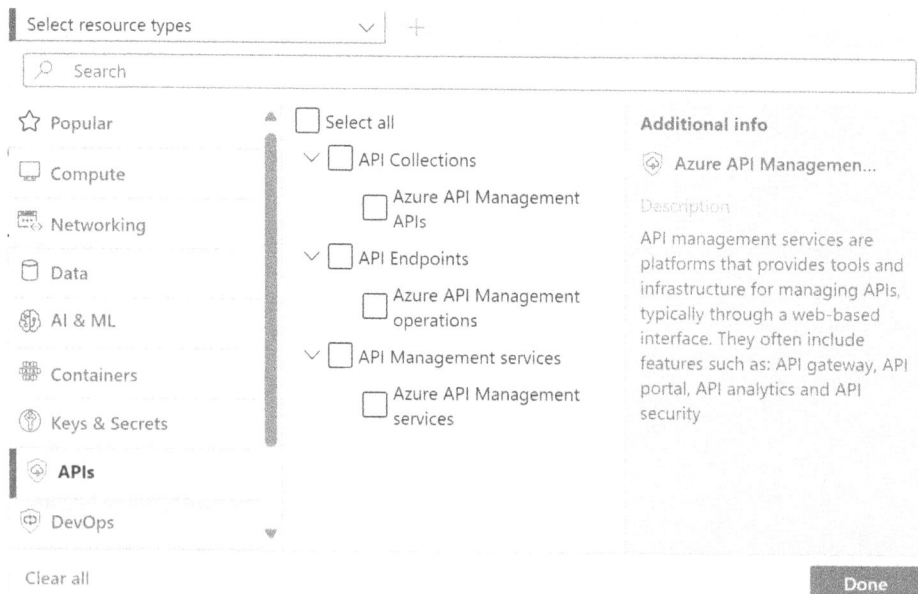

*Figure 13.10: API options available in Cloud Security Explorer*

You can also leverage the templates shown in *Figure 13.11* if you want to perform a quick query without having to think about all the parameters that are necessary to obtain this information:

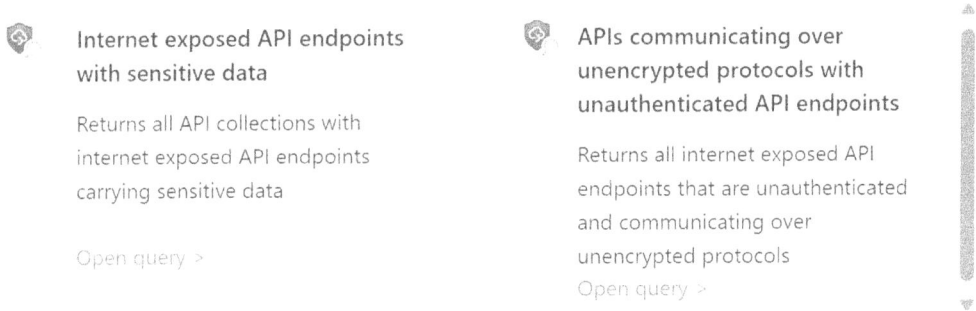

Internet exposed API endpoints with sensitive data

Returns all API collections with internet exposed API endpoints carrying sensitive data

Open query >

APIs communicating over unencrypted protocols with unauthenticated API endpoints

Returns all internet exposed API endpoints that are unauthenticated and communicating over unencrypted protocols

Open query >

*Figure 13.11: API query templates*

## Managing APIs

After the APIs are onboarded and Defender for APIs is monitoring them, you can use the **API Security** page to visualize more details about these APIs. To access the **API Security** page, click **Workload protection** under **Cloud Security** in the Defender for Cloud dashboard. There, you will see a tile called **API Security**, located at the bottom of the page, as shown in *Figure 13.12*:

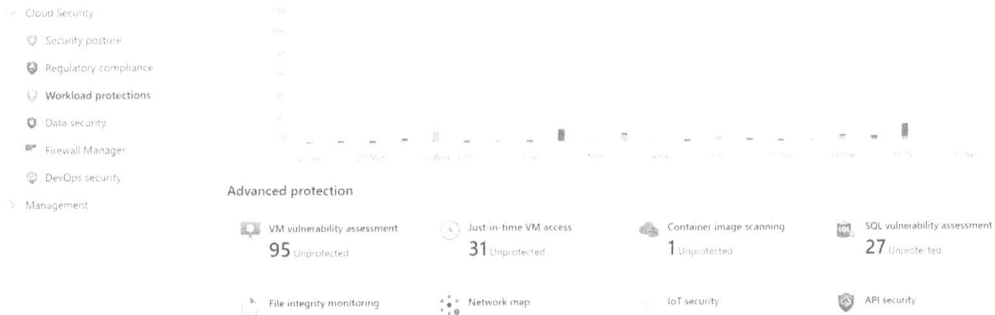

*Figure 13.12: Accessing the API security page (for better visualization, refer to https://packt.link/gbp/9781836204879)*

After clicking on this option, the **API Security** page appears with two main elements. At the top, there is a quick summary of the number of API collections, the number of API endpoints, and the number of API Management services.

At the bottom is the list of collections, with more details about each of them, as shown in *Figure 13.13*:

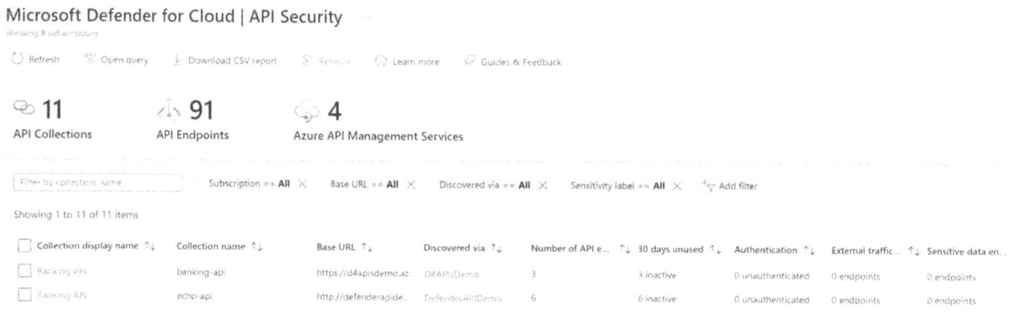

*Figure 13.13: API Security page (for better visualization, refer to https://packt.link/gbp/9781836204879)*

On this page, not only you can see the list of API collections but you can also offboard an API collection if you don't want to protect that collection anymore. All you need to do is select the API collection and click the **Remove** button. Keep in mind that by doing this, Defender for APIs will not monitor this API collection anymore.

This visualization also allows you to quickly identify the number of inactive APIs, by looking at the 30 days unused column. If you need to obtain even more details about the API collection, you can simply click on the collection that you want to see. *Figure 13.14* shows an example of the properties of an API collection:

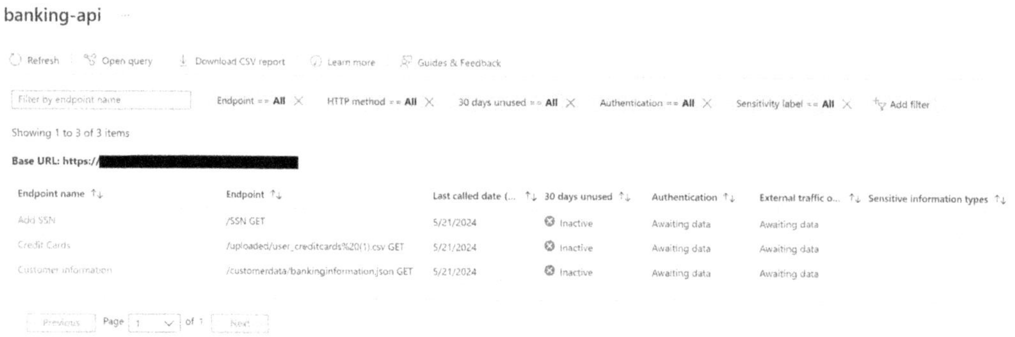

*Figure 13.14: Properties of an API collection (for better visualization, refer to https://packt. link/gbp/9781836204879)*

If you need to obtain insights from the endpoint itself, such as the recommendations that are linked to this endpoint or the potential alerts that were generated for this endpoint, you can click on the endpoint itself, and the **Resource health** page will appear, as shown in *Figure 13.15*:

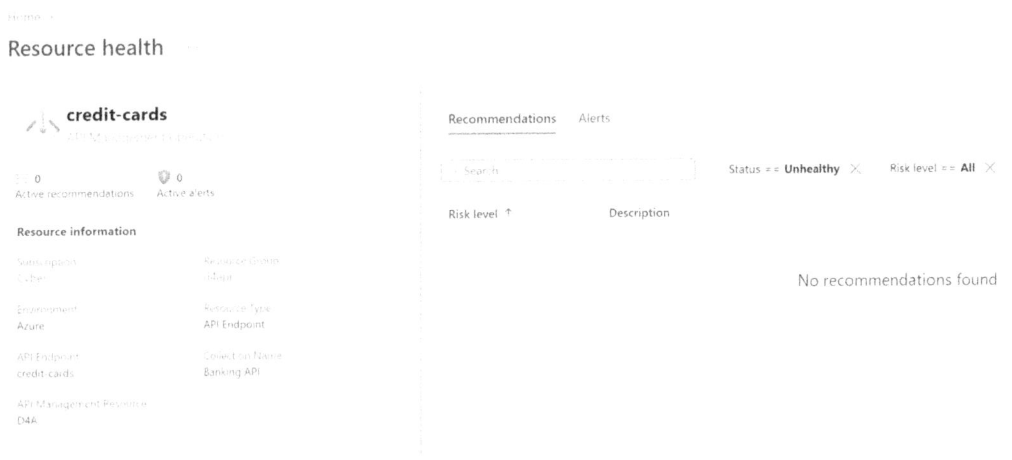

Resource health

*Figure 13.15: Resource health for the endpoint (for better visualization, refer to https://packt. link/gbp/9781836204879)*

On this page, you can see more details about the resource on the left side, and on the right side, you can access recommendations and alerts.

## Summary

In this chapter, you learned about the importance of protecting APIs, the necessary steps to prepare the environment before enabling Defender for APIs, and where Defender for APIs sits within the network architecture in a multilayered protection approach. You also learned how to enable Defender for APIs in your Azure subscription, and how to operationalize Defender for APIs. You learned about the insights that are added by Defender for APIs that can be used in Cloud Security Explorer, and the templates available to perform proactive hunting. Lastly, you learned how to manage the onboarded APIs, including how to offboard APIs.

In the next chapter, you will learn how to protect the Azure service layer.

## Notes

1. You can read the entire report at `https://bit.ly/CNAPPBook60`.

2. To learn more about this case, read `https://bit.ly/CNAPPBook79`.

3. You can read more about Azure API Management at `https://bit.ly/CNAPPBook61`.

4. More info about it at `https://owasp.org/API-Security/editions/2023/en/0xa1-broken-object-level-authorization`.

# Additional resources

- Watch this episode of *Defender for Cloud in the Field* with author Yuri Diogenes interviewing a member of the Defender for APIs team where he discusses more about the functionalities that were released at Ignite: `https://bit.ly/CNAPPBook65`
- Watch a presentation delivered by a member of the Defender for APIs team: `https://bit.ly/CNAPPBook66`

# Join our community on Discord

Read this book alongside other users. Ask questions, provide solutions to other readers, and much more.

Scan the QR code or visit the link to join the community.

`https://packt.link/SecNet`

# 14

# Protecting Service Layer

The service layer in Azure refers to the part of an application architecture that provides the business logic and core functionalities. This layer is responsible for processing requests from the presentation layer (such as web frontends or mobile apps) and interacting with the data layer (such as databases). One key element of this architecture is **Azure Resource Manager** (**ARM**), which is the deployment and management service for Azure. It provides a management layer that allows you to create, update, and delete resources in your Azure account.

Threat actors are aware of the importance of the Azure service layer, and they know that by exploiting ARM, they can perform various operations. Threat actors can target misconfigurations, leverage excessive permissions, or take advantage of vulnerabilities within the ARM framework and its components.

**Platform-as-a-service** (**PaaS**) services such as Azure App Service and Azure Key Vault are also common targets for threat actors, hence the importance of having a security layer of protection that can identify malicious and suspicious operations against these services.

Defender for Cloud has plans available to improve the Azure service layer security posture as well as to trigger different types of alerts that will be generated according to the analytics available for each one of these plans.

This chapter covers:

- Defender for Resource Manager
- Defender for App Service
- Defender for Key Vault

# Defender for Resource Manager

Up to now, you were guided to enable plans where you had the correlated workload to protect. For example, if you don't have **virtual machines (VMs)** to protect, there is no need to enable Defender for Servers. In the case of Defender for Resource Manager, you should always enable it regardless of the types of workloads that you have. In Azure, you have many different workloads (like VMs, databases, networks, etc.) that need to work together to run your applications. ARM is like the conductor that manages and coordinates these resources. It makes sure that they are created, updated, and deleted in a controlled and organized way, regardless of whether you are using the Azure portal, Azure CLI, Azure REST APIs, or other programmatic clients. This means that ARM is always there and should be protected from the get-go.

Threat actors targeting ARM are likely to use toolkits like **Microburst** (`https://github.com/NetSPI/MicroBurst`) to identify weak configurations and carry out post-exploitation activities, such as credential dumping. Defender for Resource Manager leverages internal ARM logs as a source as well as Azure Activity logs as a data source and applies advanced security analytics to detect and trigger alerts for suspicious operations.

These suspicious activities can be the identification of unusual resource management operations, such as connections from suspicious IP addresses, disabling of antimalware, and the execution of dubious scripts within VM extensions. Monitoring the Azure management layer, or, to say it more simply, monitoring ARM, is also important for identifying potential attempts to perform a lateral movement from the Azure management layer to the Azure resources data plane. The Azure data plane refers to the layer of an Azure service responsible for handling the actual data operations, such as reading, writing, processing, or storing data.

You need to be the Security Admin in the subscription that you want to enable Defender for Resource Manager for, and this is the type of plan that can only be enabled at the subscription level. To enable this plan in the Azure subscription, open the Defender for Cloud dashboard, click **Environment settings** under the **Management** section, click the subscription that you want to enable Defender for Resource Manager for, click the **Defender plans** option, and click **On** beside the **Resource Manager** plan, as shown in *Figure 14.1*:

| Resource Manager | $5/Subscription/Month Details > | Full | Off On |

*Figure 14.1: Enabling Defender for Resource Manager (for better visualization, refer to https://packt.link/gbp/9781836204879)*

After enabling, click the **Save** button to commit the changes. That's pretty much it; nothing else needs to be done. At this point, Defender for Resource Manager is monitoring ARM and will trigger alerts if necessary. You can see a list of potential alerts that can be generated by Defender for Resource Manager at https://bit.ly/CNAPPBook67. *Figure 14.2* has a Defender for Resource Manager sample alert:

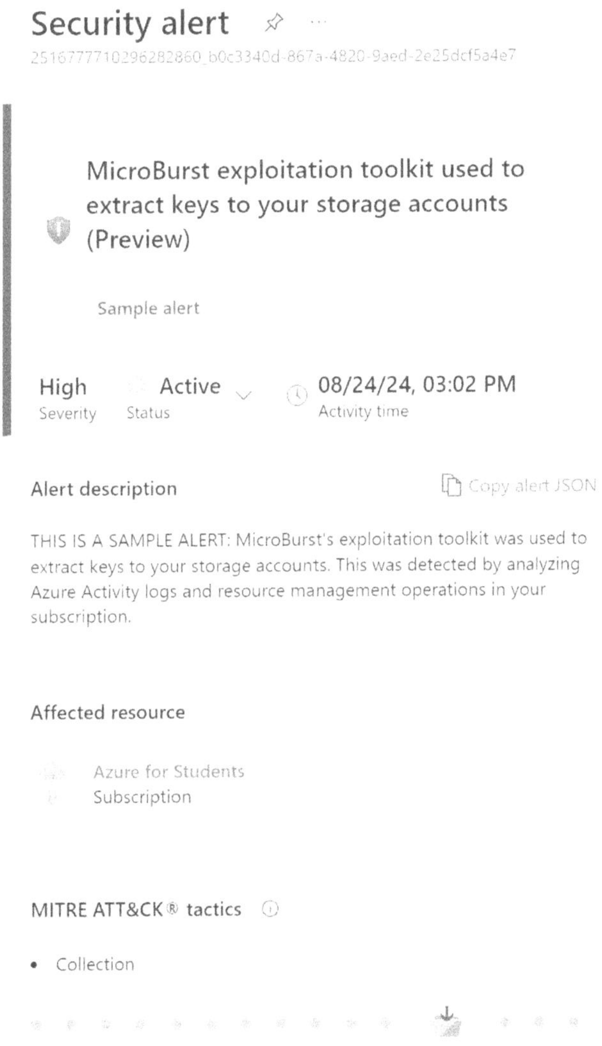

## Security alert

2516777710296282860_b0c3340d-867a-4820-9aed-2e25dcf5a4e7

### MicroBurst exploitation toolkit used to extract keys to your storage accounts (Preview)

Sample alert

| **High** | **Active** ˅ | ⏱ **08/24/24, 03:02 PM** |
|----------|--------------|--------------------------|
| Severity | Status | Activity time |

**Alert description**                    📋 Copy alert JSON

THIS IS A SAMPLE ALERT: MicroBurst's exploitation toolkit was used to extract keys to your storage accounts. This was detected by analyzing Azure Activity logs and resource management operations in your subscription.

**Affected resource**

Azure for Students
Subscription

**MITRE ATT&CK® tactics** ⓘ

- Collection

*Figure 14.2: Microburst sample alert*

As you can see in the alert's description, this type of alert is generated based on data gathered from the Azure Activity log. The other elements of this alert follow the same standards as other alerts. You can use the guidelines from this article to investigate Defender for Resource Manager alerts: `https://bit.ly/CNAPPBook68`.

## Enabling at scale

Defender for Cloud plans can be enabled at scale using Azure Policy. You can deploy the policy to the subscription as the main scope, or to a management group. The relevant part of Defender for Resource Manager for the policy is shown in the sample JSON code below:

```
{
  "policyRule": {
    "if": {
      "field": "type",
      "equals": "Microsoft.Resources/subscriptions"
    },
    "then": {
      "effect": "deployIfNotExists",
      "details": {
        "type": "Microsoft.Security/pricings",
        "name": "ResourceManager",
        "existenceCondition": {
          "field": "Microsoft.Security/pricings.pricingTier",
          "equals": "Standard"
        },
        "deployment": {
          "properties": {
            "mode": "incremental",
            "parameters": {},
            "template": {
              "$schema": "https://schema.management.azure.com/
schemas/2019-04-01/deploymentTemplate.json#",
              "contentVersion": "1.0.0.0",
              "resources": [
                {
                  "type": "Microsoft.Security/pricings",
                  "apiVersion": "2020-01-01-preview",
```

```
                    "name": "ResourceManager",
                    "properties": {
                       "pricingTier": "Standard"
                    }
                  }
                ]
              }
            }
          }
        }
      }
    },
    "parameters": {}
}
```

You also can deploy at scale using an API. For that, you need to leverage Microsoft Defender for Cloud's REST API. The overall steps are shown below:

1.  Obtain an access token.

2.  Send a PUT request to the Microsoft Defender for Cloud API, as shown in the example below (replace subscriptionID with the subscription ID that you need to use):

    ```
    PUT https://management.azure.com/subscriptions/{subscriptionId}/
    providers/Microsoft.Security/pricings/ResourceManager?api-
    version=2020-01-01-preview
    ```

3.  The request body should include the pricing tier you want to apply. The entire request looks similar to the example below:

    ```
    curl -X PUT \
      https://management.azure.com/subscriptions/{subscriptionId}/
    providers/Microsoft.Security/pricings/ResourceManager?api-
    version=2020-01-01-preview \
      -H "Authorization: Bearer <access_token>" \
      -H "Content-Type: application/json" \
      -d '{
           "properties": {
             "pricingTier": "Standard"
           }
        }'
    ```

> The same approach used to enable Defender for Resource Manager at scale can also be used to enable the other plans that will be covered in this chapter.

# Defender for App Service

Azure App Service is a PaaS offering available in Microsoft Azure that allows developers to build, deploy, and scale web apps, mobile backends, and RESTful APIs. It supports multiple programming languages and frameworks, which fits perfectly as a solution for hosting and managing applications in the cloud. Despite the built-in security features available in Azure App Service, threat actors actively scan web applications to uncover and exploit vulnerabilities. This means you still need an extra layer of security to monitor against active attempts to compromise this service.

Requests to Azure App Service pass through multiple gateways before reaching the intended environment. At these gateways, the requests are thoroughly inspected and logged. Defender for App Service analyzes this data to identify potential exploits and attackers, and to learn new attack patterns, which enhances future threat detection capabilities, including the capability to identify dangling DNS[1] types of attacks. You can see a list of security alerts available for Defender for App Service at `https://bit.ly/CNAPPBook69`.

Defender for App Service can be enabled in Azure subscriptions that have any of the following Azure App Service plans:

- Free plan
- Basic service plan
- Standard service plan
- Premium v2 Service Plan
- Premium v3 Service Plan
- App Service Environment v1
- App Service Environment v2
- App Service Environment v3

You need to be the Security Admin in the subscription where you want to enable Defender for App Service, and this is the type of plan that can only be enabled at the subscription level. To enable this plan in the Azure subscription, open the Defender for Cloud dashboard, click **Environment settings** under the **Management** section, click the subscription that you want to enable Defender for App Service for, click the **Defender plans** option, and click **On** beside the App Service plan, as shown in *Figure 14.3*:

*Figure 14.3: Enabling Defender for App Service (for better visualization, refer to https://packt. link/gbp/9781836204879)*

After enabling, click the **Save** button to commit the changes. At this point, you can also see the enablement status by accessing the App Service itself in the Azure App Service dashboard. There, you can click **Microsoft Defender for Cloud** in the left navigation pane, as shown in *Figure 14.4*, and you will see a summary of the findings.

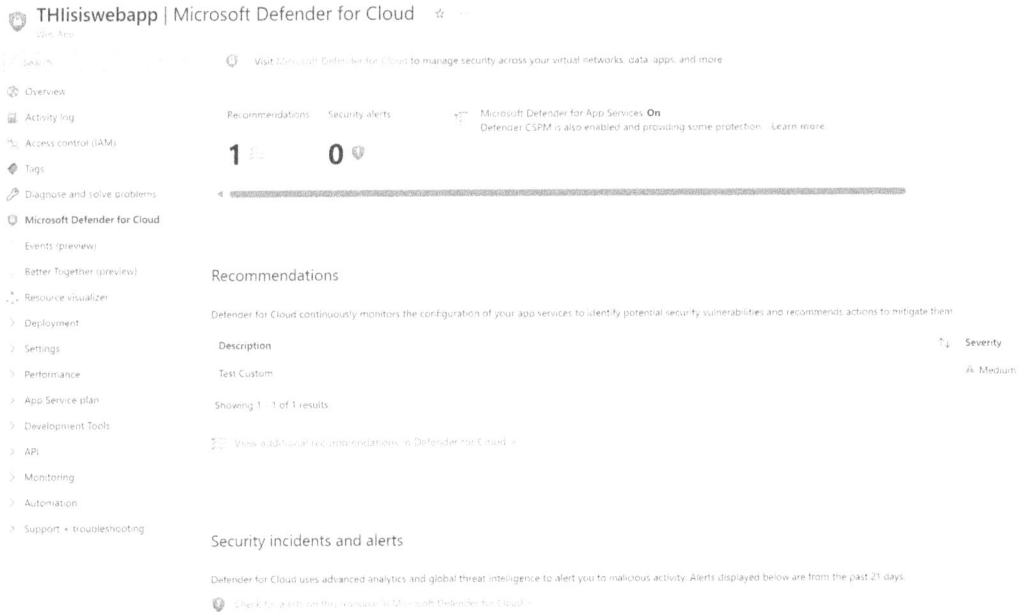

*Figure 14.4: The Defender for Cloud blade within the Azure App Service dashboard (for better visualization, refer to https://packt.link/gbp/9781836204879)*

As part of the posture management improvement, security recommendations[2] will be generated based on your existing Azure App Service settings. The example shown in *Figure 14.5* has a low-priority recommendation to enable diagnostic logs in App Service:

# Diagnostic logs in App Service should be enabled

⚥ Open query        ⊚ View policy definition    ☷ View recommendation for all resources

| Low | | contosoappainvj... | Unassigned | « |
| Risk level ⓘ | | Resource | Status | |

### Description

Enabling diagnostic logs in App Service is a crucial step for maintaining a secure and transparent system.

These logs provide a detailed record of activities, which can be invaluable for investigating security incidents or network compromises.

If diagnostic logs are not enabled, it may be difficult to identify and resolve security issues, potentially leading to prolonged system vulnerabilities and increased risk of data breaches.

| Attack Paths | Scope | Freshness |
| 0 | RO | 30 Min |

| Last change date | Owner | Due date |
| 8/24/2024 | - | - |

**Ticket ID**

-

**Risk factors** ⓘ

-

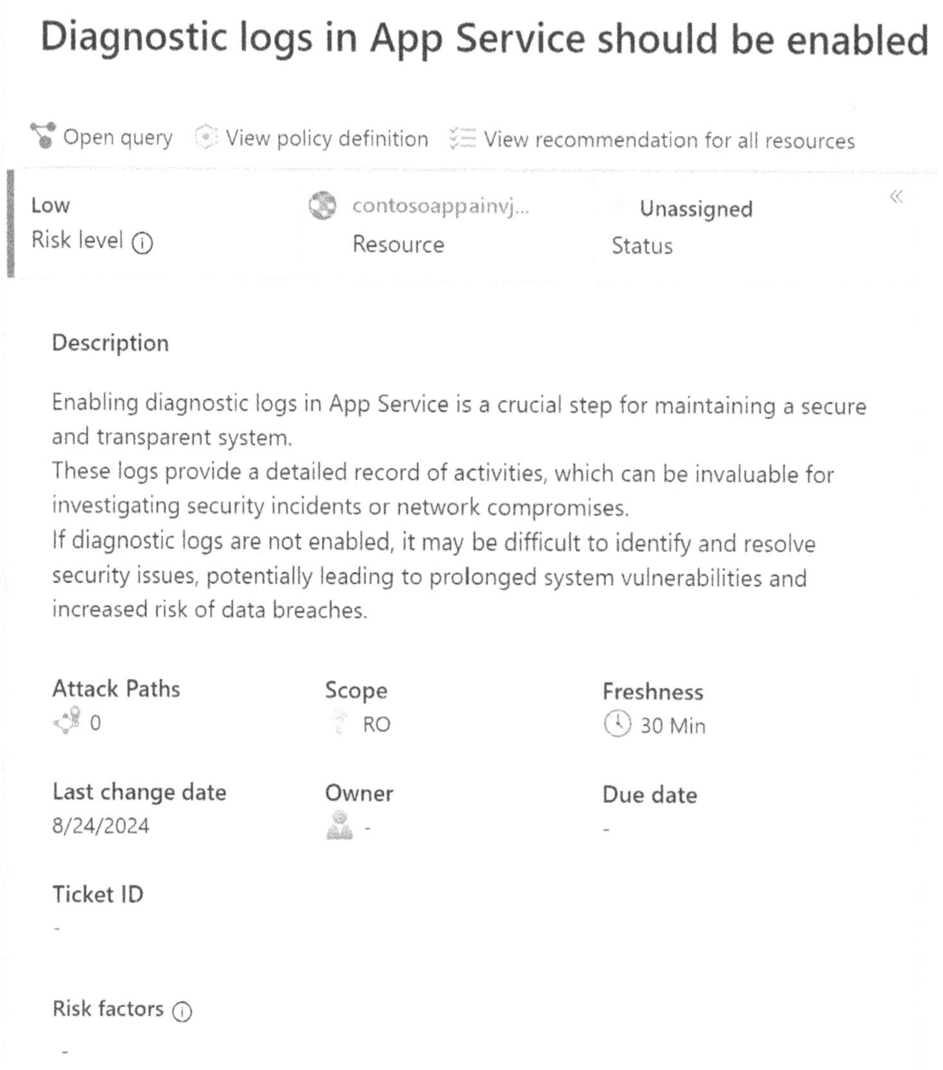

*Figure 14.5: Security recommendation for Azure App Service*

# Defender for Key Vault

Azure Key Vault is a cloud service provided by Microsoft Azure that allows you to securely store and manage sensitive information like secrets, encryption keys, and certificates. Customers can use Azure Key Vault to safeguard cryptographic keys and secrets, such as the ones used in cloud applications and services. This service provides a centralized, secure, and scalable way to manage and access this sensitive data. However, Azure Key Vault is not immune to attacks, and threat actors can leverage public scripts[3] to dump Key Vault keys via an automation account.

Microsoft Defender for Key Vault monitors and detects unusual or potentially unsafe activities aimed at accessing or exploiting your Key Vault accounts. When suspicious activities are detected, Defender for Key Vault generates alerts, which contain details about the activity and offer guidance on how to investigate[4] and address the potential threats. You can see the list of potential alerts generated by Defender for Key Vault at `https://bit.ly/CNAPPBook73`.

You need to be the Security Admin in the subscription that you want to enable Defender for Key Vault for, and this is the type of plan that can only be enabled at the subscription level. To enable Defender for Key Vault in the subscription, open the Defender for Cloud dashboard, click **Environment settings** under the **Management** section, click the subscription that you want to enable Defender for Key Vault for, click the **Defender plans** option, and click **On** beside the **Key Vault** plan, as shown in *Figure 14.6*:

*Figure 14.6: Enabling Defender for Key Vault (for better visualization, refer to https://packt.link/gbp/9781836204879)*

After enabling, click the **Save** button to commit the changes. At this point, you can also see the enablement status by accessing Key Vault itself in the Key Vault dashboard. There, you can click **Settings** in the left navigation pane, and then click the **Microsoft Defender for Cloud** option, as shown in *Figure 14.7*, and you will see a summary of the findings.

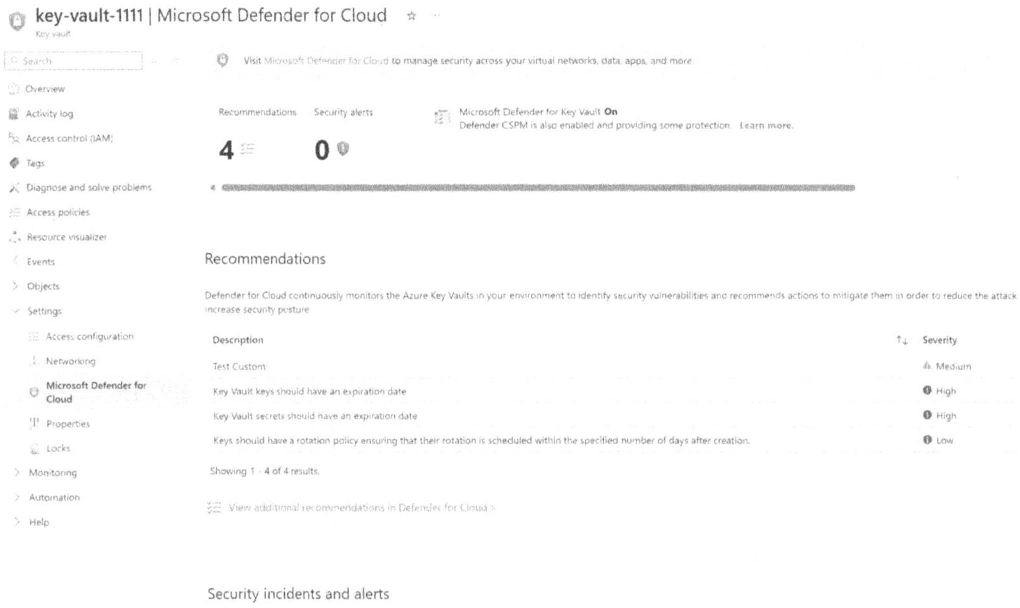

*Figure 14.7: The Defender for Cloud blade within the Key Vault dashboard (for better visualization, refer to https://packt.link/gbp/9781836204879)*

As part of the posture management improvement, security recommendations will be generated based on your existing Key Vault settings. The example shown in *Figure 14.8* has a low-priority recommendation to set an expiration date in the Key Vault keys:

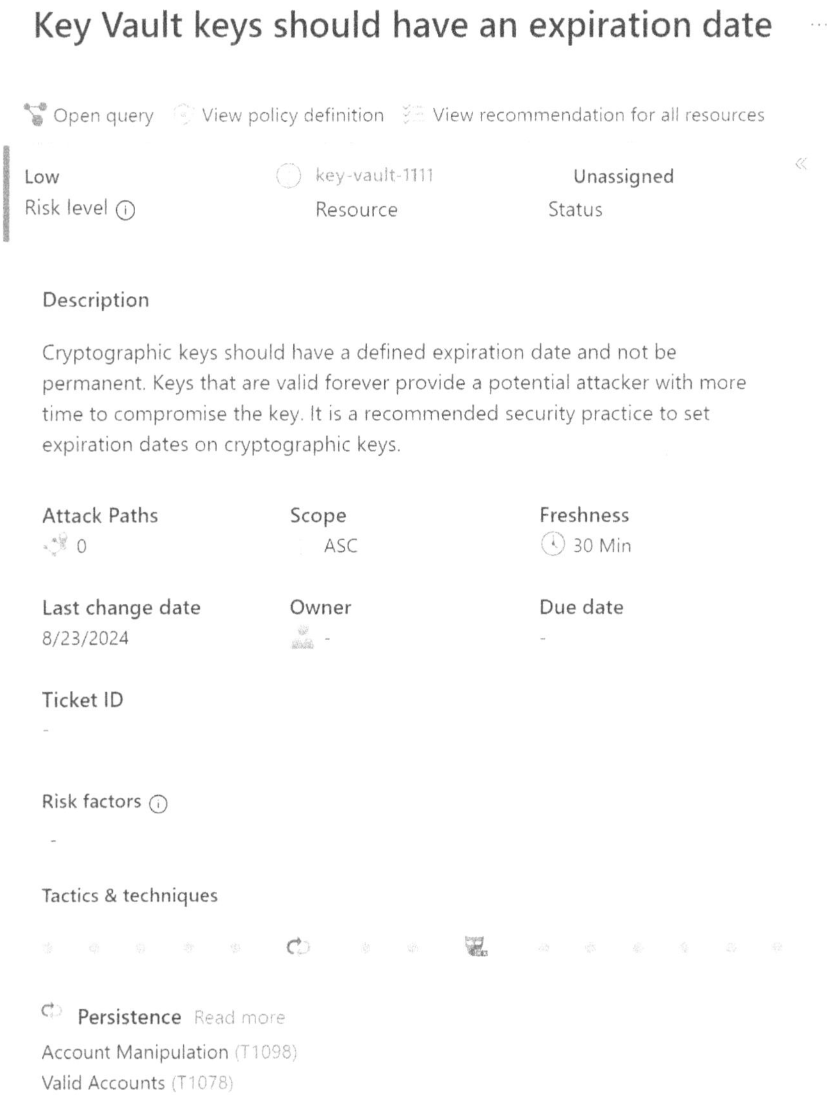

# Key Vault keys should have an expiration date  ...

◦ Open query   ◦ View policy definition   ◦ View recommendation for all resources

| Low | ○ key-vault-1111 | Unassigned | « |
| --- | --- | --- | --- |
| Risk level ⓘ | Resource | Status | |

### Description

Cryptographic keys should have a defined expiration date and not be permanent. Keys that are valid forever provide a potential attacker with more time to compromise the key. It is a recommended security practice to set expiration dates on cryptographic keys.

| Attack Paths | Scope | Freshness |
| --- | --- | --- |
| ⋅ 0 | ASC | ⏱ 30 Min |

| Last change date | Owner | Due date |
| --- | --- | --- |
| 8/23/2024 | ⚎ - | - |

**Ticket ID**

-

**Risk factors** ⓘ

-

### Tactics & techniques

⟳ **Persistence**   Read more

Account Manipulation (T1098)

Valid Accounts (T1078)

*Figure 14.8: Defender for Key Vault recommendation*

> If you want to perform controlled validation and see how alerts are triggered, you can follow the instructions from this article: `https://bit.ly/CNAPPBook75`.

## Summary

In this chapter, you learned about the importance of protecting ARM and how Defender for Resource Manager monitors and detects threats against ARM. You learned how to enable Defender for Resource Manager in your Azure subscription. You learned about the importance of protecting Azure App Service, and how Defender for App Service can help to protect your platform. You also learned about Defender for Key Vault, the use case scenario, and capabilities.

In the next chapter, you will learn how to protect the Azure service layer.

## Notes

1. You can read more about dangling DNS at `https://bit.ly/CNAPPBook70`, and how Defender for App Service helps to detect this type of attack at `https://bit.ly/CNAPPBook71`.

2. You can see the entire list of recommendations available in Defender for App Service at `https://bit.ly/CNAPPBook72`.

3. An example of a public script to get Azure Key Vault keys can be found at `https://bit.ly/CNAPPBook77`.

4. For guidelines to investigate alerts about Defender for Key Vault, visit `https://bit.ly/CNAPPBook74`.

## Additional resources

- Learn more about **Azure Resource Manager** (**ARM**): `https://bit.ly/CNAPPBook76`
- Access the Microsoft Defender for Cloud REST API library: `https://bit.ly/CNAPPBook78`

# Join our community on Discord

Read this book alongside other users. Ask questions, provide solutions to other readers, and much more.

Scan the QR code or visit the link to join the community.

`https://packt.link/SecNet`

# 15

# Incident Response

So far in this book, you have learned the value of CNAPP from the security posture perspective. The ability to leverage rich insights helps you prioritize what needs to be addressed first in your environment. However, CNAPP goes beyond posture management, as it also contains elements from **Cloud Workload Protection (CWP)** that can be leveraged by the **Incident Response (IR)** team.

To reduce the gap between an attack and a response, it is necessary to have rich analytics that are tailored to each workload that you are monitoring. The goal is to empower the IR team to have insightful information about an attack and to use this information to help during the investigation.

The analytics that are generated by the different Defender for Cloud plans are automatically streamed to Microsoft Defender **XDR (eXtended Detection and Response)** platform and can also be configured to stream to Microsoft Sentinel, which is the Microsoft **Security Information and Event Management (SIEM)** solution, or a third-party SIEM solution. These are the tools that are usually used by the **Security Operations Center (SOC)** and leveraged by the IR team.

If the organization is not using XDR or SIEM, it is still possible to leverage built-in capabilities in Defender for Cloud to help investigate incidents and automate incident response.

This chapter covers:

- Incident Response using Defender for Cloud
- Integration with Microsoft XDR
- Integration with Microsoft Sentinel

# Incident Response using Defender for Cloud

According to the **National Institute of Standards and Technology (NIST)** SP 800-61 Revision 3[1], the incident response lifecycle is composed of three phases: Detect, Respond, and Recover. The analytics generated by Defender for Cloud fit into the Detect phase and add value to the response phase.

By now, you already know the concept of alerts, and that each Defender for Cloud plan brings a set of analytics that were created specifically for the workload that it is protecting. Many of the features that were covered in *Chapter 10*, such as alert correlation and alert suppression, can be used to help the IR team to investigate alerts. For example, alert correlation generates a Security Incident that aggregates the alerts that are associated with a potential attack, as shown in *Figure 15.1*:

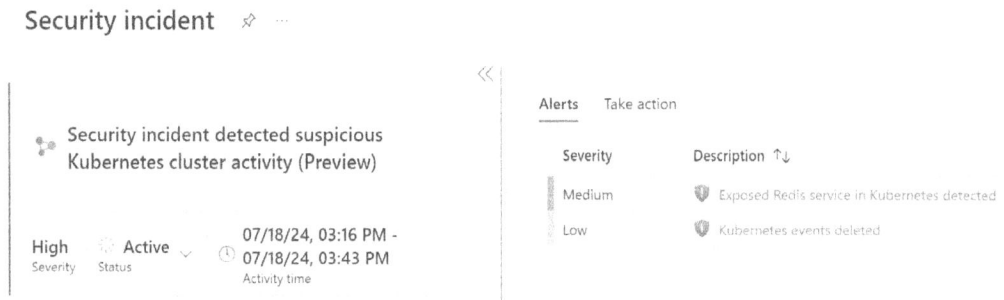

Security incident  ⚲  ⋯

Security incident detected suspicious
Kubernetes cluster activity (Preview)

| | | |
|---|---|---|
| High | Active ⌄ | 07/18/24, 03:16 PM - |
| Severity | Status | 07/18/24, 03:43 PM |
| | | Activity time |

Alerts    Take action

| Severity | Description ↑↓ |
|---|---|
| Medium | Exposed Redis service in Kubernetes detected |
| Low | Kubernetes events deleted |

*Figure 15.1: Security Incident in Defender for Cloud (for better visualization, refer to https://packt.link/gbp/9781836204879)*

The automatic correlation of alerts helps to save time during an investigation since there is no need to perform an exhaustive manual correlation between alerts. One important point when you are investigating an alert is the status of the alert, which can be adjusted by clicking on the **Status** option (in the example of *Figure 15.1*, the status is **Active**), and from there, change to the appropriate state. For example, if the alert was already addressed, you can change it to **Resolved**.

The insights provided by each alert will assist the IR team in understanding what happened, when it happened, and which resource was affected by the attack. These insights appear in the **Alert details** tab in the alert itself as shown in *Figure 15.2*:

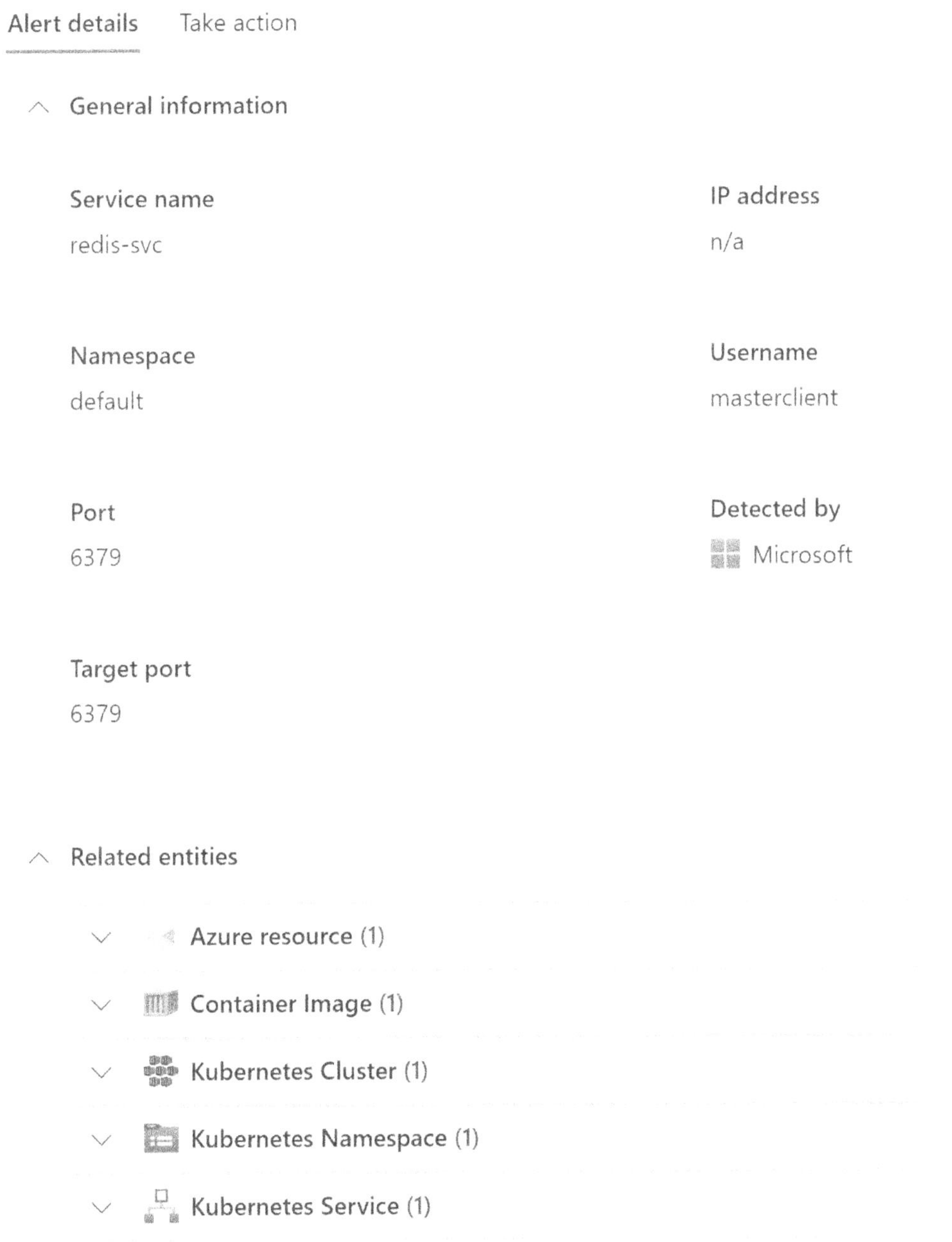

Alert details    Take action

∧  General information

Service name                          IP address
redis-svc                             n/a


Namespace                             Username
default                               masterclient


Port                                  Detected by
6379                                  ▦ Microsoft


Target port
6379


∧  Related entities

   ∨    ◁  Azure resource (1)

   ∨    ▥  Container Image (1)

   ∨    ⊞  Kubernetes Cluster (1)

   ∨    ▤  Kubernetes Namespace (1)

   ∨    ⊡  Kubernetes Service (1)


*Figure 15.2: Alert details tab*

While having these insights is important, some IR teams may not have the right playbook for each alert. In other words, it may be a new alert that the IR team has never experienced in the past, which means they will need extra guidance on how to handle it (Response phase). For that, Defender for Cloud provides a set of mitigation steps for each alert to help IR respond to that alert. These mitigation steps are located in the **Take action** tab, as shown in *Figure 15.3*:

Alert details    Take action
                 ‗‗‗‗‗‗‗‗‗

∧   🔍 Inspect resource context

    Start with examining the resource logs around the time of the alert.

    [ Open logs ]

∧   ✚ Mitigate the threat

    1. Find the external IP address of the service with the command: kubectl get service [service name] -n

    [namespace]. The service name and namespace are in the alert details.

    2. Check whether the service is exposed to the internet by trying to access to its IP address with the port listed in

    the alert details.

    3. If the service is exposed to the internet and is not protected by an authentication mechanism, escalate the alert

    to your security information team.

    You have 14 more alerts on the affected resource. View all > >

∧   🛡 Prevent future attacks

    Your top 3 active security recommendations on 🔅 woodgrove-k8-MDC:

    | High | ☰ Kubernetes clusters should be accessible only over HTTPS |
    | High | ☰ Azure running container images should have vulnerabilities resolved |
    | High | ☰ Running containers as root user should be avoided |

    Solving security recommendations can prevent future attacks by reducing attack surface.

    View all 20 recommendations > >

∨   ⦂⦂ Trigger automated response

∨   👁 Suppress similar alerts

∨   ✉ Configure email notification settings

*Figure 15.3: Take action tab (for better visualization, refer to https://packt.link/ gbp/9781836204879)*

Some alerts may also have extra insights that can be used to track down what caused the generation of this alert, and this is in the **Inspect resource context** section. This section may or may not have supporting logs; again, it will vary according to the alert and workload. The **Mitigate the threat** section contains suggestions to respond to this alert, which also vary according to the alert/workload.

The **Prevent future attacks** section helps during the lessons-learned exercise that happens after the incident gets resolved. The suggestions from this section will be used as input to improve the security posture, which, in the case of the example shown in *Figure 15.3*, means to remediate three high-severity recommendations.

In the lower part of the page, you have the capability to link an Azure Logic Apps automation for the alert, which can be done by selecting the automation in the **Trigger automated response** section. If the alert is considered a false positive, you can suppress the alert by using the **Suppress similar alerts** option. Lastly, you can adjust who will get emails regarding security alerts on this subscription by making the change using the **Configure email notification settings** option.

While investigating these alerts from Defender for Cloud is an option, you can also use Microsoft Defender XDR for a richer experience since there will be data coming from different data sources.

# Integration with Microsoft Defender XDR

Microsoft XDR is an integrated security solution designed to provide holistic visibility and protection across an organization's various digital environments. Microsoft Defender XDR is delivered through Microsoft Defender and Microsoft Sentinel, enabling better detection and investigation. Microsoft Defender XDR uses AI-driven detection mechanisms across endpoints, IoT, identities, apps, email, and cloud environments to identify and alert potential threats. Microsoft Defender XDR has insights coming from Microsoft Defender for Cloud, Microsoft Defender for Endpoint, Microsoft Defender for Office 365, Microsoft Defender for Identity, Microsoft Defender for Cloud Apps, and Microsoft Sentinel.

To access Microsoft Defender XDR, you need to visit `security.microsoft.com` and sign in with your privileged account credentials. Once you open the dashboard, you will see the navigation bar on the left side. The **Incidents & alerts** section (*Figure 15.4*) has the alerts and incidents generated by Defender for Cloud.

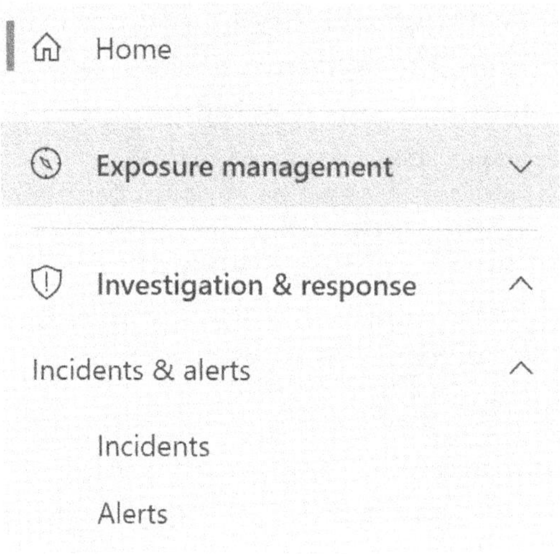

*Figure 15.4: Incidents and alerts in XDR*

When you click the **Alerts** option, you will see a list of all alerts and the product that generated this alert (**Product name** column) as shown in *Figure 15.5*. This consolidation view of all alerts helps IR teams have a better understanding of the bigger picture and have a more detailed insight into what happened and what to do to respond to an alert.

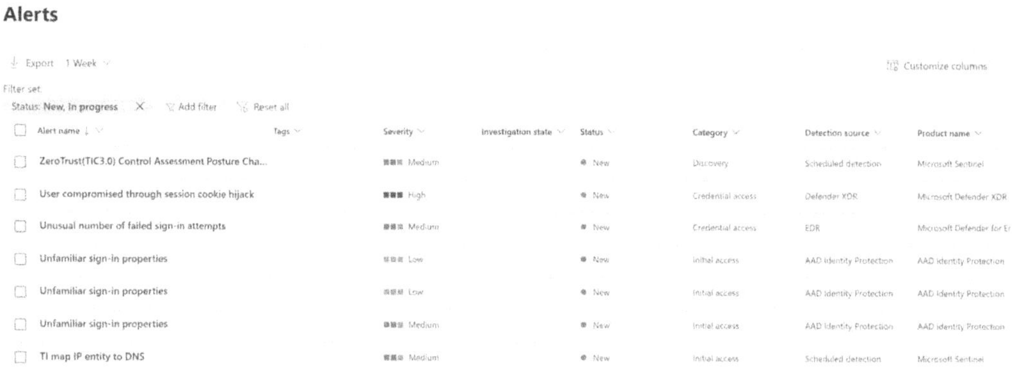

*Figure 15.5: List of alerts coming from different products (for better visualization, refer to https://packt.link/gbp/9781836204879)*

Upon opening an alert, you will also have a different look and feel, where the first part is an explanation of the story behind the alert, and the ability to switch back to the Defender for Cloud dashboard using the option **View alert page in MDC**, as shown in *Figure 15.6*:

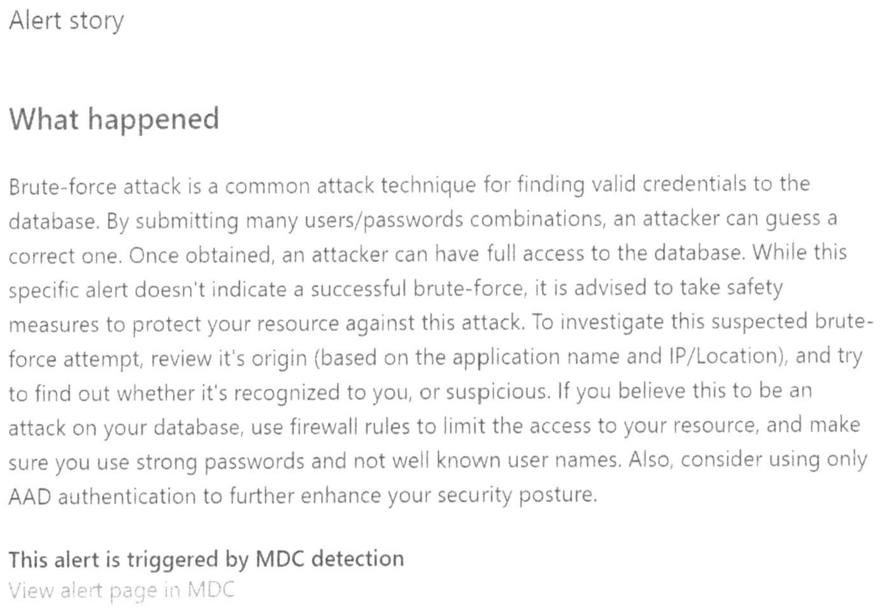

Alert story

## What happened

Brute-force attack is a common attack technique for finding valid credentials to the database. By submitting many users/passwords combinations, an attacker can guess a correct one. Once obtained, an attacker can have full access to the database. While this specific alert doesn't indicate a successful brute-force, it is advised to take safety measures to protect your resource against this attack. To investigate this suspected brute-force attempt, review it's origin (based on the application name and IP/Location), and try to find out whether it's recognized to you, or suspicious. If you believe this to be an attack on your database, use firewall rules to limit the access to your resource, and make sure you use strong passwords and not well known user names. Also, consider using only AAD authentication to further enhance your security posture.

**This alert is triggered by MDC detection**
View alert page in MDC

*Figure 15.6: Alert story*

After this field, you have more details about the compromised resource and potential cause, as shown in *Figure 15.7*:

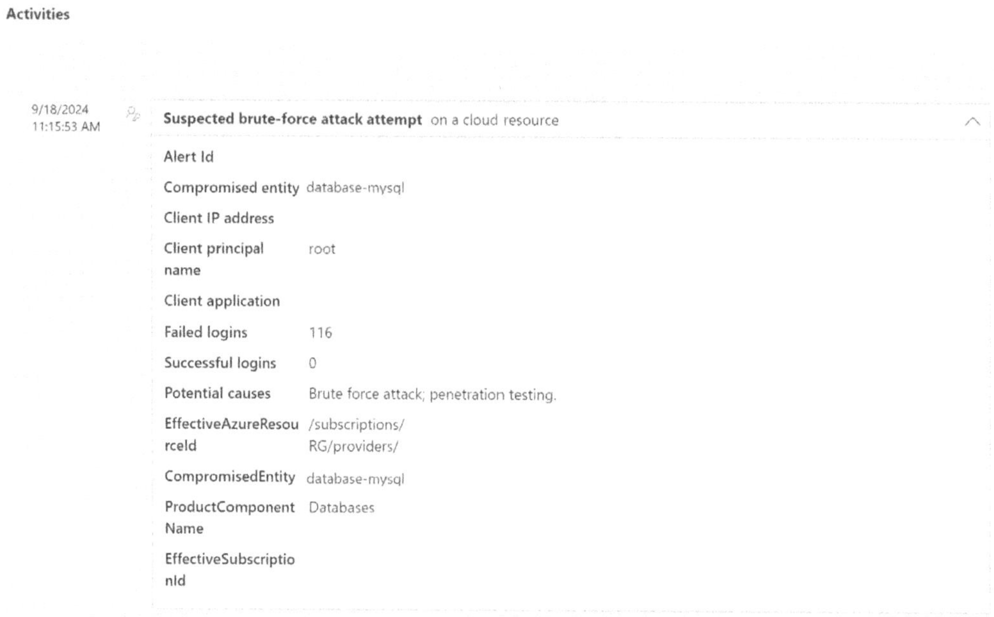

Activities

9/18/2024
11:15:53 AM    **Suspected brute-force attack attempt** on a cloud resource

Alert Id

Compromised entity  database-mysql

Client IP address

Client principal     root
name

Client application

Failed logins        116

Successful logins    0

Potential causes     Brute force attack; penetration testing.

EffectiveAzureResou  /subscriptions/
rceId                RG/providers/

CompromisedEntity    database-mysql

ProductComponent     Databases
Name

EffectiveSubscriptio
nId

*Figure 15.7: Alert activity (for better visualization, refer to https://packt.link/gbp/9781836204879)*

In *Figure 15.7*, some fields are blank due to private information about the environment. The idea is that after reading the alert story and becoming more familiar with what happened, you can then use the information from the **Activities** section to help you identify the affected workload. The alert criticality and status are available on the right side, as shown in *Figure 15.8*:

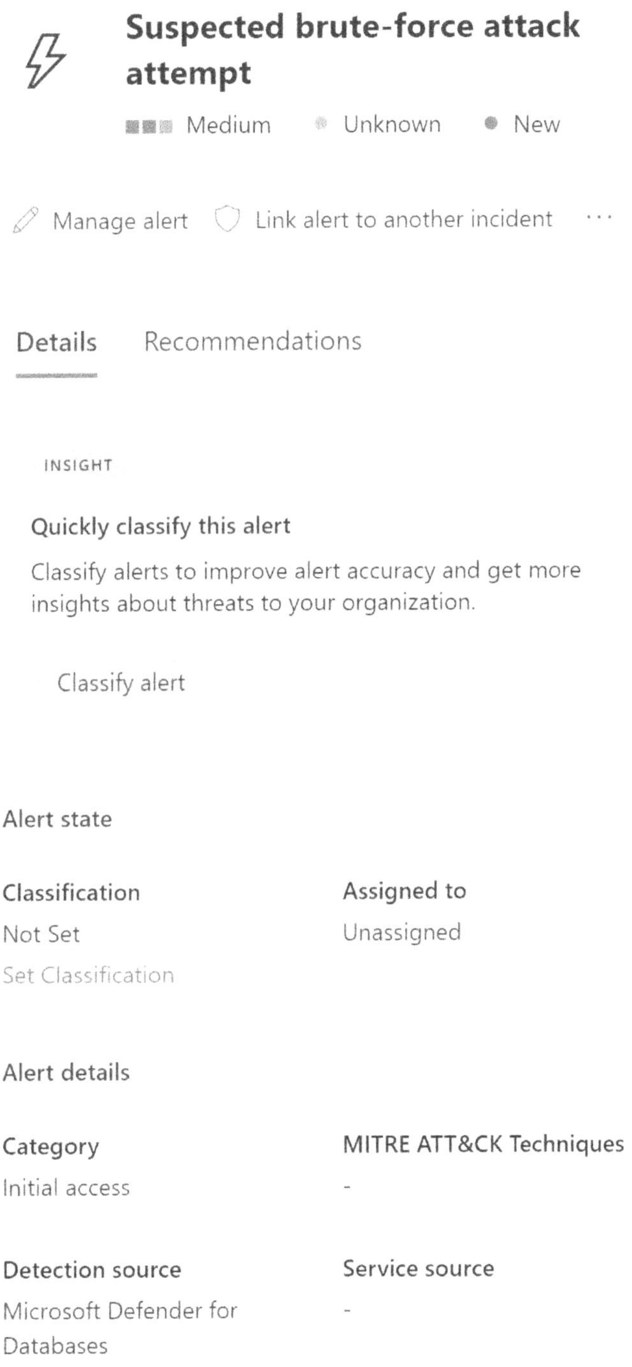

# Suspected brute-force attack attempt

■■▫ Medium　　◦ Unknown　　● New

✎ Manage alert　　🛡 Link alert to another incident　　· · ·

## Details　　Recommendations

INSIGHT

### Quickly classify this alert

Classify alerts to improve alert accuracy and get more insights about threats to your organization.

Classify alert

## Alert state

**Classification**

Not Set

Set Classification

**Assigned to**

Unassigned

## Alert details

**Category**

Initial access

**MITRE ATT&CK Techniques**

-

**Detection source**

Microsoft Defender for Databases

**Service source**

-

*Figure 15.8: Alert criticality and state*

All this data is coming from Defender for Databases, as shown in *Figure 15.8*, via the native integration of Defender for Cloud with Microsoft Defender XDR. Aside from the aggregation of all alerts across the Microsoft Security suite of products, Microsoft Defender XDR also has a hunting capability that enables you to use **Kusto Query Language** (**KQL**) to perform threat hunting, which is very useful for the IR team while investigating an incident.

# Hunting

To access this capability, click on the **Hunting** section, and then click the **Advanced hunting** option. *Figure 15.9* has an example of the **Advanced hunting** page:

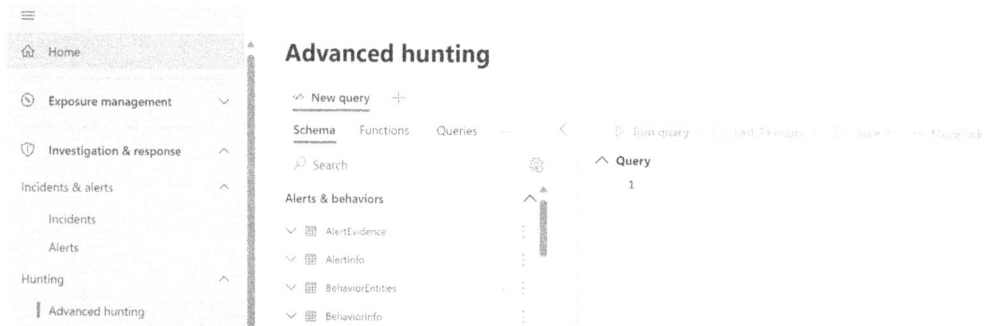

*Figure 15.9: Advanced hunting page (for better visualization, refer to https://packt.link/ gbp/9781836204879)*

The left side of the **Advanced hunting** page has the schema[2] for each product that is integrated with the platform. This helps you understand the tables that are available for you to query. Make sure to follow the guidelines from the *Advanced hunting query best practices*[3] article when you start building your queries.

To start hunting, you need to type the query and click the **Run query** button; a sample result is shown in *Figure 15.10*. I recommend you visit this GitHub repository at `https://bit.ly/ CNAPPBook104` for a rich list of sample queries for different threat-hunting scenarios and an in-depth explanation of KQL.

*Figure 15.10: Query result (for better visualization, refer to https://packt.link/gbp/9781836204879)*

You can use this page to also save your query (with the **Save** button) or share your query with someone else (with the **Share link** button).

While Microsoft Defender XDR provides this comprehensive aggregation of data coming from different Microsoft security solutions, some organizations may be using a SIEM, such as Microsoft Sentinel, to aggregate data coming from other data sources that are not Microsoft. For those scenarios, the IR team may prefer to use the SIEM platform to help during the investigation, since it contains data coming from all data sources available in the company, which may include firewall and other devices.

# Integration with Microsoft Sentinel

Microsoft Sentinel uses a Log Analytics workspace to store the data that is consumed from different data sources. Defender for Cloud is part of the free data sources, which means there is no extra cost in ingesting data from Defender for Cloud into Microsoft Sentinel's workspace. In addition to Defender for Cloud security alerts, the following other Microsoft data sources are also free of cost to ingest:

- Azure activity logs
- Microsoft Sentinel Health
- Office 365 audit logs, including all SharePoint activity, Exchange admin activity, and Teams
- Security alerts, including alerts from the following sources:

  - Microsoft Defender XDR
  - Microsoft Defender for Office 365
  - Microsoft Defender for Identity
  - Microsoft Defender for Cloud Apps
  - Microsoft Defender for Endpoint

To configure the Defender for Cloud connector in Microsoft Sentinel, open the Microsoft Sentinel dashboard. On the left navigation pane, click the **Configuration** section, and then click the **Get these data connectors** button on the **Data connectors** page as shown in *Figure 15.11*:

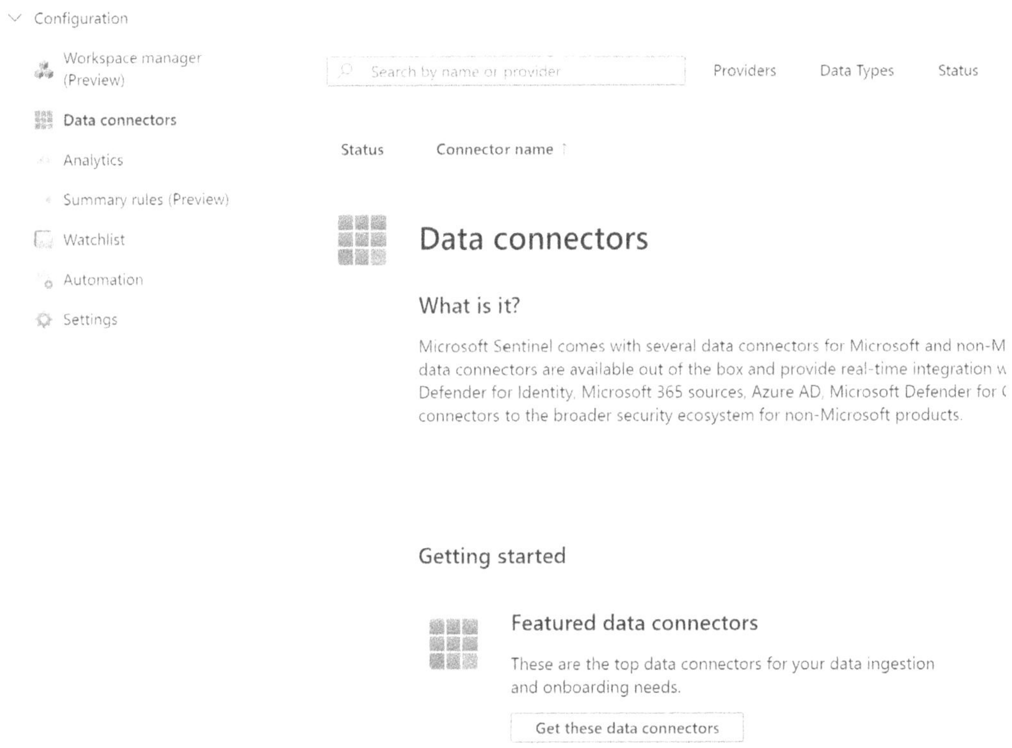

Configuration

Workspace manager
(Preview)

Data connectors

Analytics

Summary rules (Preview)

Watchlist

Automation

Settings

Search by name or provider            Providers      Data Types      Status

Status        Connector name ↑

# Data connectors

## What is it?

Microsoft Sentinel comes with several data connectors for Microsoft and non-M
data connectors are available out of the box and provide real-time integration w
Defender for Identity, Microsoft 365 sources, Azure AD, Microsoft Defender for (
connectors to the broader security ecosystem for non-Microsoft products.

## Getting started

### Featured data connectors

These are the top data connectors for your data ingestion
and onboarding needs.

Get these data connectors

*Figure 15.11: Accessing the connectors (for better visualization, refer to https://packt.link/
gbp/9781836204879)*

On the **Content hub** page, click **Microsoft Defender for Cloud** from the list, as shown in *Figure 15.12*:

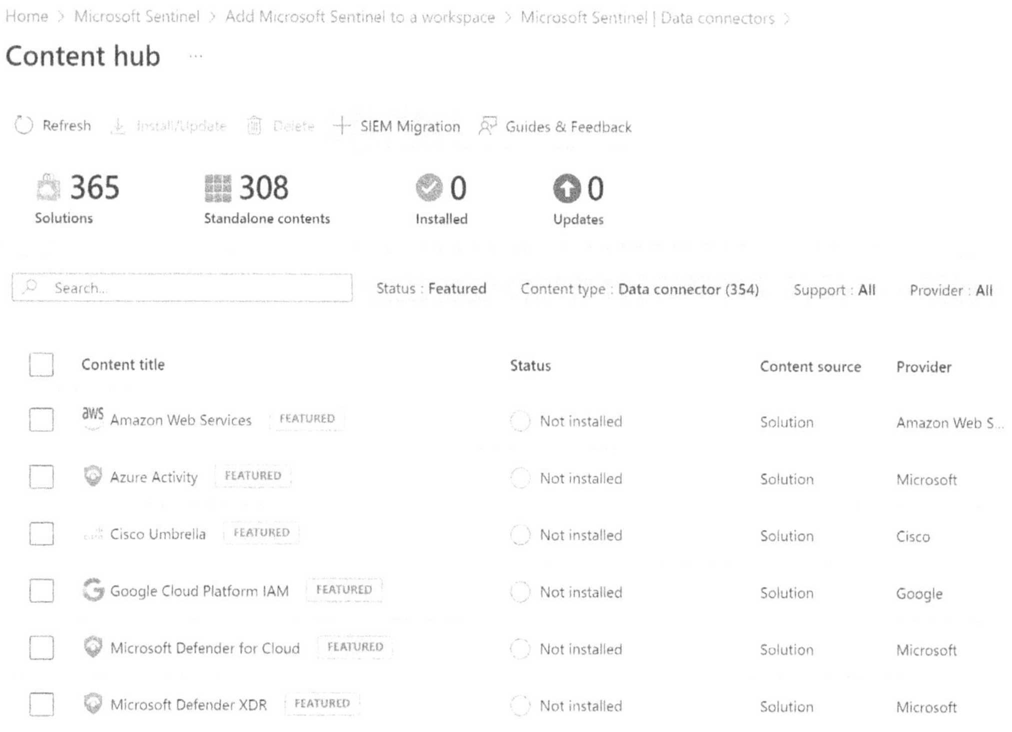

Home > Microsoft Sentinel > Add Microsoft Sentinel to a workspace > Microsoft Sentinel | Data connectors >

## Content hub ...

◯ Refresh   ⤓ Install/Update   🗑 Delete   ＋ SIEM Migration   📖 Guides & Feedback

🛍 **365**          ▦ **308**                    ⊘ **0**              ⬆ **0**
Solutions          Standalone contents       Installed          Updates

| ⌕ Search... | Status : **Featured** | Content type : **Data connector (354)** | Support : **All** | Provider : **All** |

| | Content title | Status | Content source | Provider |
|---|---|---|---|---|
| ☐ | aws Amazon Web Services  FEATURED | ◯ Not installed | Solution | Amazon Web S... |
| ☐ | 🌐 Azure Activity  FEATURED | ◯ Not installed | Solution | Microsoft |
| ☐ | Cisco Umbrella  FEATURED | ◯ Not installed | Solution | Cisco |
| ☐ | G Google Cloud Platform IAM  FEATURED | ◯ Not installed | Solution | Google |
| ☐ | 🌐 Microsoft Defender for Cloud  FEATURED | ◯ Not installed | Solution | Microsoft |
| ☐ | 🌐 Microsoft Defender XDR  FEATURED | ◯ Not installed | Solution | Microsoft |

*Figure 15.12: Content hub (for better visualization, refer to https://packt.link/ gbp/9781836204879)*

Once you select **Microsoft Defender for Cloud** from the list, a blade will open on the right side with more details about the connector. Click the **Install** button as shown in *Figure 15.13*:

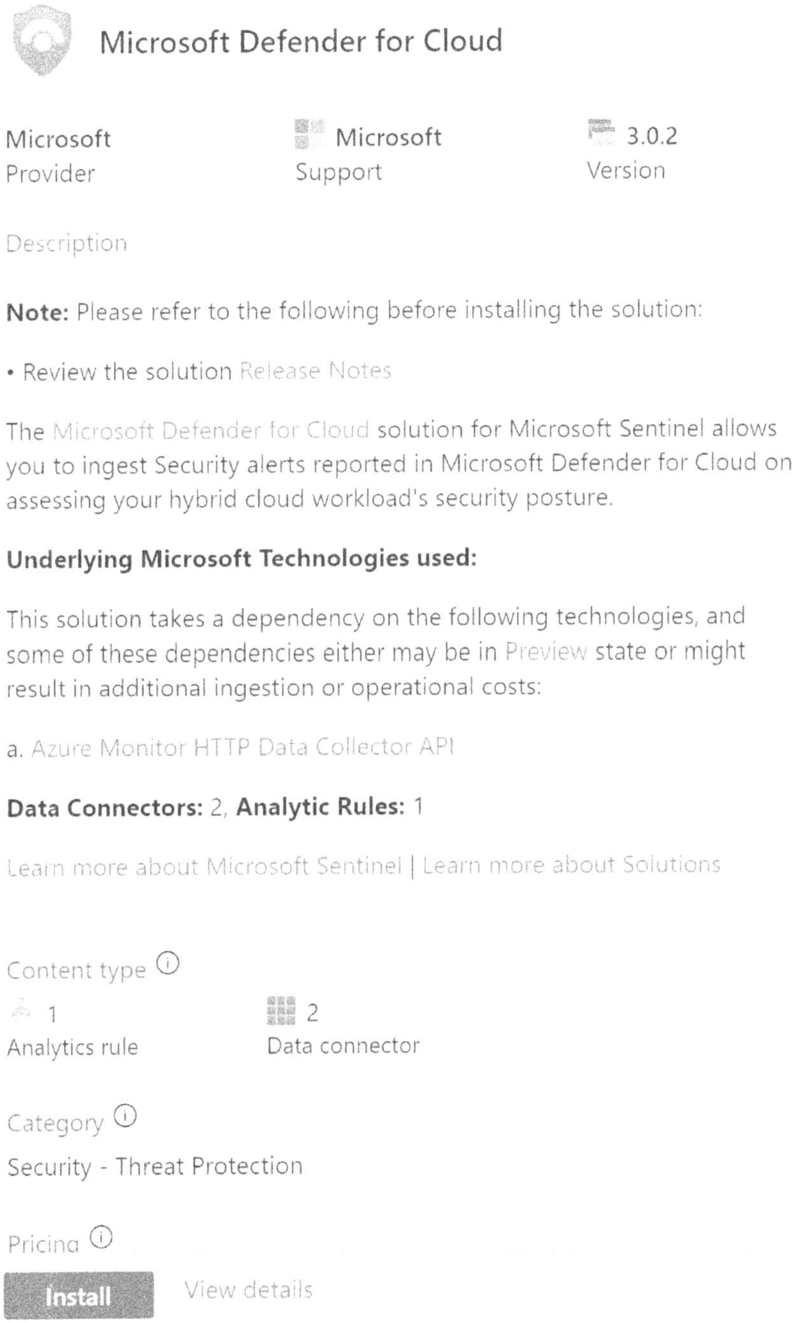

## Microsoft Defender for Cloud

| Microsoft | Microsoft | 3.0.2 |
|---|---|---|
| Provider | Support | Version |

Description

**Note:** Please refer to the following before installing the solution:

• Review the solution Release Notes

The Microsoft Defender for Cloud solution for Microsoft Sentinel allows you to ingest Security alerts reported in Microsoft Defender for Cloud on assessing your hybrid cloud workload's security posture.

**Underlying Microsoft Technologies used:**

This solution takes a dependency on the following technologies, and some of these dependencies either may be in Preview state or might result in additional ingestion or operational costs:

a. Azure Monitor HTTP Data Collector API

**Data Connectors:** 2, **Analytic Rules:** 1

Learn more about Microsoft Sentinel | Learn more about Solutions

Content type ⓘ

1
Analytics rule

2
Data connector

Category ⓘ

Security - Threat Protection

Pricing ⓘ

Install    View details

*Figure 15.13: Defender for Cloud connector blade*

After the installation process finishes, the **Install** button becomes the **Manage** button. Click on it and the **Microsoft Defender for Cloud** page appears as shown in *Figure 15.14*:

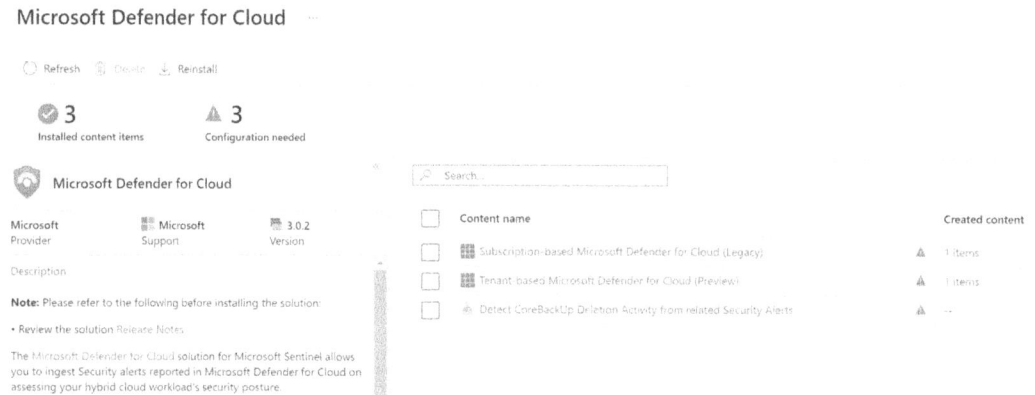

*Figure 15.14: Defender for Cloud connector page (for better visualization, refer to https:// packt.link/gbp/9781836204879)*

The most important items are on the **Content name** list, which are the available options to perform the connection with Defender for Cloud. You can connect to Microsoft Sentinel using the legacy connector, which only connects to one subscription at a time, or you can use the connector for the tenant level (which was in public preview at the time this chapter was written), which will ingest Defender for Cloud alerts from all subscriptions that have Defender for Cloud in the target tenant. For the purpose of this example, select the checkbox for **Subscription-based Microsoft Defender for Cloud (Legacy)**, and in the blade that appears on the right side (*Figure 15.15*), click the **Open connector page** button.

# ▦ Subscription-based Microsoft Defender for Cloud (Legacy)

**Disconnected**
Status

✎ **Microsoft**
Provider

🕒 --
Last Log Received

Last data received

--

Content source ⓘ
Microsoft Defender for Cloud

Version
1.0.0

Author
Microsoft

Supported by
Microsoft Corporation | Email

Data received                                              Go to log analytics

SecurityAlerts
**0**

**Open connector page**

*Figure 15.15: Connector blade*

On the bottom of the right side of the page that appears, select the subscription and toggle the status to **Connected** as shown in *Figure 15.16*:

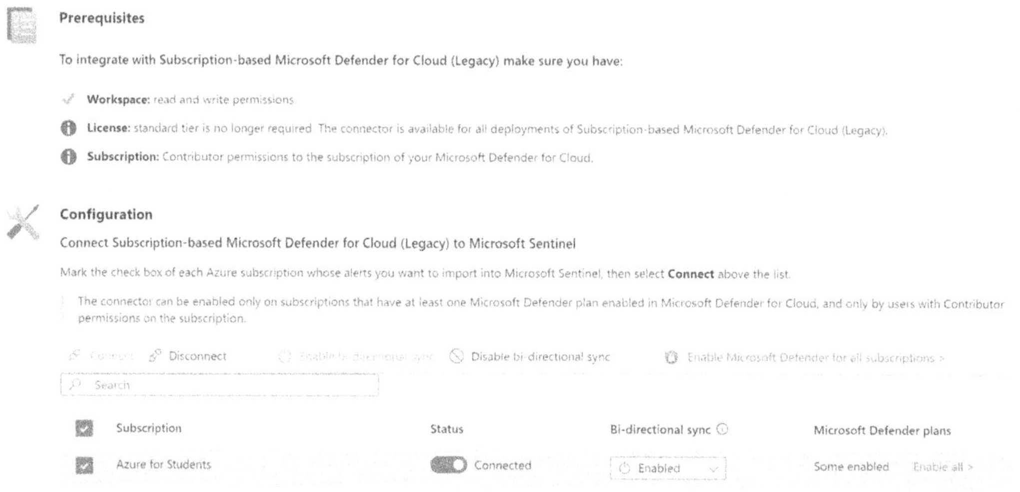

**Prerequisites**

To integrate with Subscription-based Microsoft Defender for Cloud (Legacy) make sure you have:

✓ **Workspace:** read and write permissions.

ℹ **License:** standard tier is no longer required. The connector is available for all deployments of Subscription-based Microsoft Defender for Cloud (Legacy).

ℹ **Subscription:** Contributor permissions to the subscription of your Microsoft Defender for Cloud.

**Configuration**

Connect Subscription-based Microsoft Defender for Cloud (Legacy) to Microsoft Sentinel

Mark the check box of each Azure subscription whose alerts you want to import into Microsoft Sentinel, then select **Connect** above the list.

The connector can be enabled only on subscriptions that have at least one Microsoft Defender plan enabled in Microsoft Defender for Cloud, and only by users with Contributor permissions on the subscription.

| | Subscription | Status | Bi-directional sync ⓘ | Microsoft Defender plans |
|---|---|---|---|---|
| ☑ | Azure for Students | ⬤ Connected | ⟳ Enabled ∨ | Some enabled   Enable all > |

*Figure 15.16: Changing the connection status (for better visualization, refer to https://packt. link/gbp/9781836204879)*

Notice that, the bi-directional sync is enabled by default. Keep this setting as is since the goal is to automatically sync the alert status between Defender for Cloud and Microsoft Sentinel. The status of the connector will change to **Connected**, as shown in *Figure 15.17*:

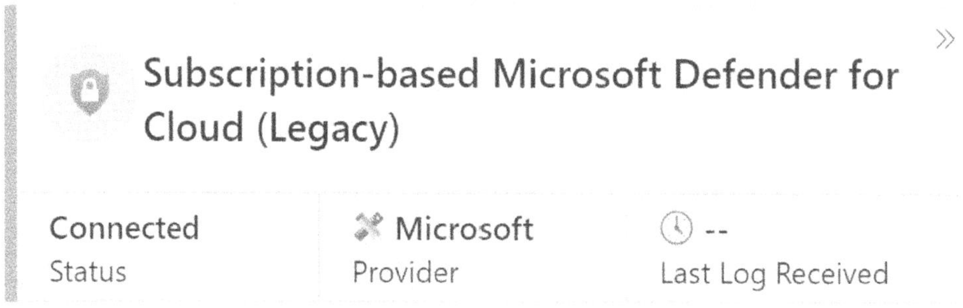

# Subscription-based Microsoft Defender for Cloud (Legacy)

| Connected | Microsoft | ⏱ -- |
|---|---|---|
| Status | Provider | Last Log Received |

*Figure 15.17: Connector status*

Notice that the **Last Log Received** is still blank. Allow some time for the flow to occur. If you don't have new alerts flowing in, you can also validate using sample alerts. Go to the Defender for Cloud dashboard, open the **Security alerts** dashboard, and click **Sample alerts**. Trigger some sample alerts for some plans, and this should be streamlined to Microsoft Sentinel.

To verify if the security alert is already ingested in Microsoft Sentinel, open the Microsoft Sentinel dashboard, click **Logs** under the **General** section, type the query below, and click the **Run** button:

```
SecurityAlert
    | where ProductName == "Azure Security Center"
```

*Figure 15.18* has an example of the result with the sample alerts at the bottom. This means that the flow between Defender for Cloud and Microsoft Sentinel is working properly.

*Figure 15.18: Accessing Defender for Cloud alerts in Microsoft Sentinel (for better visualization, refer to https://packt.link/gbp/9781836204879)*

# Summary

In this chapter, you learned about the use of Defender for Cloud for Incident Response and the different alert insights provided by Defender for Cloud to empower IR teams to do better investigations. You also learned about the integration of Defender for Cloud with Microsoft Defender XDR and the alert experience in the Microsoft Defender XDR portal. You received an introduction to the advanced hunting capability in Microsoft Defender XDR. Lastly, you learned how to configure the Defender for Cloud connector in Microsoft Sentinel to enable the ingestion of Defender for Cloud security alerts in Microsoft Sentinel.

In the next chapter, you will learn how to leverage AI to improve your security posture.

# Notes

1. You can read more about this document at `https://csrc.nist.gov/Projects/incident-response`.

2. To learn more about the different schemas, visit `https://bit.ly/CNAPPBook103`.

3. Read more about advanced hunting best practices at `https://bit.ly/CNAPPBook102`.

# Additional resources

- Learn more about the investigation workflow using Microsoft XDR: `https://bit.ly/CNAPPBook105`

- Use the guidelines to investigate alerts about Defender for Key Vault: `https://bit.ly/CNAPPBook74`

# Join our community on Discord

Read this book alongside other users. Ask questions, provide solutions to other readers, and much more.

Scan the QR code or visit the link to join the community.

`https://packt.link/SecNet`

# 16

# Leveraging AI to Improve Your Security Posture

**Artificial Intelligence (AI)** is here to stay and will continue to revolutionize the industry. The opportunities that are already unfolding with the use of AI are nothing short of amazing. One of the opportunities that became clear was the advantage of using AI to improve cloud security posture management. AI can enable security posture management teams to use natural language to ask important questions, and rapidly obtain the answer. In a matter of seconds, posture management teams can understand the most important risks to remediate and ask for assistance to remediate security recommendations at scale by writing a script.

This scenario is also known as **AI for Security**, which means the use of AI to help with information security. However, it is not only the "good guys" that are using AI to improve security posture. Threat actors are also already using AI to perform malicious operations. This means that it is important to also have Security for AI—in other words, having the right visibility of your GenAI applications and models to ensure that they are secure.

Defender for Cloud covers both scenarios. AI for Security is covered with the integration with Copilot for Security, and Security for AI is covered with AI posture management, which is part of Defender CSPM.

This chapter covers:

- Defender for Cloud integration with Copilot for Security
- AI posture management

# Defender for Cloud integration with Copilot for Security

At Ignite 2023, Microsoft announced the integration of Defender for Cloud with Copilot for Security and Copilot for Azure[1]. While Copilot for Security in Defender for Cloud is not bound to any of the paid plans, you will get more insights from it if you have Defender for CSPM enabled in your subscription. Since Defender CSPM is responsible for all the artifact scanning that helps to bring insights to the attack path, there is a lot of data collection that is missing if Defender CSPM is not enabled. This means that Copilot for Security in Defender for Cloud may have limited visibility to provide a better answer. The prerequisites to have Copilot for Security in Defender for Cloud are the following ones:

- Defender for Cloud must be enabled
- Access to Azure Copilot

    - Copilot in Azure is initially accessible to all users within a tenant by default. However, Global Administrators can control and regulate access for their organization[2].

- You must have **security compute units (SCUs)**[3] available and assigned to Copilot for Security

    - Copilot for Security is offered through a provisioned capacity model and is charged on an hourly basis. You can allocate SCUs and adjust the amount as needed. Billing is determined hourly, with a minimum charge of one hour.

## Exploring recommendations

The first scenario in which Copilot for Security can help security posture management teams to better understand their current risks is risk exploration. In this scenario, Copilot for Security will leverage the insights from the risk-based recommendations (when Defender CSPM is enabled) and enable the cloud security administrator who is interacting with the platform, via prompts, to continue asking more questions about risk.

To access Copilot for Security in Defender for Cloud, to perform risk exploration, open Defender for Cloud dashboard and click on **Recommendations** in the left navigation pane. In the top-right corner, click the **Analyze with Copilot** option. Copilot for Security will suggest some prompts, as shown in *Figure 16.1*:

Figure 16.1: Initial prompts

While you can get started with your risk exploration by using those predefined prompts, you can also type your question directly in the prompt that appears in the bottom part of this page, as shown in *Figure 16.2*:

Figure 16.2: Prompt for open question

For the purpose of this example, let's use the predefined question **Show risks for resources with sensitive data**. Once you click on it, you will notice that Copilot for Security will go through a series of steps before providing the answer, as shown in *Figure 16.3*:

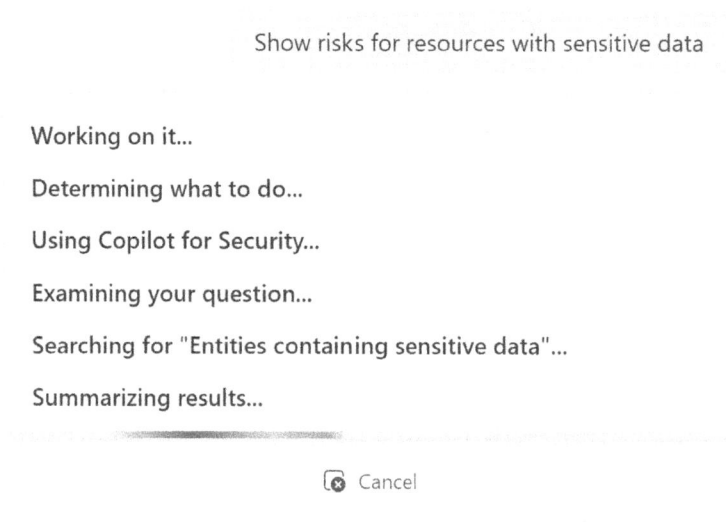

Show risks for resources with sensitive data

Working on it...

Determining what to do...

Using Copilot for Security...

Examining your question...

Searching for "Entities containing sensitive data"...

Summarizing results...

Cancel

*Figure 16.3: Steps taken before providing the answer*

After performing the steps shown in *Figure 16.3*, Copilot for Security will summarize an answer, as shown in *Figure 16.4*:

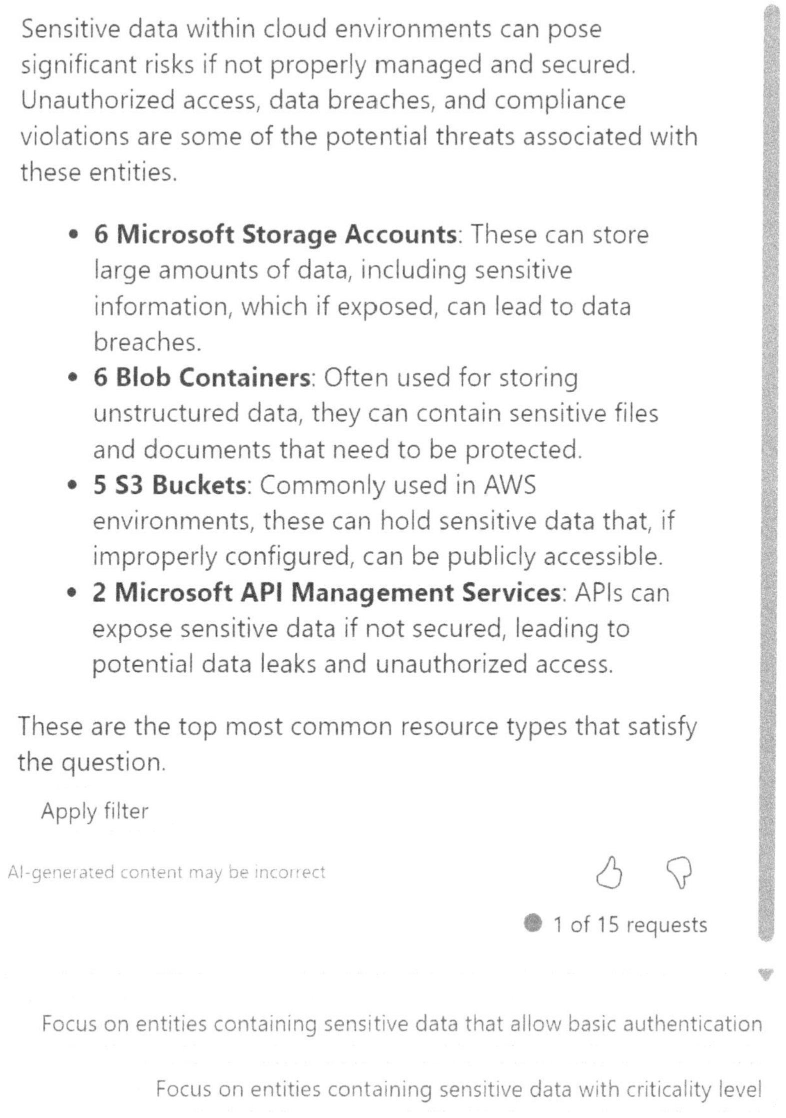

Sensitive data within cloud environments can pose significant risks if not properly managed and secured. Unauthorized access, data breaches, and compliance violations are some of the potential threats associated with these entities.

- **6 Microsoft Storage Accounts**: These can store large amounts of data, including sensitive information, which if exposed, can lead to data breaches.
- **6 Blob Containers**: Often used for storing unstructured data, they can contain sensitive files and documents that need to be protected.
- **5 S3 Buckets**: Commonly used in AWS environments, these can hold sensitive data that, if improperly configured, can be publicly accessible.
- **2 Microsoft API Management Services**: APIs can expose sensitive data if not secured, leading to potential data leaks and unauthorized access.

These are the top most common resource types that satisfy the question.

Apply filter

AI-generated content may be incorrect

1 of 15 requests

Focus on entities containing sensitive data that allow basic authentication

Focus on entities containing sensitive data with criticality level

*Figure 16.4: Summarization of the answer*

Copilot for Security gives an overview of the number and type of resources that have sensitivity data. While this is a good start, you can continue to explore the risk by using the predetermined follow-up questions (the two at the bottom of *Figure 16.4*). You can also click the **Apply filter** button to automatically create a filter on the **Recommendations** page that reflects the open recommendations for these resources highlighted in the answer.

Once you find the recommendation that you need to address and open it, you will also be able to use Copilot for Security to summarize the recommendation. The **Summarize with Copilot** button will be available on the right side of the recommendation, as shown in *Figure 16.5*:

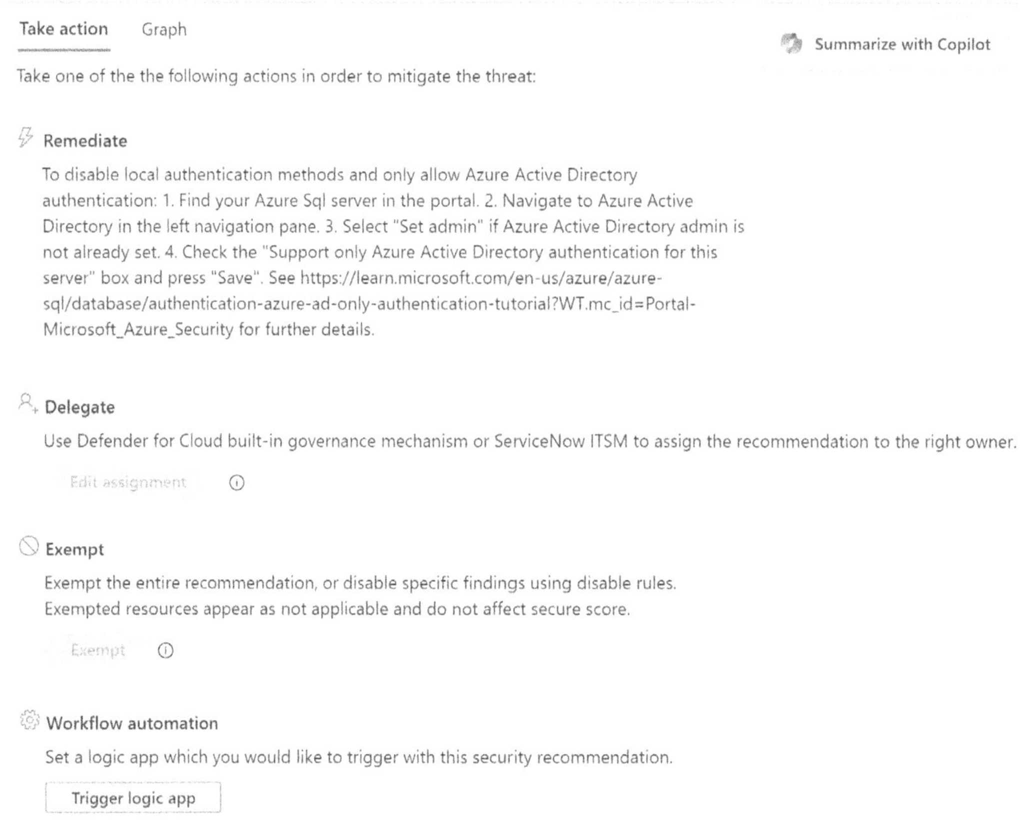

Take action    Graph                                                       Summarize with Copilot

Take one of the the following actions in order to mitigate the threat:

⚡ **Remediate**

To disable local authentication methods and only allow Azure Active Directory authentication: 1. Find your Azure Sql server in the portal. 2. Navigate to Azure Active Directory in the left navigation pane. 3. Select "Set admin" if Azure Active Directory admin is not already set. 4. Check the "Support only Azure Active Directory authentication for this server" box and press "Save". See https://learn.microsoft.com/en-us/azure/azure-sql/database/authentication-azure-ad-only-authentication-tutorial?WT.mc_id=Portal-Microsoft_Azure_Security for further details.

🧑‍🦱 **Delegate**

Use Defender for Cloud built-in governance mechanism or ServiceNow ITSM to assign the recommendation to the right owner.

    Edit assignment    ⓘ

🚫 **Exempt**

Exempt the entire recommendation, or disable specific findings using disable rules. Exempted resources appear as not applicable and do not affect secure score.

    Exempt    ⓘ

⚙️ **Workflow automation**

Set a logic app which you would like to trigger with this security recommendation.

    Trigger logic app

*Figure 16.5: Summarize with Copilot button (for better visualization, refer to https://packt.link/gbp/9781836204879)*

When you click this button, Copilot for Security will fetch the result and create a summary, as shown in *Figure 16.6*:

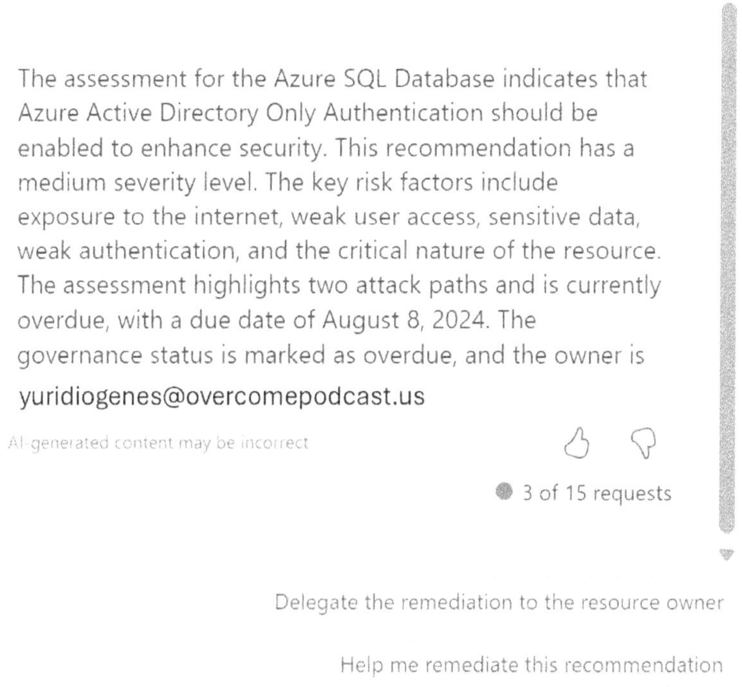

The assessment for the Azure SQL Database indicates that Azure Active Directory Only Authentication should be enabled to enhance security. This recommendation has a medium severity level. The key risk factors include exposure to the internet, weak user access, sensitive data, weak authentication, and the critical nature of the resource. The assessment highlights two attack paths and is currently overdue, with a due date of August 8, 2024. The governance status is marked as overdue, and the owner is yuridiogenes@overcomepodcast.us

AI-generated content may be incorrect

● 3 of 15 requests

Delegate the remediation to the resource owner

Help me remediate this recommendation

*Figure 16.6: Recommendation summary*

Notice that one of the follow-up questions suggested is **Help me remediate this recommendation.** Copilot for Security will provide the steps to remediate the recommendation, which can be similar to the steps that you already have in the recommendation itself (under the **Take action** tab). However, you can ask for help to create a script.

For example, while exploring the recommendation **Storage account public access should be disallowed**, you can ask for direct assistance to create a script, as shown in the example of *Figure 16.7*:

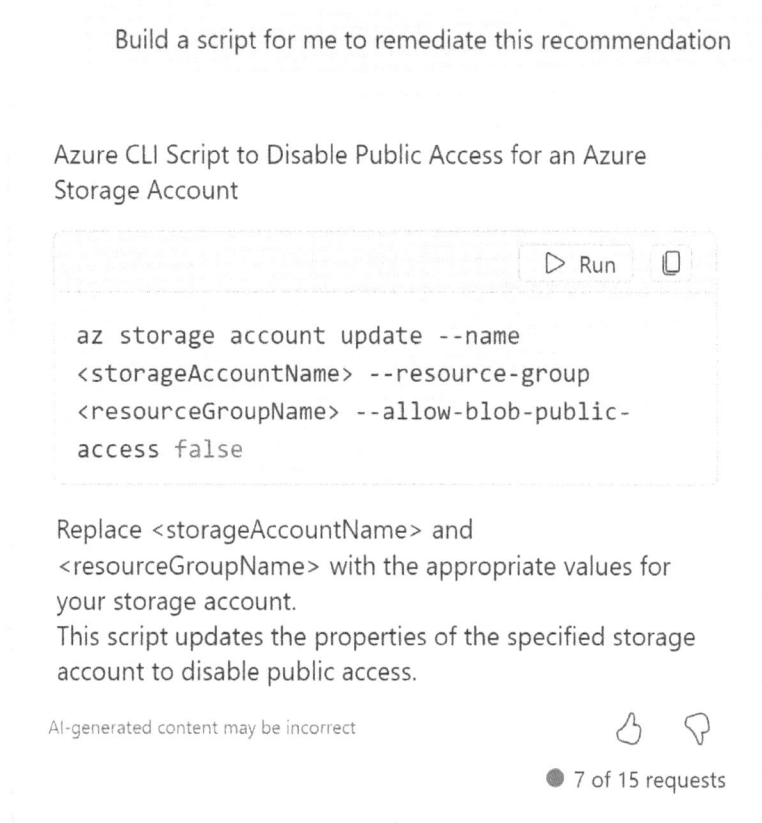

Build a script for me to remediate this recommendation

Azure CLI Script to Disable Public Access for an Azure Storage Account

▷ Run    🗍

```
az storage account update --name
<storageAccountName> --resource-group
<resourceGroupName> --allow-blob-public-
access false
```

Replace <storageAccountName> and <resourceGroupName> with the appropriate values for your storage account.
This script updates the properties of the specified storage account to disable public access.

AI-generated content may be incorrect                              👍    👎

● 7 of 15 requests

*Figure 16.7: Assistance to build a script*

While this is not a "ready to run" type of script, it gives you the template that you just need to replace with the final values before executing. Also, the answer may vary according to the recommendation, so keep in mind that this is just one example of assistance to build a script to remediate resources.

While this integration will help you to leverage AI to improve your security posture, now you need to look at how to protect GenAI apps from potential attacks, and that's where AI posture management features will assist you.

# AI posture management

According to the **First annual generative AI study: business rewards vs. security risks** 2023[4], the top concern for business leaders and cybersecurity professionals was the leakage of sensitive data by staff using AI. While companies feel the need to empower their employees to leverage AI, the concern around security is not only valid but also important to address.

The first step in securing AI workloads is to understand which workloads are available, in other words, visibility. Defender CSPM continuously discovers GenAI apps deployed across Azure OpenAI Service, Azure Machine Learning, and Amazon Bedrock. In addition to that, it will also discover vulnerabilities within generative AI library dependencies such as **TensorFlow**, **PyTorch**, and **LangChain**. This is done by the DevOps security capabilities built into Defender CSPM that will be used to scan source code for **Infrastructure as Code** (**IaC**) misconfigurations and container images for vulnerabilities.

Once it finds the AI workloads, it will perform an assessment, and the result of this assessment may generate a security recommendation, such as the one shown in *Figure 16.8*:

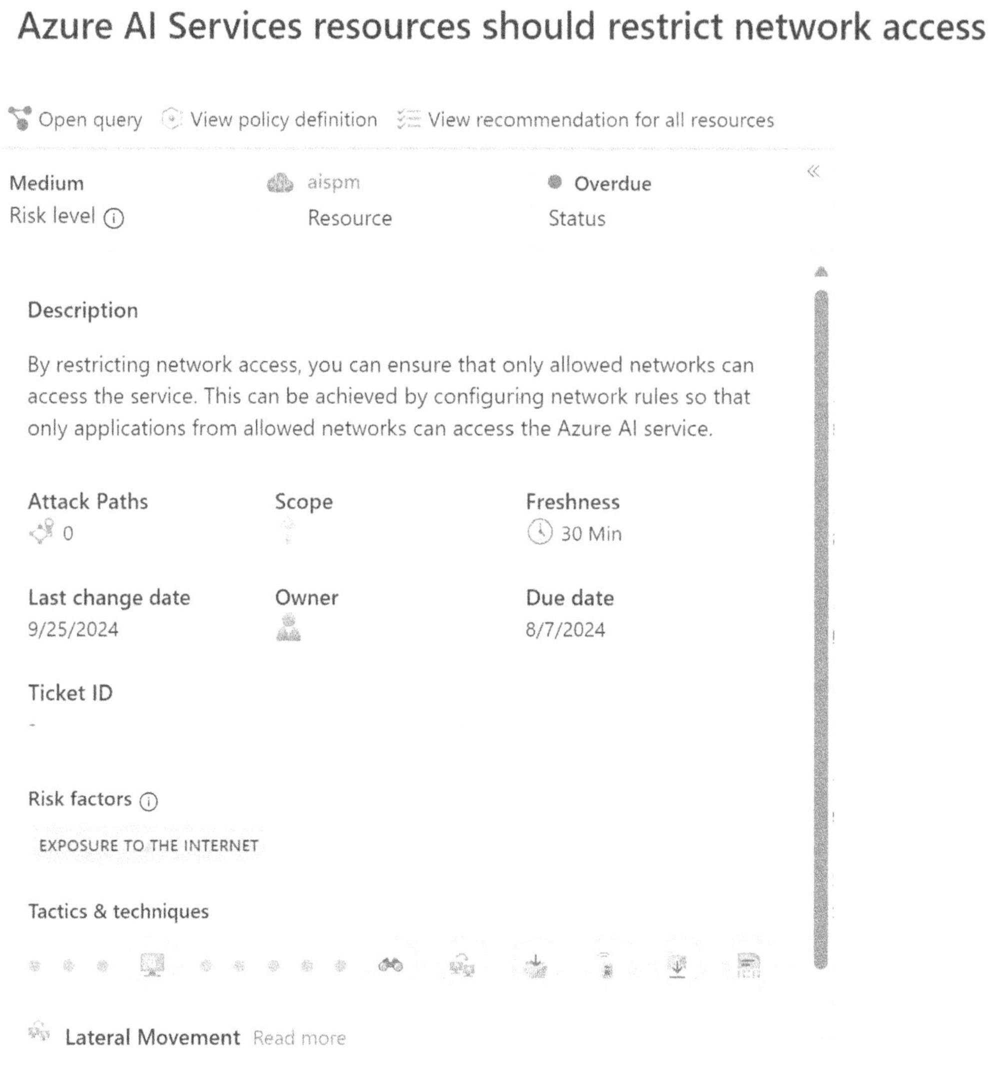

# Azure AI Services resources should restrict network access

🕸 Open query    ⟳ View policy definition    ⋮≡ View recommendation for all resources

| Medium | aispm | ● Overdue | « |
|---|---|---|---|
| Risk level ⓘ | Resource | Status | |

### Description

By restricting network access, you can ensure that only allowed networks can access the service. This can be achieved by configuring network rules so that only applications from allowed networks can access the Azure AI service.

| Attack Paths | Scope | Freshness |
|---|---|---|
| ⫟ 0 | | 🕐 30 Min |

| Last change date | Owner | Due date |
|---|---|---|
| 9/25/2024 | 🧍 | 8/7/2024 |

Ticket ID

-

Risk factors ⓘ

EXPOSURE TO THE INTERNET

Tactics & techniques

⊡ Lateral Movement  Read more

*Figure 16.8: AI security recommendation*

In addition to recommendations, Defender CSPM is also able to identify lateral movements from other workloads (such as VMs) to Azure OpenAI. For example, the attack path **Internet exposed Azure VM with high-severity vulnerabilities allows lateral movement to Azure OpenAI** provides visibility about a scenario where the threat actor can exploit vulnerabilities in a VM and leverage managed identity to access an Azure AI service.

The details of this attack path are shown in *Figure 16.9*:

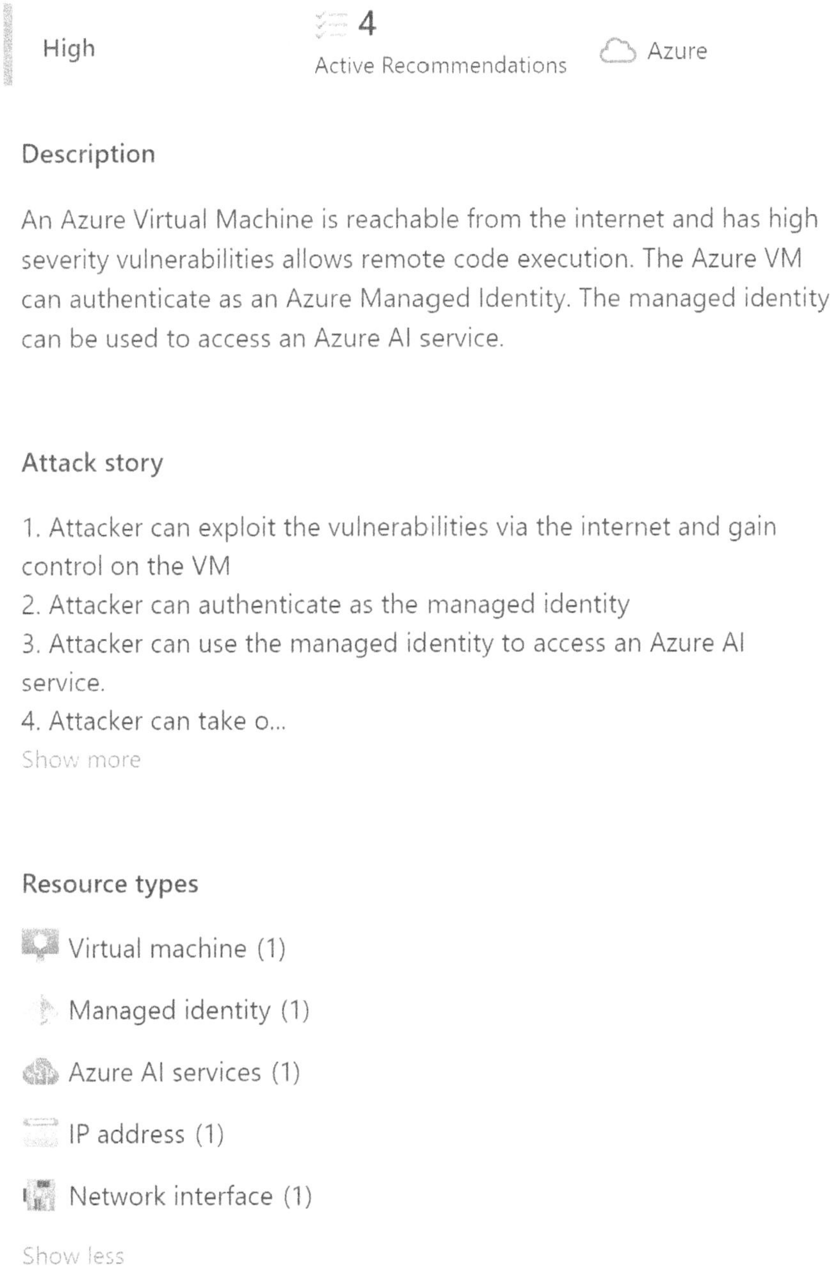

High

**4**
Active Recommendations

Azure

## Description

An Azure Virtual Machine is reachable from the internet and has high severity vulnerabilities allows remote code execution. The Azure VM can authenticate as an Azure Managed Identity. The managed identity can be used to access an Azure AI service.

## Attack story

1. Attacker can exploit the vulnerabilities via the internet and gain control on the VM
2. Attacker can authenticate as the managed identity
3. Attacker can use the managed identity to access an Azure AI service.
4. Attacker can take o...
Show more

## Resource types

Virtual machine (1)

Managed identity (1)

Azure AI services (1)

IP address (1)

Network interface (1)

Show less

*Figure 16.9: Attack path that leads to Azure OpenAI service*

The combination of risk-based recommendations tailored for AI and the possibility to disrupt potential attacks targeting AI Services with the attack path, are tools that will help to improve your AI security posture. However, just like you learned in *Chapter 9*, it is always important to perform proactive hunting to ensure that there are other areas that can be improved before a threat actor's exploit.

To gain visibility into all code repositories within your environment that contain known GenAI vulnerabilities and were provisioned in Azure OpenAI, the AI posture management capability added two new query templates in Cloud Security Explorer, as shown in *Figure 16.10*:

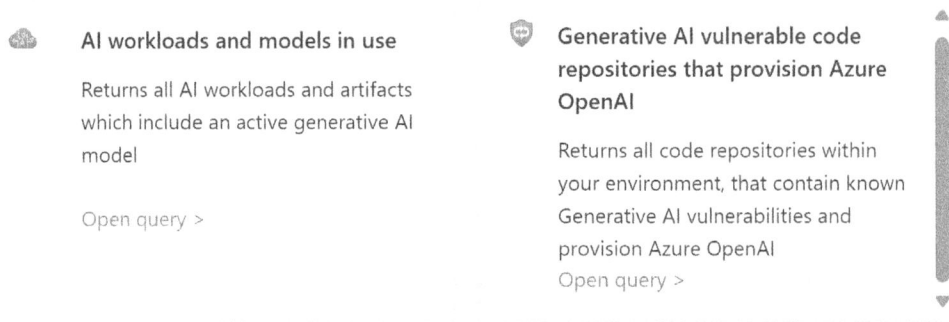

**AI workloads and models in use**

Returns all AI workloads and artifacts which include an active generative AI model

Open query >

**Generative AI vulnerable code repositories that provision Azure OpenAI**

Returns all code repositories within your environment, that contain known Generative AI vulnerabilities and provision Azure OpenAI

Open query >

*Figure 16.10: Cloud Security Explorer templates*

These templates will enable you to perform queries using Cloud Security Explorer and obtain useful information about the AI model in use, the deployment model, and potential AI-vulnerable code.

# Summary

In this chapter, you learned about the Defender for Cloud integration with Copilot for Security. You learned how to perform risk exploration by using Copilot for Security embedded experience in Defender for Cloud. You learned how to summarize recommendations and how to ask Copilot for Security to generate a remediation script. You also learned about AI posture management, the importance of having security recommendations tailored for AI scenarios, how Defender CSPM takes into consideration Azure AI as part of the attack path, and the AI queries available in Cloud Security Explorer.

In the next chapter, you will learn about Security Exposure Management.

## Notes

1. To see how Microsoft Copilot for Security integrates with other Microsoft solutions, review this diagram: `https://bit.ly/CNAPPBook110`.

2. For more information about how to manage user access to Microsoft Copilot in Azure, visit `https://bit.ly/CNAPPBook108`.

3. To learn more about SCUs, visit `https://bit.ly/CNAPPBook109`.

4. You can download this report from `https://bit.ly/CNAPPBook111`.

## Additional resources

- You can watch this episode of *Defender for Cloud in the Field* where the author (*Yuri Diogenes*) interviews the PM responsible for the AI Posture Management capability in Defender for Cloud: `https://bit.ly/CNAPPBook112`

- Use the guidelines to investigate alerts about Defender for Key Vault: `https://bit.ly/CNAPPBook74`

## Leave a Review!

Thank you for purchasing this book from Packt Publishing—we hope you enjoy it! Your feedback is invaluable and helps us improve and grow. Once you've completed reading it, please take a moment to leave an Amazon review; it will only take a minute, but it makes a big diff erence for readers like you. Scan the QR or visit the link to receive a free ebook of your choice.

`https://packt.link/NzOWQ`

# 17

# Security Exposure Management

In today's evolving digital landscape, most of the organizations are more exposed to risks than ever, bringing more areas and expanding out the landscape creates a lot of growth and amazing opportunities but also significant risks. The main challenges are prioritization and the ability to focus on the areas that are considered as the most critical of the business and likely named as crown jewels. The overgrowing amounts of threats, vulnerabilities, control misconfigurations, overprivileged access, and cyber threats becomes to be more challenging mainly due lack of human resources, this can lead to situation when the strategy will be to handle incidents after they already happen which is not the best way to go. Issues like data breaches can lead to financial losses, reputational damage, and, ultimately, a loss of trust.

Protecting and remediating the full threat landscape is challenging, time-consuming and costly. To effectively manage these risks, organizations must create visibility across their landscape and identify potential threats before they will harm and create damage on their crown jewels. Implementing proactive measures to prevent risks before they will become to be security incidents is the key. However, achieving this level of visibility and control could be very challenging. Usually most of the organizations today implements wide array of security tools to manage different areas in their security posture and the CWPP (cloud workload protection platform), resulting in fragmented insights and limited capabilities to mitigate risks holistically.

MSEM (Microsoft Security Exposure Management) is the evolving name for XSPM (Extended Security Posture Management), and the meaning of the name is to extend the security posture area by providing a unified view of security posture across the landscape. By integrating Defender tools from Microsoft and the current existing security tools MSEM helps to better understand the security landscape and make the life of the decision makers and the practitioners much easier.

MSEM is trying to change the legacy way and to go beyond traditional vulnerability management. By collecting all the asset information with security context such as criticality (crown jewels) it can helps to identify, assess, and prioritize security risks. This enables organizations to proactively manage their attack surfaces by reducing the number of choke points, protect critical assets that are marked with issues, and mitigate exposure risks. The result is a more resilient security posture, enabling businesses to stay ahead of evolving threats.

This chapter covers:

- Understanding unified security management
- Onboarding Microsoft Security Exposure Management
- Operationalizing unified exposure management
- Addressing the attack surface

# Understanding unified security management

To be able to manage a rapidly growing environment there is a critical need for a unified solution to assist and manage security across all the infrastructure. The traditional methods of securing such "patching everything" and often involve solutions that don't communicate well with one another, are no longer relevant. Implementing strategies to identify, prioritize, and mitigate risks effectively is a must to have.

MSEM is exactly the tools that can assist here, by providing a comprehensive unified view of the security posture across the entire digital estate. It collects and enriches asset information with critical security context combining that with data that could assist and allow organizations to proactively manage attack surfaces, protect critical assets, and explore and mitigate exposure risks.

## Integration with Microsoft Defender for Cloud

One of the key strengths of MSEM is its integration with Microsoft Defender for Cloud, a comprehensive cloud-native security solution that provides unified security management and advanced threat protection across hybrid cloud environments. The main idea behind this integration is to get a "bird's eye view" of the security cloud coverage that is protected by Microsoft Defender for Cloud for both CSPM and CWPP.

MSEM's integration with Defender for Cloud brings together critical cloud security initiative and providing metrics and recommendations to your cloud infrastructure. This combination of Cloud Security Posture Management (CSPM) and Cloud Workload Protection (CWP) delivers a score mechanism and holistic view approach to assess and remediate the environment.

By leveraging the integration with MDC (Microsoft Defender for Cloud) and the DCSPM plan more capabilities such as cloud attacks paths coverage can be achieved. The main idea of enhancing visibility into one area cross security posture, better understand potential attack vectors, and reviewing recommendations to remediate about how to protect their cloud and on-premises assets made a real impact.

To be able to manage exposure risks more effectively MSEM's Oversight which involves proactive approach such as initiatives and CAP (critical asset protection) enables organizations to achieve the target. By highlighting security gaps, misconfigurations, and multiple areas businesses can build a plan and implement a remediation plan to remediate issues before they become an incident. The combination with criticality that allow us to know which asset is defined as critical and the data that that enriching the graph both the decision makers and the practitioners can prioritize risks based on their potential impact, ensuring that security teams focus on the most pressing issues first.

# Onboarding Microsoft Security Exposure Management

Before onboarding MSEM, you need to ensure that you have tenant-level permissions. This involves providing access at the tenant level to allow users to fully utilize MSEM features. It's essential that user roles have the necessary access to all Defender for Endpoint device groups1. Users that configured with device groups restriction will not be able to access critical information related to other device groups and related attack paths. This restriction also will affect access to the attack surface area and schema tables in advanced hunting, to overcome this restriction users should not be scoped.

In addition to tenant-level permissions, you need to assign Role-Based Access Control (RBAC) permissions to the administrators responsible for creating the MSEM workspace. This MSEM workspace will be used for managing as the central hub for collecting security-related data and insights, ingesting asset information with criticality level that will assist to proactively reduce and manage initiatives and attack surfaces

To create and manage the MSEM workspace, your tenant must include at least one Global Admin or Security Admin. For users to gain full access to the MSEM environment, they must be assigned one of the following Microsoft Entra ID roles2: Global Admin for complete read and write permissions, Global Reader for read-only access, Security Admin for security-specific read and write permissions, Security Operator for read and limited write permissions, or Security Reader for read-only access. These roles ensure that the right personnel have the appropriate levels of access to manage security tasks efficiently.

It's crucial to verify that all roles and permissions aligned with your security policies and oper-ational requirements, ensuring that MSEM (Microsoft Security Exposure Management)  will be able to operate otherwise some areas such as exposed entities and attack paths will be hidden.

After creating a user with a Global Reader role, test the access by navigating to `https://security.` `microsoft.com/` and expanding the Exposure management menu to select **Overview**, as shown in *Figure 17.1*:

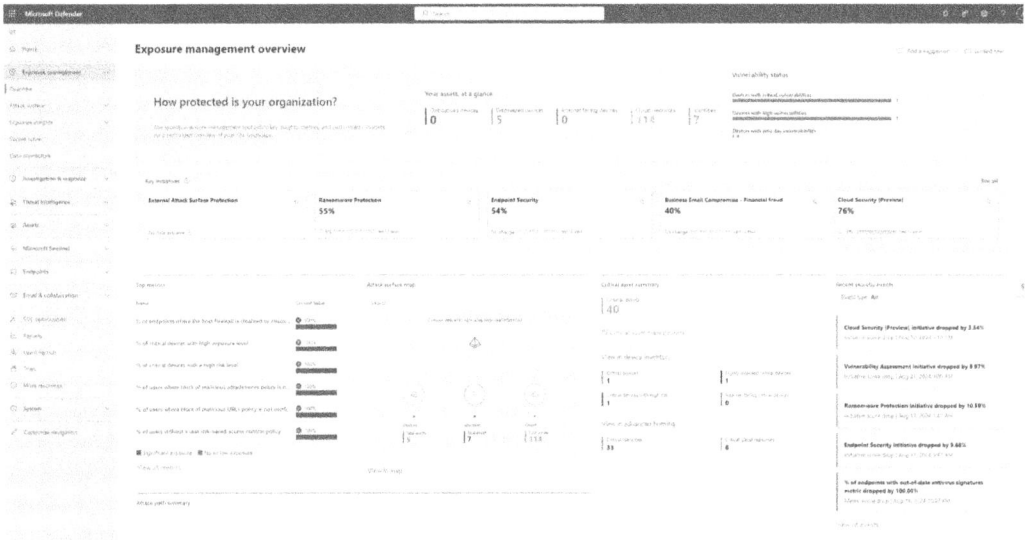

*Figure 17.1: Microsoft Security Exposure management overview page (for better visualization, refer to https://packt.link/gbp/9781836204879)*

The first thing you need to define and configure after accessing the unified security portal is crit-ical assets for your organization. This is an important task and will be explained in more detail in the next section.

# Critical asset validation

As part of the overall functionality to help you focus on critical assets in your infrastructure, there are built-in classification rules provided by MSEM. These predefined classifications include a variety of entities, such as cloud resources, identities, and devices. Some examples of these pre-defined rules are:

- IT admin workstations
- ADFS (Active Directory Federation Services)
- Domain controllers
- Exchange servers

To view the complete list of predefined rules, you can navigate to Microsoft Defender XDR under the **System settings** menu. This will provide you with a comprehensive list of all the classification rules available, as shown in *Figure 17.2*:

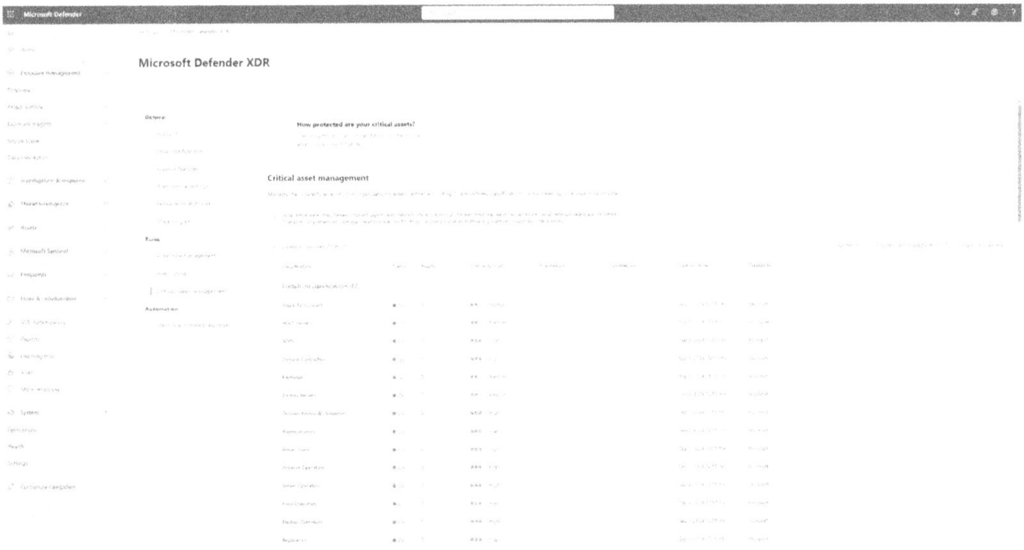

*Figure 17.2: Critical asset management view (for better visualization, refer to https://packt.
link/gbp/9781836204879)*

After reviewing that list, you should validate that these critical assets are relevant to your infra-structure. This list should be monitored and tracked over time since new devices or assets that will be populated will be in a pending approval state. To make the final determination, you will need to approve these devices to be treated as critical.

For the onboarding stage, it is essential to review the predefined rules and validate the device and the asset list to ensure they are critical. In a situation where you do not want to use the predefined rule, you can turn it off, as shown in *Figure 17.3*:

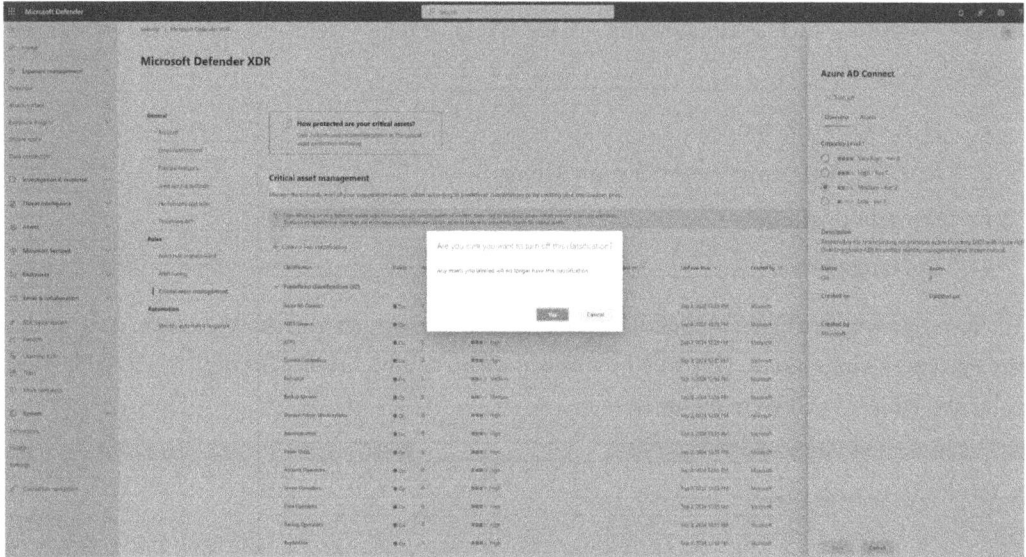

*Figure 17.3: Critical asset management—turning off a rule (for better visualization, refer to https://packt.link/gbp/9781836204879)*

This approach ensures that rules for critical assets are identified accurately while providing flexibility where there is uncertainty.

Although the customization of critical assets via classification rules is not mandatory, it is recommended since every organization has different business needs. In some cases, devices will not be marked as critical assets as part of the automated discovery process, so you will need to create custom classification rules. One example could be devices in the DMZ network that have not detected the predefined classification rules. In that case, you will need to create a custom classification rule. Another example could be a group of servers that are part of a critical application, but they do not have a role that makes them appear as a critical asset, and therefore, a custom predefined rule needs to be created. To create a custom classification, use the **Create a critical asset classification** wizard, as shown in *Figure 17.4*:

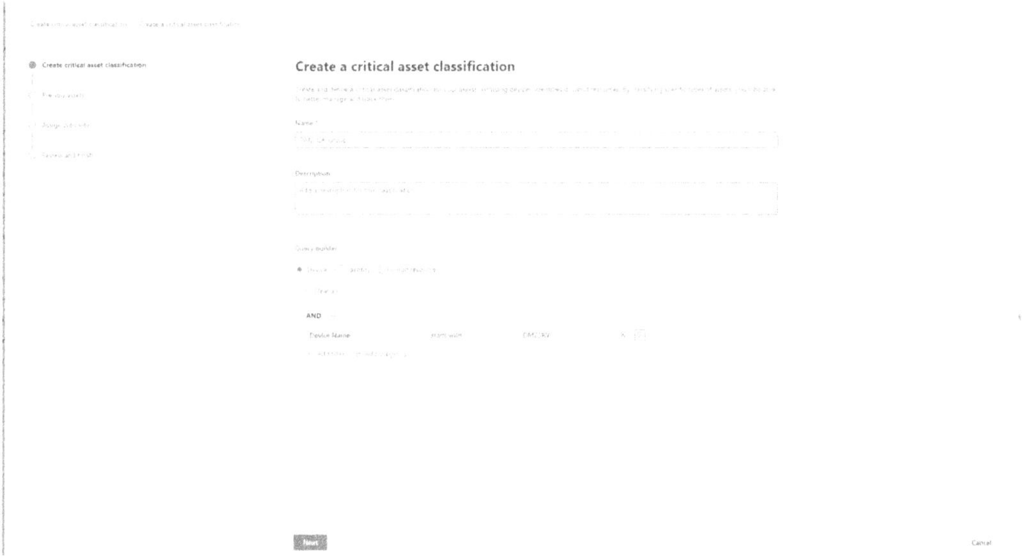

*Figure 17.4: Critical asset management rules—set the name of the rule and use the Query builder operators (for better visualization, refer to https://packt.link/gbp/9781836204879)*

In this wizard, you can review the assets list result based on your criteria, set the criticality level, and click the **Save** button to commit the changes. After completing the steps, you should be able to see the new predefined classification rule, as shown in *Figure 17.5*:

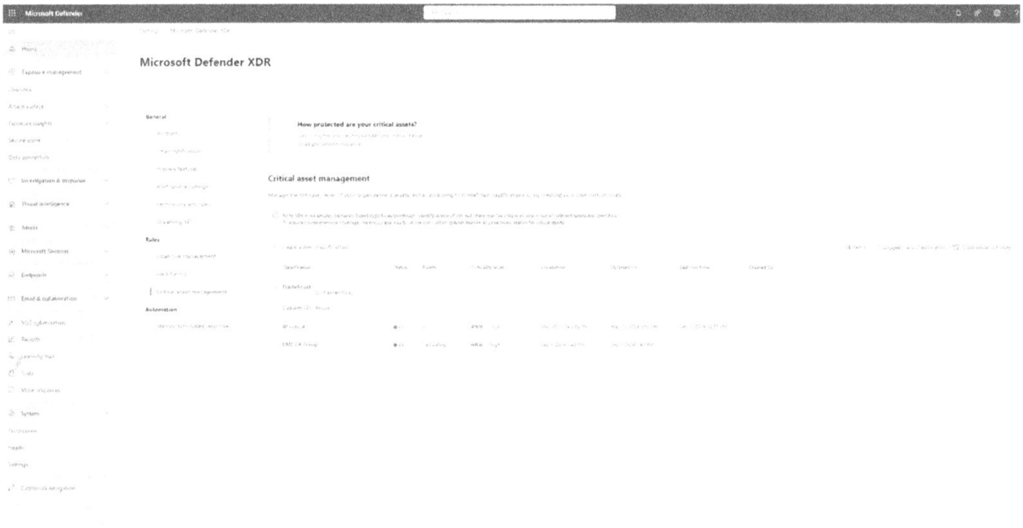

*Figure 17.5: Critical asset management—review the predefined rule (for better visualization, refer to https://packt.link/gbp/9781836204879)*

To finish the onboarding, you must connect first-party and third-party solutions. The first-party solutions are from the Microsoft Defender family, which includes:

- Microsoft Defender for Endpoint
- Microsoft Defender for Identity
- Microsoft Defender for Cloud Apps
- Microsoft Defender for Office
- Microsoft Defender for IoT
- Microsoft Secure Score
- Microsoft Defender Vulnerability Management
- Microsoft Defender for Cloud
- Microsoft Entra ID
- Microsoft Defender External Attack Surface Management (EASM)

If you have only some of those workloads, that should be fine, but from the onboarding perspective, the more you have, the broader the insights and benefits you will receive from the product. The last part of the onboarding process is connecting third-party connectors if you have different security solutions. As of the time of writing this chapter, this feature only allows a subset of connectors.

# Operationalizing unified exposure management

As you learned at the beginning of this chapter, one of the critical elements of exposure management is unified exposure management.

To be able to manage security exposure, there are two focus areas: Key initiatives and Top metrics. From my experience working with a wide range of customers, these discussions are always at the forefront. The target goal is to allow both the decision make and the practitioner to be able to decide on which security programs(initiatives) he would like to focus and identifying which ones to prioritize, whether it is based on the areas they want to invest in or those needing an overall improvement. By achieving an improving score over time and reducing critical crown jewels assets that are marked as critical could be a great jump start for this journey.

## Reviewing key initiatives

The **Key initiatives** option provides a quick view of initiatives, marked as priority or newly recommended. From discussions with customers, I have found that selecting the right initiatives is crucial for success. Organizations can start by focusing on areas they are already investing in or areas that need improvement.

Whether the goal is to enhance security in specific departments or to reduce risks across the board, key initiatives serve as a roadmap.

By identifying and prioritizing key initiatives, organizations can reduce their exposed entities that are marked as critical and focused on the areas that will have the most significant impact, as shown at Figure 17.6:

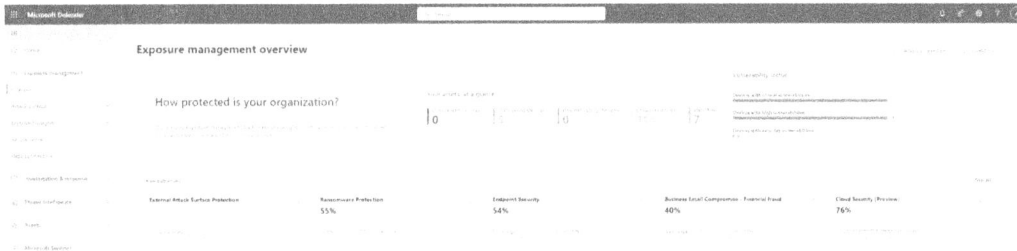

*Figure 17.6: Key initiatives—reviewing key initiatives (for better visualization, refer to https:// packt.link/gbp/9781836204879)*

# Reviewing top metrics

Alongside initiatives, the Top metrics section is an essential tool for identifying the security areas with the most exposure. I often advise customers to combine their focus on reducing exposed assets with tracking top security metrics. These metrics provide real-time data on the current state of exposure and trends over time.

Focusing on metrics with the highest current exposure values can allows proactively mitigate risks. By targeting the areas with the highest negative scores and remediating the associated recommendations to be marked as compliant can reduce short-term risks and improves overall security hygiene, as shown at *Figure 17.7*:

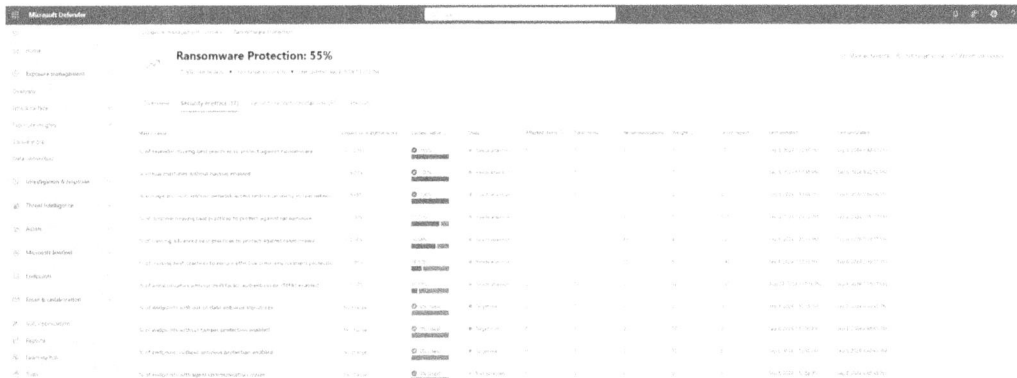

*Figure 17.7: Security metrics tab (for better visualization, refer to https://packt.link/ gbp/9781836204879)*

# Combining initiatives with metrics for proactive security

In my experience working with enterprise large scale of organizations, the best results could arrive from both strategies that align key initiatives with top security metrics and a well-designed remediation plan. By combining those both and educating the teams to be able mitigating and remediating those recommendations which will lead to overall score improvement can be a very meaningful achievement. This proactive journey investment not only mitigates immediate risks but also helps with communicating to the c-level and providing a great visibility around the impacts combining the long-term stability and security.

The recommendations feature is accessible by going to the initiative itself and choosing the Security recommendations tab, as shown in *Figure 17.8*. This page is designed to offer organizations an actionable list of steps that should be taken to remediate security risks.

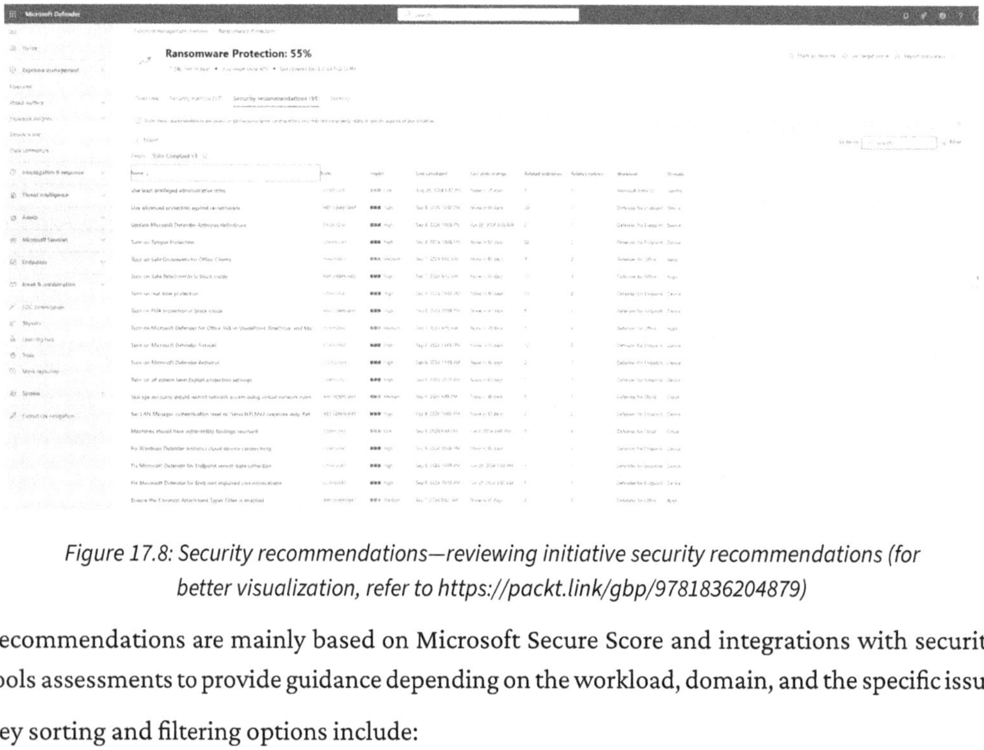

*Figure 17.8: Security recommendations—reviewing initiative security recommendations (for better visualization, refer to https://packt.link/gbp/9781836204879)*

Recommendations are mainly based on Microsoft Secure Score and integrations with security tools assessments to provide guidance depending on the workload, domain, and the specific issue.

Key sorting and filtering options include:

1. Name: The name of the recommendation.

2. State:  Can be Compliant or Not Compliant.

3. Impact: The impact rating (high, medium, or low).

4. Workload: The workload name such as Defender for Cloud, Defender for Endpoint and similar).

5.    Domain: The domain area, such as devices, apps, data, or identity.

Once the recommendations have been reviewed, security teams can proceed with the remediation process by using the following steps:

1.    Select the specific recommendation from the **Recommendations** page.

2.    Navigate to the **Remediation steps** tab, where you will find a detailed guide on resolving the issue.

To be able to Remediate effectively it will often require coordination between various teams that owns specific areas within the organization. For example, if the recommendation is related to cloud security, practitioners will be needed to perform actions in Microsoft Defender for Cloud or Microsoft Defender Vulnerability Management, as shown in *Figure 17.9*:

**Turn on Tamper Protection**

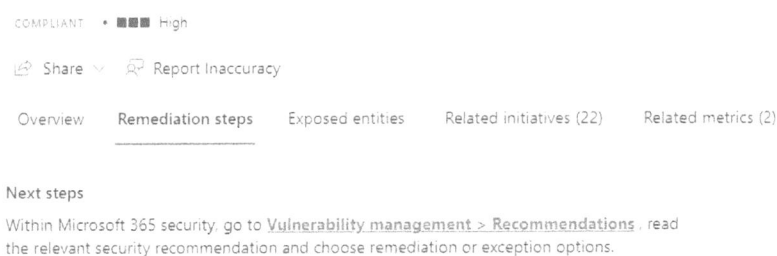

COMPLIANT   •  ■■■  High

🖉 Share ∨   🔎 Report Inaccuracy

Overview      Remediation steps      Exposed entities      Related initiatives (22)      Related metrics (2)

Next steps

Within Microsoft 365 security, go to <u>Vulnerability management > Recommendations</u> , read
the relevant security recommendation and choose remediation or exception options.

*Figure 17.9: Security recommendations—reviewing the remediation steps (for better visualization, refer to https://packt.link/gbp/9781836204879)*

It is very important that recommendations are reviewed in a timely manner and the history graph could assist here to make sure everything is aligned. Effective remediation reduces the needs of mitigating, but we always need to ensure that compliance and security goals are consistently met.

# Proactive security recommendations

In my experience, organizations that take a proactive approach to implementing security recommendations tend to see the best results in terms of minimizing incidents. By acting on recommendations early, these organizations are better positioned to handle emerging threats and prevent potential breaches.

MSEM Security events provides great view to monitor our overall security posture. Reviewing security events to better track significant fluctuations in the scores of key initiatives or security metrics could be very insightful.

Mainly these events provide insights into changes that impact areas such as the initiatives and the metrics, scores trends such as increasing or decreasing can be very helpful and provides reasoning answers to security changes in our environment.

MSEM offers organizations an opportunity to experiment with tracking and interpreting security events. The product is still evolving, and some features may be adjusted before its official release. To effectively monitor and review security events in **Microsoft Defender**, you can use two different sets of steps to access the relevant data:

> **Option 1:** In the **Overview** dashboard, navigate to the right side of the page, where you will find a section titled **Recent security events.** Click on it, and you will see a detailed view of the most recent incidents affecting your organization's security posture, as shown in *Figure 17.10*:

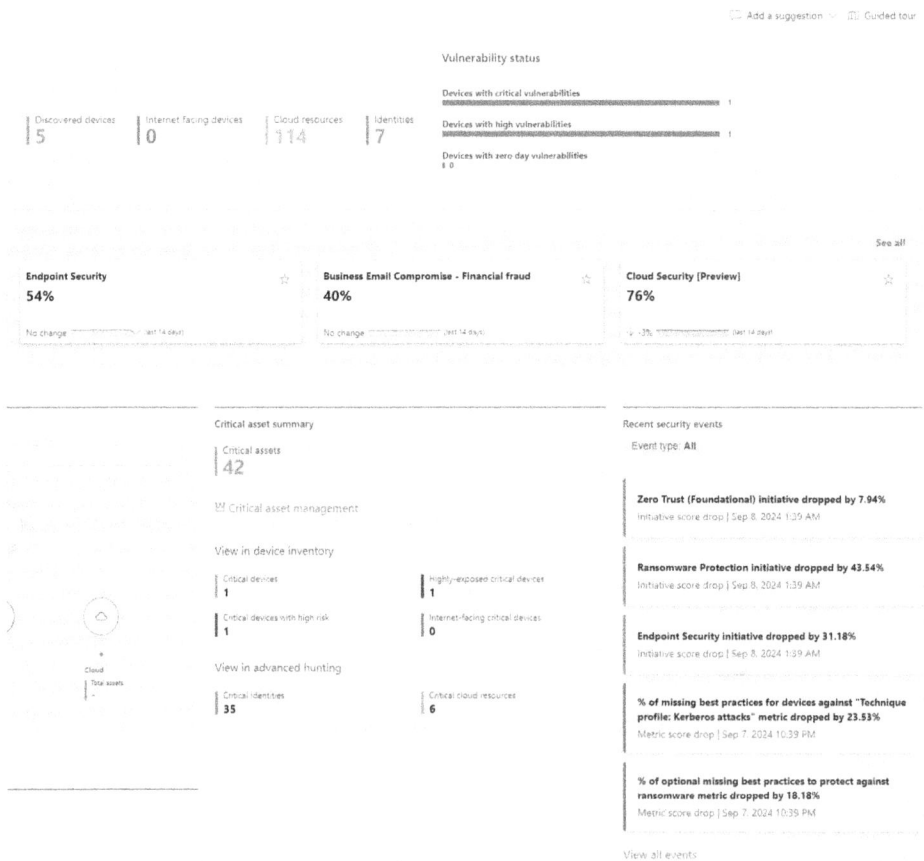

*Figure 17.10: Security events from the Overview page (for better visualization, refer to https://packt.link/gbp/9781836204879)*

**Option 2:** You can also access security events by expanding the **Exposure insights** menu on the left-hand navigation pane. From there, select **Events** to open the events page, where you can filter and sort by different parameters, such as initiative or metric score drops. *Figure 17.11* shows an example:

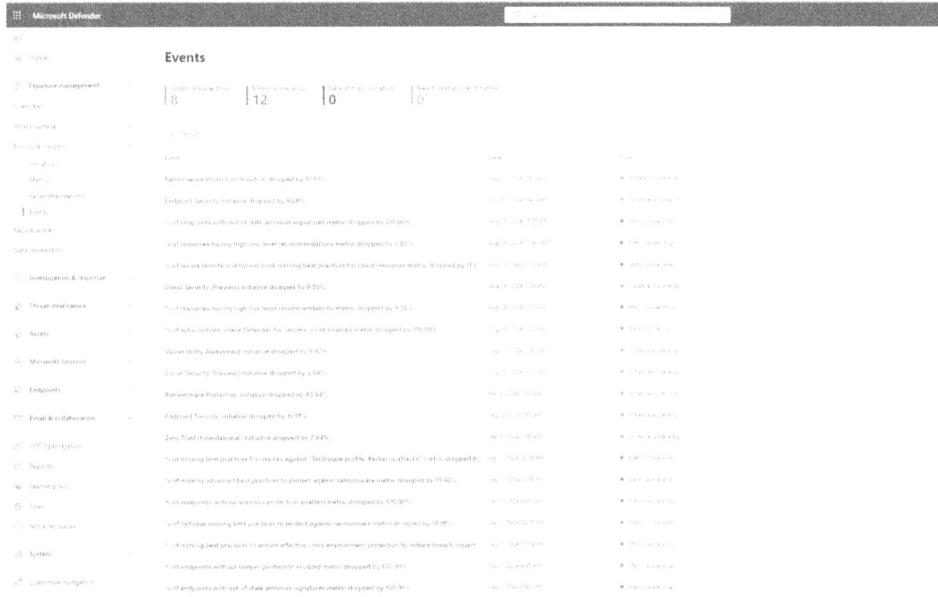

*Figure 17.11: Security events from the Exposure insights page (for better visualization, refer to https://packt.link/gbp/9781836204879)*

Each event provides a detailed look into areas where the organization's security posture has experienced notable changes. By selecting an event, you can explore its details and see how it connects to a broader initiative or metric. You can access more in-depth information from this view by navigating to the relevant **Initiatives** or **Metrics** page.

By Monitoring the security events organizations can better understand trend with regards to the relevant security area. As an example, a drop in initiative or metric scores signals a shift in the security landscape, potentially indicating misconfiguration that needs attention. By reviewing and acting quickly when these events occur, mitigating can be implemented and reduce the risks before they lead to more serious security incidents.

# Attack surface

MSEM attack surface provides tools for managing the organization's attack surface by attack surface map and visualizing attack paths. These attack paths, combined with the attack surface map, allow decision makers and practitioners to gain deeper insights into the assets and better understanding the potential risks and remediate them before they become a critical incident.

An attack path is a simplified way of graphically visualizing bad actors can use to navigate your on-prem and cloud environments. Attackers can leverage these different "paths" to access sensitive data and exploit a vulnerable configuration or resource.

**MSEM** automatically generates these attack paths based on data from integrated assets and workloads such as Microsoft Defender for Cloud, simulating potential attack scenarios—the more comprehensive the data, the more accurate and insightful the attack path analysis. However, if critical assets or integrated workloads are not fully defined, the generated paths may not fully represent your organization's actual risk landscape.

## Identifying and addressing attack paths

MSEM automatically collects data from assets and workloads to generate attack paths. These paths simulate potential attack scenarios and pinpoint vulnerabilities that attackers could exploit. The **Attack paths graph view** visually represents the attack paths, showing how attackers could move through the network. Hovering over nodes and connectors reveals detailed information about each component in the path. For example, an attack could start from a virtual machine containing TLS/SSL keys and move through permissions to access critical storage accounts.

The enterprise exposure map enhances visibility into attack paths by showing multiple potential paths and identifying critical choke points—nodes where multiple attack paths converge. Securing these choke points can disrupt several attack vectors at once.

MSEM also provides actionable recommendations to help mitigate potential attack paths. These insights offer clear, practical steps organizations can take to reduce exposure and secure vulnerable assets. *Figure 17.12* has an example of an attack path:

*Figure 17.12: Attack paths—reviewing attack paths and choke points (for better visualization, refer to https://packt.link/gbp/9781836204879)*

A critical part of managing attack paths is identifying and securing **choke points**—locations where multiple attack paths intersect to a critical asset. These bottlenecks represent high-risk areas and securing them can significantly reduce the number of potential attack paths.

MSEM allows you to group and visualize these choke point nodes. By focusing mitigation efforts on these critical areas, you can simultaneously reduce the risk of multiple attack paths. Strengthening security around choke points ensures that even if an attacker gains access to part of the network, their ability to reach valuable assets is severely limited.

Securing these nodes offers a strategic advantage, allowing security teams to concentrate on areas with the highest potential impact.

In addition to understanding attack paths, organizations must manage the potential entry points and vulnerabilities. The **Attack surface map** feature in MSEM offers a powerful visualization tool for exploring the connections between assets and identifying where your organization is most vulnerable.

*Figure 17.13* has an example of this capability:

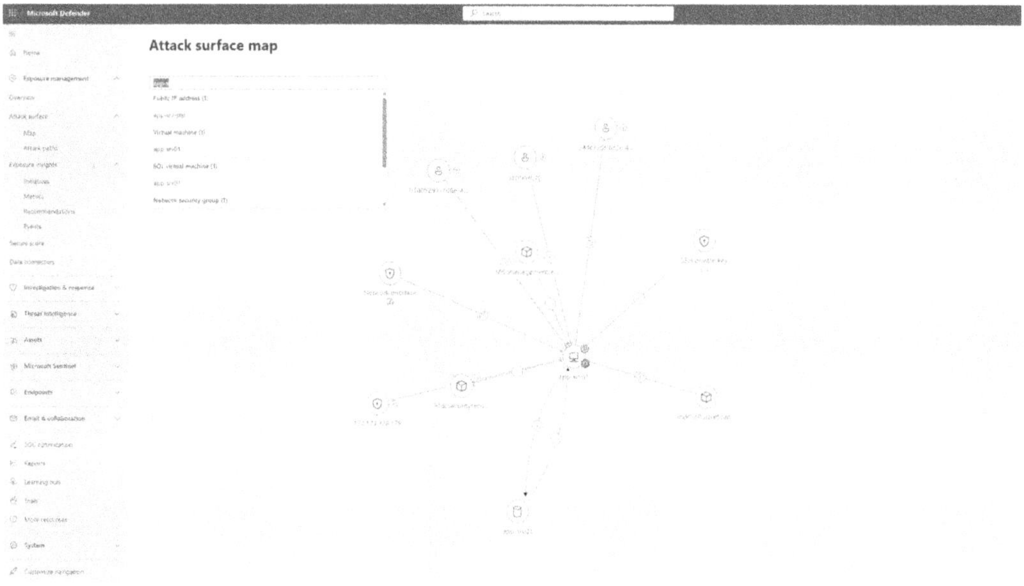

*Figure 17.13: Attack surface map—reviewing Attack surface map (for better visualization, refer to https://packt.link/gbp/9781836204879)*

Attack surface map provides a detailed visual representation of asset connections within the network, offering insights into how assets are linked and where vulnerabilities might exist:

1.  **Visual indicators**: The map uses icon indicators to represent node and edge types, such as a **high-criticality crown** to signify critical assets or a **vulnerability bug** to highlight at-risk assets. These visual indicators allow security teams to identify areas that need attention quickly.

2.  **Expandable groups**: Similar assets can be grouped together to help with organization. You can expand these groups for deeper exploration or collapse them for a more streamlined view. This flexibility allows security teams to focus on the most critical areas of the attack surface while avoiding cluttered views.

3.  **Interactive features**: Hovering over nodes and edges provides additional information about the connections and vulnerabilities. You can also click on assets to view further details or explore connected assets using the plus sign or contextual menu.

The **Attack surface map** feature allows you to zoom in on specific assets or connections, focusing on areas of particular interest. This is like the graph view used for exploring individual attack paths. By refocusing the map on a specific node, security teams can investigate its vulnerabilities and connections in greater detail.

Additionally, the search feature allows you to find specific assets by node type, making it easy to locate devices, identities, or cloud assets within the network. Filters can refine your search to focus on specific asset types or vulnerabilities.

## Summary

In this chapter, you learned more about Microsoft Security Exposure Management and how to enable it in your organization to manage the security posture of your workloads proactively. You learned the importance of incorporating **key initiatives, top metrics,** and **security recommendations** into a cohesive strategy. You also learned the importance of using **security events** to track how specific incidents and score drops affect an organization's security landscape. Lastly, you learned about **Attack surface map** and **attack paths** to visualize your environment and identify potential vulnerabilities.

Throughout this book, you've learned the entire lifecyle of adopting a CNAPP solution using Microsoft Defender for Cloud, starting from the foundational concepts of CNAPP, the planning, all the way to the implementation of each Microsoft Defender for Cloud plan. The book closed with this introduction to Microsoft Security Exposure Management, which gives you an extra component to add to your security posture management journey. I hope you had a great time reading this book, and that you can implement the insights shared in this book in your production environment.

## Notes

1. For more information about how to manage devices in MDE, visit `https://bit.ly/CNAPPBook106`.

2. For more information about Entra roles, visit `https://bit.ly/CNAPPBook107`.

## Additional resources

- Watch the author of this chapter, Shay Amar, talking about how to stay ahead of threats with proactive security:

    - Part 1 of the video: `https://bit.ly/CNAPPBook91`

    - Part 2 of the video: `https://bit.ly/CNAPPBook92`

- Visit the official Microsoft Security Exposure Management documentation: `https://bit.ly/CNAPPBook93`

# Join our community on Discord

Read this book alongside other users. Ask questions, provide solutions to other readers, and much more.

Scan the QR code or visit the link to join the community.

```
https://packt.link/SecNet
```

# Leave a Review!

Thank you for purchasing this book from Packt Publishing—we hope you enjoyed it! Your feedback is invaluable and helps us improve and grow. Please take a moment to leave an Amazon review; it will only take a minute, but it makes a big difference for readers like you.

*https://packt.link/r/1836204876*

Scan the QR code below to receive a free ebook of your choice.

*https://packt.link/NzOWQ*

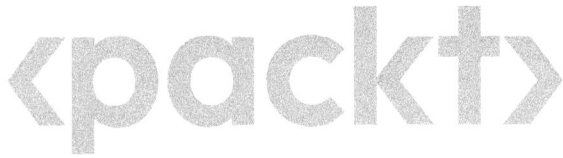

# Other Books You May Enjoy

If you enjoyed this book, you may be interested in these other books by Packt:

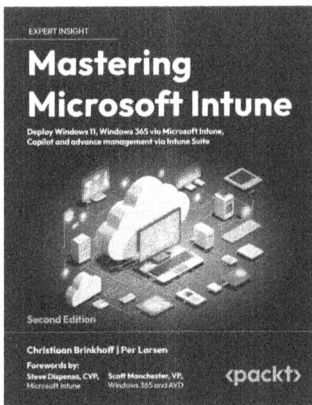

**Mastering Microsoft Intune**

Christiaan Brinkhoff, Per Larsen

ISBN: 9781835468517

- Simplify the deployment of Windows in the cloud with Windows 365 Cloud PCs
- Deliver next-generation security features with Intune Suite
- Simplify Windows Updates with Windows Autopatch
- Configure advanced policy management within Intune
- Discover modern profile management and migration options for physical and Cloud PCs
- Harden security with baseline settings and other security best practices
- Find troubleshooting tips and tricks for Intune, Windows 365 Cloud PCs, and more
- Discover deployment best practices for physical and cloud-managed endpoints

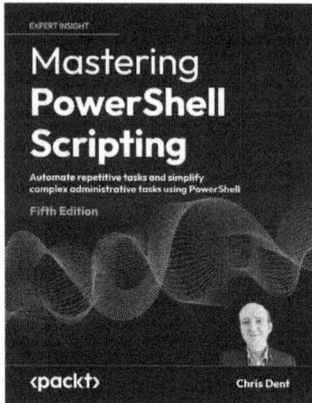

**Mastering PowerShell Scripting**

Chris Dent

ISBN: 9781805120278

- Create scripts that can be run on different systems
- PowerShell is highly extensible and can integrate with other programming languages
- Discover the powerful command-line interface that enables users to perform various operations with ease
- Create reusable scripts and functions in PowerShell
- Utilize PowerShell for various purposes, including system administration, automation, and data processing
- Integrate PowerShell with other technologies such as .NET, COM, and WMI
- Work with common data formats such as XML, JSON, and CSV in PowerShell
- Create custom PowerShell modules and cmdlets to extend its functionality

# Packt is searching for authors like you

If you're interested in becoming an author for Packt, please visit authors.packtpub.com and apply today. We have worked with thousands of developers and tech professionals, just like you, to help them share their insight with the global tech community. You can make a general application, apply for a specific hot topic that we are recruiting an author for, or submit your own idea.

# Index

# Download a free PDF copy of this book

Thanks for purchasing this book!

Do you like to read on the go but are unable to carry your print books everywhere?

Is your eBook purchase not compatible with the device of your choice?

Don't worry, now with every Packt book you get a DRM-free PDF version of that book at no cost.

Read anywhere, any place, on any device. Search, copy, and paste code from your favorite technical books directly into your application.

The perks don't stop there, you can get exclusive access to discounts, newsletters, and great free content in your inbox daily.

Follow these simple steps to get the benefits:

1.  Scan the QR code or visit the link below:

*https://packt.link/free-ebook/9781836204879*

2.  Submit your proof of purchase.
3.  That's it! We'll send your free PDF and other benefits to your email directly.